COMPUTER SIMULATION IN BIOLOGY
A BASIC Introduction

COMPUTER SIMULATION IN BIOLOGY
A BASIC Introduction

Robert E. Keen
Department of Biological Sciences
Michigan Technological University
Houghton, Michigan

James D. Spain
Department of Chemistry
Clemson University
Clemson, South Carolina

WILEY-LISS

A JOHN WILEY & SONS, INC., PUBLICATION
New York • Chichester • Brisbane • Toronto • Singapore

For Karen and Pat

For time and good times

Address All Inquiries to the Publisher
Wiley-Liss, Inc. 605 Third Avenue, New York, NY 10158-0012

© 1992 Wiley-Liss, Inc.

Printed in the United States of America

Apple, Applesoft, Apple II+, Macintosh, PreDOS, and Silentype are trademarks of Apple Computer, Inc. Hercules is a trademark of Hercules computer Technology. CGA, IBM, IBM-PC, and PC-DOS are trademarks of International Business Machines, Inc. GWBASIC, Microsoft, Microsoft BASIC, MS, and MS-DOS are trademarks of Microsoft Corporation. Photoready copy was prepared by Way With Words, Terre Haute, Indiana.

Recognizing the importance of preserving what has been written, it is a policy of John Wiley & Sons, Inc., to have books of enduring value published in the United States printed on acid-free paper, and we exert our best efforts to that end.

The publication of this volume was facilitated by the authors and editors who submitted the text in a form suitable for direct reproduction without subsequent editing or proofreading by the publisher.

Library of Congress Cataloging-in-Publication Data

Keen, Robert E.
   Computer simulation in biology : a BASIC introduction /
Robert E. Keen, James D. Spain.
        p.   cm.
   Rev. ed. of: BASIC microcomputer models in biology / James
D. Spain. 1982.
   Includes bibliographical references and index.
   ISBN 0-471-50971-X
   1. Biology—Computer simulation.   2. BASIC (Computer program
language)   I. Spain, James D.   II. Spain, James D.   BASIC
microcomputer models in biology.   III. Title.
QH323.5.K45 1991
574'01'13—dc20   91-24063
                                                              CIP

PREFACE

We have written this book because we are convinced that computer simulation based on quantitative models is a fundamental biological skill. Simulation studies pervade the recent literature of almost every biological field. Biologists with simulation skills find them to be useful in comprehending biological models quickly, separating sense from nonsense. Simulation techniques ought to be as familiar to practicing biologists as are data analytical techniques, which have been part of biological training for decades. There are many general biometry textbooks for teaching experimental design and analysis to biologists. However, as far as we are aware, this is the only general textbook of biological simulation techniques.

There are several excellent texts that cover computer simulation in particular disciplines of biology. The previous edition of this book, titled *BASIC Microcomputer Models in Biology*, was the first textbook to introduce simulation techniques with models from many biological fields. The similarity of simulation techniques across all of biology argues strongly for the usefulness of a general text.

The book is structured around the premise that students (and teachers) cannot understand the power and the limits of computer simulation unless they actually write computer programs based on biological models. For this reason, numerous exercises are integrated with the descriptive material of each chapter, and in places the book resembles a laboratory manual. When students complete an exercise, they begin to see the modeled system from new perspectives. In fact, their insights into a system seem to come more from writing the simulation program than from examining the simulation results. Students are always more enthusiastic about developing their own simulation programs than they are about working with "canned" simulation programs written by someone else to show how a model behaves. The simulation process serves to reinforce their basic understanding of quantitative principles of biology.

This book tries to take advantage of the wide application of many basic biomathematical concepts among most disciplines in biology. For example, the concepts of chemical kinetics employed in biochemistry are (for better or worse) often used in modeling physiological and ecological systems. Compartmental modeling and transport processes find applications at all levels of biological organization, from sub-cellular systems to global biogeochemistry. Each chapter of the book builds on preceding chapters in a logical, cumulative fashion. We ask our students to complete at least one or two representative exercises from each chapter. Experience with these simple simulations provides them with background to appreciate the potential role of computer simulation and modeling of biological systems, including complex models of ecosystems, of human physiology, and of subcellular metabolism.

The book is written for a diverse audience including essentially any biologist who wishes to learn techniques of biological simulation based on quantitative models. Applied mathematicians may also find it useful as an introduction to some applications of simple numerical techniques to biological problems. The material demands little in the way of prerequisites. Mathematical training through algebra and trigonometry is assumed, and beginning calculus is helpful but not essential. A first course in statistics is also useful.

The text is divided into four parts. The first part deals with simple equations that model single biological systems or components of systems. Individual chapters illustrate the ways in which equations for simple models are derived, and how they may be used to generate simulation data. Students can learn techniques of computer operation and programming while doing simple exercises. A chapter on program planning, debugging, and BASIC pitfalls is strategically placed to help students prepare for more complex programs in succeeding chapters. A critical chapter on numerical integration concludes the first part.

The second part is concerned with deterministic models of multicomponent systems. The models in these chapters generally require numerical techniques to solve multiple equations having interdependent variables. This second part is organized according to specific areas of biological application and/or modeling approaches within such areas. Each chapter first covers some concepts and techniques which are essential for understanding simulation in general, and which are needed for work in succeeding chapters. The latter sections of each chapter contain more advanced material, which permits students to pursue particular interests in some subdisciplines.

The third part of the book covers some elements of simulation with probabilistic biological models. We have argued with ourselves and students about the proper placement of the concepts and material of these chapters. An earlier placement has advantages, but some of the programming techniques require considerable practice.

The fourth section covers a variety of models that use the techniques of previous sections, applied to biological disciplines that were not covered in depth in the second part of the book. Students and instructors may be selective about their coverage of this material.

References are given throughout the text to the literature of biological modeling and simulation. This is an introductory text, so the literature review cannot be rigorous for each subject. We have attempted to cite some of the recent books on simulation and modeling in most areas of biology. These may be consulted for more comprehensive coverage, comment and evaluation of different disciplines.

Most of the exercises ask for x-y plots of simulation results. Such plots can be produced in several ways, including drawing graphs by hand from numerical output, or directing output to a software graphics package. In Appendix 2 we provide a BASIC graphing routine usable on three common personal computers; students can write their programs directly into this program. Appendix 1 is a handy reference for the elements of BASIC programming. Other appendices contain some curve-fitting programs discussed in Chapter 3.

A number of models in this book have been developed in conjunction with our own teaching, and these are included without any indicated source or citation. Many of these deal with common biological systems, so that it is likely that similar models have been described elsewhere in biological literature. We apologize in advance to those who see their pet ideas published here without proper citation and credit, and we would appreciate being informed of any such oversight.

Courses taught with this textbook may use a diversity of computer equipment. We have used time-sharing terminals connected to a large mainframe computer, laboratories of personal computers devoted completely to a bio-simulation course, and the diffuse personal computer environment that prevails currently on most North American campuses. Almost any computer system is suitable if it includes graphics-capable BASIC with some method of printing graphical output. Because most students are now familiar with personal computers, we favor their use. Students can quickly begin BASIC programming if they are given one or two pages of "how-to" instructions for operating a specific machine, and a diskette containing the appendix programs from this book. After 30 minutes of supervised instruction, students can successfully write and complete at least the initial exercises.

We have taken pains to minimize errors in the book. However, the precision of expression required for writing a successful computer simulation almost ensures that some errors will be detected. Such errors can be frustrating for both readers and authors. We assume full responsibility for these and ask that we be informed of problems.

We are heavily indebted to numerous colleagues who critically read the first edition of this book and who have offered comments on this second edition. Some of these include Martin Auer, Martin Boraas, Janice Glime, Bruce Hannon, James Horton, Kenneth Kramm, Tim Mack, Roy

Meyers, James Randall, Thomas Snyder, Carl Walters, Edwin Williams, Brian Winkel, and Lois Young. The critical students at Michigan Tech University and the University of Minnesota-Duluth especially deserve our gratitude. The major part of this edition was written while the first author was on sabbatical leave. During that period, the U.S. EPA Environmental Research Lab in Duluth and particularly Nelson Thomas provided technical and other support for the writing.

Finally, we suspect we are not the most qualified individuals to write such a textbook. We share a conviction that simulation should play a central role in biological research, but neither of us considers our mathematical background to be particularly strong. While we think this edition presents a good cross-section of simulation in biological disciplines, we both wish the really qualified authors would get on with the job.

CONTENTS

PART FOUR: SUPPLEMENTARY MODELS

INTRODUCTION

THE ROLE OF MODELING AND COMPUTER SIMULATION IN BIOLOGY

Life scientists in the future may be characterized more by their ability to use the computer for data analysis and simulation than by their ability to use the traditional microscope. This is so because biology has a greater potential for mathematical development than other sciences. Living systems are made up of interacting chemical and physical processes, all of which can be described in mathematical terms. However, these systems are so immensely complex that they have resisted mathematical analysis by the classical methods successfully used by chemists and physicists. Only within the last four decades have biologists been able to use digital computers to begin to deal with these complex systems.

Computers now are used in biological investigation for planning experiments, collecting data, searching the literature, exploring and analyzing data, and publishing results of experiments. These uses have become common in most laboratories during the past decade with the general introduction of powerful personal computers. This book is about another important role, that of simulation based on quantitative models. Such simulations are important for formulating and improving the conceptual models that biologists use in the practice of their science. In particular, simulations based on complex models can provide unexpected insight into behavior of intricate systems. Personal computers are admirably adapted for such simulations, which should also be a **part of** common laboratory work.

The overall objective of this book is to introduce and provide practice with some methods of computer simulation based on quantitative biological models. To keep the emphasis on simulation rather than on programming techniques and computer operation, only elementary methods of programming are employed or illustrated. The most effective way to learn simulation is by actually developing, programming and testing computer models. One does not learn to play the piano only by reading books about piano playing; some actual practice is needed. Similarly, one cannot become very knowledgeable about computer simulation without running a computer. This book presents a variety of biological models to introduce principles of simulation using a direct approach that avoids extensive mathematics.

A major goal of this textbook is to convince students that all biological disciplines are quantitative sciences with mathematical foundations. It should become obvious that quantitative approaches are required for developing satisfactory conceptual models in biology.

I.1 Conceptual Models

Models and other analogies have always played an important role in the thinking of scientists. The use of models is such a fundamental part of human thinking that we are usually unconscious of the important difference between models and real systems. The difference becomes quite clear if we become lost in the woods or in a city, for then we are faced bluntly with the failure of our conceptual model of location and its lack of conformity with the real system.

A model is any representation of a real system, and may deal with the structure or function of the real system. The model may involve words, diagrams, mathematical notation, or physical structures in representing the system. "Model" may have the same meaning as concept, hypothesis, or analogy. Because no model can totally represent the real system in every detail, it must always involve varying degrees of simplification or abstraction. (A road map is a useful model for travel by automobile, but it does not show stoplights or cracks in the paving.) Given this broad definition, it is obvious that essentially all science involves the formation, examination, and improvement of conceptual models about the universe. A quick review of some conceptual models will demonstrate how they contribute to current ideas about living systems.

The atom as a unit of elemental structure and the molecule as a unit of chemical reactivity were for a very long time only conceptual models. Not until recently have large molecules been observed directly with electron microscopes. We usually depend on x-ray diffraction patterns to construct physical models of molecules. This is an indirect technique for describing structures that may never be directly observed. For example, the α-helix structure now found to some degree in most protein polypeptide chains was worked out by Linus Pauling, who experimented with paper models of molecular structure until he found one with repeat distances consistent with his x-ray data. Another classic example is the double helix structure of deoxyribose nucleic acid (DNA), first proposed by Watson and Crick, based upon a molecular model which matched many of the known properties of DNA (Watson 1968). Like most good models, it triggered a burst of experimental activity which has continued to the present.

The textbook diagrams of typical animal or plant cells are based upon a composite of many observations of many kinds of cells using a variety of observation techniques (Hardin 1966). The three-dimensional structure of most organs within the body is based upon serial analysis of hundreds of two-dimensional tissue sections using the technique of stereology (Elias and Pauly 1966). The resulting three-dimensional models of organs are essential to understanding their function under normal and abnormal conditions.

Conceptual models of the gene have a long history of being proposed, discarded, modified and refined. The blending theory of heredity gave way to the concept of hereditary particles described as "beads on a string". Subsequently, the gene has been considered as that portion of chromosomal DNA which codes for a single polypeptide chain. Despite the great advances made in genetics, it is still dominated by conceptual models based largely upon the observation of genetic effects. Each new model has led to new sets of questions, which in turn have produced new and better understanding of the nature of the gene.

In ecology, the food chain and the ecosystem are important conceptual models which have prompted studies of the transport of energy and materials within the ecosystem. Organic evolution has proven to be an extremely fertile ground for the production of conceptual models for mechanisms of speciation.

In biochemistry, we deal with a variety of conceptual models about enzyme action, including enzyme attachment to the substrate and enzyme responses to changes in temperature and pH. The metabolic pathways that we call glycolysis, Krebs' cycle, and the carbon cycle of photosynthesis are really conceptual models which will continue to require modification to be consistent with new findings. These models raised many new questions besides answering older ones, and they provoked many new experiments. The models are significant too because they are used in decisions about medical treatment, drug action, toxic effects, and nutrition of plants and animals.

It is clear that models play important roles in biology. Some of the earliest models involved careful observation and drawing in great detail the morphology of systems under investigation. This technique still provides a valuable means of forcing oneself to see details about a structure which would otherwise go unnoticed. Those who have taken the time to draw accurately the details of biological structure soon realize that they have never really examined the subject before. New details of structure seem to appear suddenly, and interesting relationships become apparent. This illustrates an important point: that the greatest benefit of a model often comes as a direct result of the thinking required to develop it. From this viewpoint, modeling has been significant in understanding of most biological systems.

The life sciences involve more than just describing and understanding natural systems. Manipulating the environment seems to be a fundamental human characteristic, and one of the objectives of biological study is to understand enough about complex living systems to manipulate them "beneficially". Whether the system manipulator is a physician deciding on the course of treatment for an ill patient or a political appointee deciding on bag limits for a game bird, decisions are made on the basis of conceptual models of the manager. Improving conceptual models should lead to better management.

Conceptual models, by themselves, are generally lacking in rigor. They can be imprecise and interpreted differently by different persons. To avoid this problem, the conceptual model should be given a more precise form that is unambiguous and can be evaluated and validated. One form of conceptual model that provides these advantages is the mathematical model.

I.2 The Mathematical Model

A mathematical model of a biological system may be as simple as a single equation relating one variable to another, or it may be a multicomponent model involving the interaction of many equations having several mutually dependent variables. Mathematical models may deal with rates of change of systems of cells, organisms, populations, or molecules with time and under the impact of environmental factors such as light, temperature, pH, etc. The simpler mathematical models may be developed in various ways; most either are derived in a straightforward way from conceptual models, or are obtained empirically from analysis of experimental data. Hall and Day (1977) refer to these two approaches as "mechanistic" and "descriptive". The development of multicomponent models with many equations will be discussed below.

An equation or group of equations by themselves may not contribute much to the understanding of a particular phenomenon unless the reader is gifted or practiced mathematically. For this reason, it is usually necessary to solve the equation for some representative values of the independent variable (for example, time, pH or temperature) and to present the resulting information in graphical form. Because of the amount of calculating involved, this is best done by implementing the mathematical model on a computer and examining the graphical output.

I.3 Computer Simulation

In simplest form, computer simulation involves implementing a mathematical model on a computer to produce simulation data. In this way, the output of the mathematical model may be compared readily with experimental data from the real system to evaluate the model. Hand calculations may be sufficient for very simple models, but computer simulation is almost essential for understanding multicomponent models and their complex interrelationships.

Implementing a simulation model on a computer involves programming the mathematical expressions, and assigning various rate constants and coefficients. This may be accomplished either with general purpose algebraic languages, or with special simulation languages. The simulation is allowed to proceed, resulting in an output of simulation data. The simulation data may be examined to correct the implementation of the model, or to correct the model itself.

In many ways, simulation resembles a game that one plays with the computer. The computer keeps track of all the rules of the game, follows through each play in correct sequence, and provides any random numbers that may be needed to simulate effects of chance. Persons working with the simulation will have various decisions to make just as they would playing a computerized ballgame, where, depending on conditions, they would select certain strategies to optimize the prospects of scoring. In the same way, a person working with a simulation of population growth would have an opportunity to determine initial population densities, growth rates, presence of predators, and the impact of predation. The simulation would be allowed to run so that its behavior might be observed. As a follow-up, the person may decide to alter conditions in an attempt to stabilize the system or to observe the effects of different rates of growth. The outcome of such a "game" will depend on the assumptions made in designing the simulation and on the selection of initial constant values.

Clearly, a computerized ballgame cannot replicate totally the real game in complexity, chance, and the multiplicity of necessary decisions. Likewise, a biological simulation can only approach the real system to varying degrees, depending upon the complexity of the mathematical model on which it is based. However, it is possible to simulate systems which would be impossible to investigate experimentally because of the amount of time and space involved. For example, investigation of a real predator-prey system could involve population estimates taken over a 10- to 50-year period, and a 10- to 100-square-kilometer area. Even population experiments in small closed systems may require intense study over many weeks to provide meaningful results. A simulation of these same systems could be completed in seconds on the computer.

I.4 Relationship of Modeling and Simulation to the Research Process

Some of the concepts presented above are summarized in a diagram relating modeling and simulation to the overall process of research and management (Figure I.1).

The focus of our interest is the Real System. It must be understood that any biological system must always remain to some extent a "black box". No matter how much information we have about a particular biological entity, we must always remain on the outside looking in. For this reason, the Real System is distinguished from the other components of the diagram of Figure I.1 as a circle. All of the boxes represent forms of information that are derived in one way or another from the Real System. As such, they are reflections or perceptions of the Real System. The arrows of the diagram are processes by which the information is obtained and manipulated.

The primary concern of science is production of a conceptual model which most nearly reflects the Real System. The conceptual model is our mental picture of the Real System based on available information interpreted in the light of experience with similar systems. Science usually becomes aware of a new biological system as a result of investigating some other system. However, even the most casual observation generally results in the formulation of at least a crude conceptual model. Initially this usually takes the form of an analogy to known systems. A good conceptual model will suggest hypotheses that can be tested experimentally to corroborate the model. Through experiments, new data are obtained to throw new light on the original observation or on its interpretation. In one way or another, this process generally results in the improvement of the conceptual model. The sequence describes the classical research loop.

A mathematical model results from a formalization of the conceptual model in quantitative terms, usually as an equation describing the response of a system to some variable such as time or temperature. The mathematical model may also be derived from statistical analysis of the experimental data, or by a combination of these two approaches. In any case, the thinking required to formulate the mathematical model generally improves the conceptual model. This improved model will in turn suggest further experimentation. Repeated excursions through the loop of quantitative research in Figure I.1 can result in further improvements to both the mathematical model and the conceptual model.

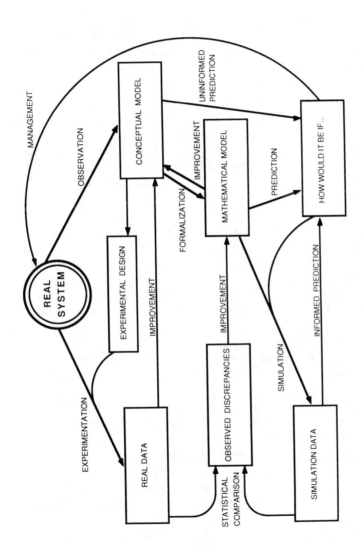

Figure I.1. A diagram showing the role of quantitative modeling and simulation within the process of research. Based in part on a diagram by Gérardin (1968).

The process of simulation expands the power of the quantitative research loop by using the mathematical model to generate simulation data which may be compared to real data to show differences. Assuming that the model was properly programmed for the computer, these differences indicate problems with the validity of the assumptions used in formulating the mathematical model, and hence flaws in the conceptual model. The errors may be so large as to require an entirely new conceptual model, or so small as to require only slight modifications of coefficients in the mathematical model. In any case, new questions are raised and new experiments may be needed. The fine tuning of a simulation can produce further improvements in both the conceptual model and the mathematical model.

Thus, simulation permits one to take advantage of a more powerful research loop. Its chief advantage is that it provides a clear and unbiased test of the thinking involved in the formulation of the mathematical model, and therefore of the conceptual model. Simulation can therefore play a key role in understanding biological systems. No other technique can contribute such a powerful addition to developing and testing concepts.

If this were the only value of simulation and modeling in biological research, it would be sufficient justification for their use. However, simulation provides another important benefit: it allows "experiments" on system models that lie outside the range of normal possibility. It allows us to ask questions about "what would happen to the system if ...?". This process is what we might label the enlightened management loop. By simulating various management strategies on the mathematical model, we are able to make intelligent predictions about their success or failure on the Real System. Of course, there are less effective management loops. For example, the farthest right-hand loop in Figure I.1 might describe politically expedient management in which observation leads to a conceptual model and then directly to predictive management. In ecology, this produces what Fretwell (1972) has described as the "shoot-the-chicken-hawk" school of wildlife management.

Computer simulation is particularly valuable for systems which involve multiple, nonlinear interactions. This is especially true in areas of biology such as physiology and ecology which often involve control or management of multicomponent systems. Such systems usually contain numerous variables interrelated by positive and negative feedback loops. Forrester (1969) has shown that human decisions based on intuition about such systems often produce effects that are exactly the opposite of those desired. It is essential therefore that management decisions be based on something more than simple conceptual models. Watt (1970) gives this as one of the most important reasons for simulation in the area of environmental management. It is clear that in many circumstances experiments with the Real System are impossible. On the other hand, it is desirable that the control strategy be correct because populations of economically or socially important organisms may be involved. In such cases, computer simulation provides a logical base for informed management decisions.

Modeling and simulation are not able to provide insight into all questions about biological systems because some questions are not yet reducible to quantitative terms and mathematical models. Some questions will never be reduced to these terms. In most systems, however, it seems evident that modeling and simulation can be extremely powerful aids to understanding and management.

I.5 Development of Multicomponent Models

At first it appears terribly difficult to produce models of complex systems and to verify such models. The problem is not as great as it seems at first sight, because multicomponent models are made up of clusters of submodels with multiple dependencies. Each submodel may be developed independently and validated by comparing its simulation output with that of the appropriate real subsystem. The submodels may be merged into a single model with the potential for simulating the total system. This reductionist approach to whole-system modeling permits the most complex simulations to be manageable.

The essence of this approach is the idea that systems are in fact hierarchies of organization (Walters 1971). Each component of the system embodies sub-components which in turn contain sub-sub-components, etc. (Figure I.2). Fortunately it is not necessary to have a conceptual, mechanistic understanding of all the sub-components in order to produce workable simulations. Numbers of the components and sub-components may be lumped into one or more sub-models which define the net effect of these sub-components on the system as a whole. This process of lumping reduces the resolution of the model, but may have the effect of focusing the modeler's attention on primary controlling factors.

When we are modeling an intermediate level in the hierarchy, higher levels are usually considered to be external to the system of interest. The behavior of an external variable is usually approximated and then incorporated into the model as an environmental input. If it later becomes obvious that the approximate input is important to the behavior of the component model, the model may have to be expanded to include these other levels.

It is a mistake to think that simulation data must duplicate exactly the data obtained by experimentation with the Real System. By definition,

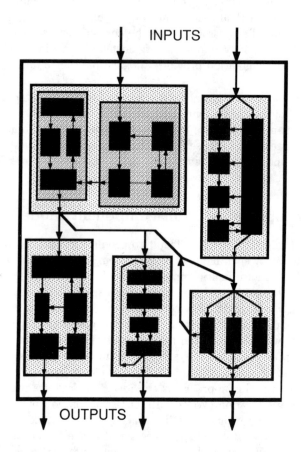

Figure I.2. A diagram showing the organizational hierarchy of systems within any living system. Based in part on a diagram by C. J. Walters (1971).

a simulation model is not the real system, and it is not reasonable to expect simulation data to match that of the real system point by point (Meadows et al. 1972). It is usually sufficient if the model is able to show general correspondence with the behavior of the Real System. How far the simulation data may depart from the actual data and still produce confidence in the model will depend upon the function of the model in the research process and the system being modeled. It is also possible to make a simulation so complex that it contributes little or nothing to the understanding of the system being modeled. The major factors that control the system may be hidden if the model is overly complex.

The results of simulations produced by multicomponent models of complex systems are sometimes controversial. These models are most dependent upon the computer's ability to handle large amounts of data precisely. They are also the most likely to produce results which we do not expect. It is only through experience working with more familiar simple model systems that we are able to understand the capabilities and limitations of computer simulations. With this experience, we gain confidence in the results of more complex simulations even when they behave in a manner that runs against our intuition.

The techniques of biological simulation are still quite new and being developed actively. Some pioneering simulation modelers overstated the possible developmental rates of their field. Others put too much emphasis on predictive capability without recognizing that the principal value of simulation is the ability to improve conceptual models. Because of the complexity of the systems involved, it is almost certain that biological predictability will develop slowly.

Biologists who intend to perform any quantitative research should be trained equally in both statistical analysis and simulation techniques. Simulation should play an active role in most quantitative research programs, being used as a tool in developing and testing concepts and models. There are, of course, biologists who specialize in techniques of simulation, just as there are those who specialize in biometry and data analytical methods. However, the basic techniques of simulation modeling need to be understood by all quantitative biologists.

I.6 Simulation Modeling: Languages and Machines

The fundamental ideas of biological modeling and simulation do not depend on particular computing machines or languages. For example, many of the simulations described in the following chapters were first implemented on analog computers that are now largely obsolete. Biological simulations can be programmed using any of the general purpose algebraic languages and most of the specialized simulation languages. The specific emphasis in this book is on the development of modeling skills with the BASIC language on personal computers.

The BASIC language is the most practical for teaching elementary simulation. A person trained in one of the specialized simulation languages can produce results of a complex simulation in a remarkably short time. However, learning to use one of these languages almost presupposes some experience in modeling, such as that provided in this text. None of these languages are as widely available as the popular general-purpose languages, and translations among them are rather more difficult than among BASIC, Pascal and FORTRAN, for example.

The recent technological development of the personal computer has made available to practically every biologist and biology student a convenient tool for developing and implementing a wide range of complex simulations. The graphical capability of most personal computers is particularly attractive for many simulations. BASIC is the most widely used and understood computer language. It was developed specifically as an easily learned language, and has no equals for this. The generic BASIC used in this text is designed for easy implementation on most machines using most dialects of the language.

PART ONE

SIMPLE MODEL EQUATIONS

The critical assumption behind all biological simulation is that an equation may serve as an analog or model of a simple biological process. The assumption seems reasonable because almost any biological process may be described by a cause-effect or stimulus-response curve (Figure P1.1). The curve will relate the intensity or amount of some causative or stimulating agent to the intensity or amount of a biological response or effect. Traditionally, the measure of the agent, called the independent variable x, is recorded along the horizontal axis of a graph, and the measure of the biological effect or response, called the dependent variable y, is expressed along the vertical axis. It is possible to arrange the axes differently, but we will try to retain the conventional arrangement as far as possible.

The response of a biological system may be studied experimentally under different conditions, and the resulting data may be used to construct

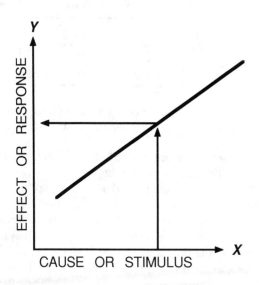

Figure P1.1. Graph of the simple cause-and-effect relationship.

a cause-effect curve. The curve will show the quantitative performance of the system under a given set of conditions. It is therefore a diagrammatic model of the system. Curves are also used to show the behavior of equations that relate one variable to another. Thus it is possible, in one way or another, to find an equation which will generate output data that resemble mathematically the data produced by the biological system. The equation, therefore, also becomes a model of the biological process. The equation can stand in place of or substitute for the biological system in terms of its relationship to other components of the system involved in the process. Mathematically, the equation is interchangeable with the process, as illustrated in Figure P1.2.

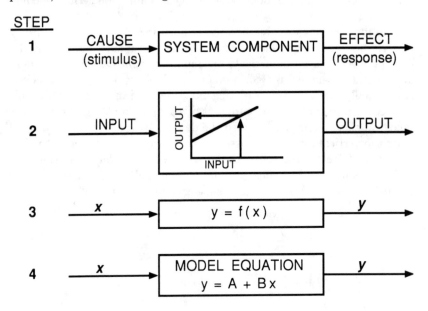

Figure P1.2. Stepwise replacement of a component of a biological system with a model equation.

The procedure for finding an appropriate model equation first involves identifying the functional relationship of the system involved in the process we are modeling. By examining the data we are usually able to set up a simple function that describes the output variable y, as a function of the input variable x. Conventionally, the general form of the function equation is $y = f(x)$. Sometimes the output variable y is determined experimentally to be a function of more than a single input variable, so that the equation has the form $y = f(x, z)$. Most of the biological processes considered in this book involve only a single independent variable other than time.

After finding an equation that effectively mimics the biological response, we still must find appropriate values for the constants, coefficients, and exponents of the model equation. These may be estimated with trial and error, or by more formal procedures such as curve fitting. With very complex models, trial and error may be the only available technique. After assigning values, it is possible to implement a simulation program on a computer. The computer program may be verified to make sure it is performing as expected. The model equation may be validated or corroborated by comparing simulation data with data obtained by observation or experiment.

Part One of this text considers the two principal techniques used to find the functional relationship between x and y. The analytical technique gives rise to theoretical or mechanistic model equations, while the empirical technique results in descriptive or empirical equations.

Empirical models are obtained using statistical methods to fit one of several generalized equations to experimental data. Such models do not depend upon any insight the modeler may have about the workings of the biological system or process. In contrast, analytical or mechanistic models are based on equations that are derived from conceptual models of the biological process. The distinction is not completely clear, because empirical techniques may play a role in the initial formation of the conceptual model. The theoretical basis for an analytical model is corroborated when experimental data and simulation data correspond more or less closely.

In the five chapters making up Part One of this book, you will be introduced to several of the more common mathematical models in biology, and to methods that are used in almost all biological simulations. You should be able to accomplish the following:

(1) Learn how to write BASIC programs to produce simulation data from simple model equations;

(2) Learn how to produce graphs of simulation data;

(3) Know the steps used to produce mechanistic equations from simple conceptual models;

(4) Understand the usefulness and derivation of models produced from assumptions of stable systems;

(5) Become familiar with the form of curves produced by several function equations that are found frequently in biological work;

(6) Know how to find values for constants of both theoretical and empirical model equations, using curve fitting techniques;

(7) Learn about programming flowcharts and about some simple techniques for testing computer programs;

(8) Know how to program some elementary Euler techniques for numerical integration of simple equations.

When you have accomplished these goals, you should be ready to proceed to Part Two of the book, which involves application of these methods to a number of models from a variety of biological fields.

CHAPTER 1

ANALYTICAL MODELS
BASED ON
DIFFERENTIAL EQUATIONS

Analytical models are often expressed as differential equations that define a rate of change of some dependent variable with respect to some independent variable. In biological models, the independent variable is usually time, distance or concentration. To show how models may be developed using differential equations, we will look at a model for biological growth, with time as the independent variable and growth as the dependent variable. Several other simple analytical models based on differential equations will also be presented as further illustrations. These simple models can be used in writing short computer programs for simulating biological processes. Even though they are brief, these programs will let you become familiar with techniques used for the remainder of the book.

1.1 A Model of Biological Growth

A major reason for using differential equations to develop models is that these equations are easily obtained from common sense "function equations". For the purpose of developing this growth model, we will be interested in a population of cells, perhaps cells in a tissue culture dish, or yeast cells or bacteria cells in a culture flask. We observe that cell growth rate (number of cells added per hour) depends on the number of cells already present. That is, if we have one culture with 10 cells and another with 100, the culture with 100 cells will produce more new cells in an hour than the culture with just 10 cells. Likewise, we note that a culture with 0 cells will not produce any new cells. From these general observations, we can write a simple function equation for growth of cell numbers:

$$G = f(N) \tag{1.1}$$

where G is the growth rate and N is number of cells.

If we assume that growth is a direct function of N (i.e. that growth depends directly on N, or that it is directly proportional to N), and if we also assume no other factors are involved, then the growth rate equation will take this form:

$$G = kN \qquad (1.2)$$

where k is a constant of proportion. If you were to graph this equation, showing N (population number) on the x-axis and G (growth rate) on the y-axis, the result would be a straight line with a slope of k. That is, as N increases, G would increase in direct proportion.

The equation has limited value in this form, because just now we are interested in population numbers during the time of growth, rather than in the rate of growth. We need to convert our equation to a form giving this information. To do this, first we define growth rate G as dN/dt. This new expression symbolizes the instantaneous rate of change of number with respect to time. Our equation is now written as

$$\frac{dN}{dt} = kN \qquad (1.3)$$

Equations 1.2 and 1.3 have identical meanings, but Equation 1.3 is in terms of the two variables we want, N and t. Now we can find the equation for growth of the population by integrating Equation 1.3. Mathematicians know how to perform integrations as a result of their experience with the reverse process of differentiation. From their experience they have developed a large number of integration rules, which are found in most textbooks of calculus.

If we use these rules, several useful forms of the equation will be obtained following integration:

$$\ln N_t - \ln N_0 = kt \qquad (1.4)$$

and

$$\frac{N_t}{N_0} = e^{kt} \qquad (1.5)$$

and

$$N_t = N_0 e^{kt} \qquad (1.6)$$

The intermediate steps in this integration may be found in most textbooks of calculus and of population ecology (e.g. Hutchinson 1978). For biological purposes, Equation 1.6 is extremely useful and will occur in many different contexts. Here it describes the number of cells in the population at any time (N_t), based on the initial population size (N_0) and the growth constant (k). The base of natural logarithms is given as e. The form of the curve of numbers over time is shown in Figure 1.1. Over a limited range of population sizes and time periods, this equation may be a useful model to describe growth of bacteria, rabbits, people, money, and other quantities.

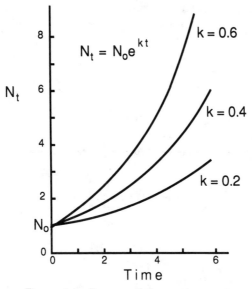

Figure 1.1. Exponential growth curves.

Exercise 1-1: When Equation 1.6 is used to describe growth of bacterial populations, it is often given in the form

$$N_t = N_0 e^{\mu t} \tag{1.6A}$$

where μ is the specific growth constant.

Write a BASIC computer program that uses this equation to simulate unlimited growth of a bacterial culture. Assume for a very rapidly multiplying bacterium that $\mu = 0.092$ min^{-1}, and that initial density $N_0 = 2$ bacteria ml^{-1}. To find powers of e, you may either use the EXP(X) function which is built into BASIC, or raise 2.71828 to the appropriate powers. Appendix 1 provides a review of some essentials of programming in BASIC.

Implement your program on a computer. Your program should have the computer print the results in two parallel columns, one indicating time (t) and the other showing the corresponding bacterial density (N_t) at each minute, $0, 1, 2, 3, \ldots, 50$.

Exercise 1-2: Rewrite your program from Exercise 1-1 so that it uses the graphical capabilities of your computer to produce a graph

```
100 REM SIMULATION OF EXPONENTIAL GROWTH
110 REM
120 LPRINT "TIME",,"POPULATION SIZE"
130 LPRINT"   0--------5--------10--------15--------20--------25"
140 U = .1
150 NO = 2
160 FOR T=0 TO 25
170    NT = NO * EXP (U*T)
180    Y = INT(NT*60/30)
190    LPRINT T; TAB(5); "+"; TAB(Y+5); "*"
200 NEXT T
210 LPRINT"   0--------5--------10--------15--------20--------25"
220 END
```

```
TIME                      POPULATION SIZE
    0--------5--------10--------15--------20--------25
 0  +      *
 1  +      *
 2  +      *
 3  +       *
 4  +       *
 5  +        *
 6  +         *
 7  +          *
 8  +          *
 9  +           *
10  +            *
11  +             *
12  +              *
13  +              *
14  +               *
15  +                *
16  +                 *
17  +                  *
18  +                   *
19  +                    *
20  +                      *
21  +                       *
22  +                         *
23  +                          *
24  +                            *
25  +                              *
    0--------5--------10--------15--------20--------25
```

Figure 1.2. Program and output for Exercise 1-2, set up for a simple line printer. (The program was written for the IBM-Microsoft BASIC interpreter. The LPRINT statements in Lines 120, 130, 190 and 210 send output to the "line printer". The LPRINT statement is not available in all versions of BASIC, although the equivalent exists in all BASICs.)

showing bacterial density as a function of time from 0 to 120 minutes. Set up the output so that bacterial density is plotted on the y-axis and time is plotted on the x-axis. Appendix 2 gives a listing of graphical programs for some microcomputers. Figure 1.2 shows the growth model used in a sample program that produces graphical output for computers equipped with simple printers.

1.2 Exponential Decay

As a sort of converse to the growth process described above, some biological systems will show a decline in concentration of a certain substance through time, with the loss rate proportional to the concentration of the substance present at any time. Following the same procedure as above, we can arrive at a differential equation describing this process of decline:

$$\frac{dC}{dt} = -kC \tag{1.7}$$

Here, C is the concentration of the substance being used up, and k is the rate constant. The negative sign is needed to indicate the reduction in C. This equation integrates to

$$C_t = C_0 e^{-kt} \tag{1.8}$$

This is the classical model for exponential decay used to describe processes such as weight loss during starvation, excretion of drugs or a radioisotope from an organism, light absorption in a liquid, radioactive decay, and other phenomena. Note that in each of these examples the dependent variable will approach zero as t approaches infinity.

Exercise 1-3: Write a program using Equation 1.8 to simulate the decay of the radioactive isotope ^{32}Phosphorus. Begin your simulation with a specific activity of 500 μcuries, and use a value of $k = 0.04847$ day^{-1}. Produce graphical output that shows remaining activity from 0 to 100 days. Use a pencil and straightedge to estimate the isotopic half-life (time for activity to be reduced by 50%).

1.3 Distribution of Organisms

The exponential decay model has been used in a variety of biological research areas to describe the distribution of plants and animals. The organisms are assumed to have a central location of maximum concentration. Their density away from that point is assumed to follow a classical diffusion pattern, which will result in a negative exponential decline in density away from the central locus. Two such situations are given in the following pair of exercises.

Exercise 1-4: Aquatic crustaceans and immature insects that inhabit flowing water will at times release themselves from the streambed and drift downstream. McLay (1970) used the following model based on exponential decline to describe downstream densities:

$$N_x = N_0 e^{-Rx} \tag{1.9}$$

where N_0 is population density at the source of animals, N_x is the density at a distance x meters downstream from the source, and R is a constant that applies uniquely to the organism in a given stream. Write a program to simulate stream drift for larval chironomid insects in a stream, where $R = 0.13$ m^{-1}. Assume density of these animals is 1200 meter^{-2} at their source. Your graphical output should show density at each meter for a distance of 60 meters downstream from the source.

Exercise 1-5: Van Dover et al. (1987) used this same exponential model to describe distribution of a deepwater crab, *Bythograea*, that lives around hydrothermal vents in the Pacific Ocean. The animals were observed to be most abundant immediately around the vents, with a density of about 100 per unit of camera viewing area (about 845 m^2). In any direction from a vent, their density decreased, with $R = 8.56$ km^{-1}. Write and implement a program to find density of crabs as a function of distance from a hydrothermal vent. As output, produce a graph that shows their symmetrical distribution along a line running through a vent, from 600 meters on one side to 600 meters on the other.

1.4 Newton's Law of Cooling

The basic form of exponential decline given in Equation 1.8 has been modified slightly to provide the basis of numerous biological models. An example is Newton's Law of Cooling as a model for loss of heat from a cooling object. This law states that temperature of an object drops at a rate proportional to the difference between the temperature of the object and the temperature of the environment. The rate of temperature change with time is given by

$$\frac{dT}{dt} = -k(T - C) \tag{1.10}$$

where C is the environmental temperature and k is a cooling rate constant. Equation 1.10 integrates to an equation describing the temperature of a cooling object through time:

$$T_t = C + (T_0 - C)e^{-kt} \tag{1.11}$$

where T_t is the temperature of the object at time t and $(T_0 - C)$ is the difference between the initial temperature and the environmental temperature, with C held constant throughout the cooling process. The relationship of Equation 1.11 to Equation 1.8 for exponential decay is obvious when C is set to zero. (Note that Equation 1.11 also holds for "negative cooling" when C exceeds T_0.)

Exercise 1-6: Use Equation 1.11 to write a program for simulating the cooling of a human corpse with $k = 0.06$ hour^{-1}, which is the approximate value for an average clothed adult male in still air. Assume a normal body temperature of 37°C initially and a constant environmental temperature of 8°C. Set up your program to produce a graph showing body temperature during the 48-hour period following death.

1.5 Passive Diffusion Across a Membrane

An equation similar to Equation 1.10 may be used to model the process of passive diffusion. The rate of change of concentration of an internal solute of a cell, caused by passive diffusion into an environment with a constant solute concentration, is given with

$$\frac{dC}{dt} = -k(C - C_x) \tag{1.12}$$

where C is the internal concentration for a cell of unit volume and unit surface area, C_x is the environmental concentration, and k is the proportionality constant for diffusion rate. The integrated form of the equation is

$$C_t = (C_0 - C_x)e^{-kt} + C_x \tag{1.13}$$

where C_t is the concentration in the cell at time t, C_0 is the initial internal concentration, and C_x is concentration of the external environment, assumed to be constant for the duration of the diffusion process.

Exercise 1-7: Write a program using Equation 1.13 to simulate diffusion from a cell of unit volume and unit surface area having an initial internal solute concentration of 100 units per unit volume. At time zero the cell is put into an environmental solute with a concentration of 50 units per unit volume. The diffusion rate constant is 0.20 minute^{-1}. The graphical output from your program should show the concentration of internal cell solute for a period of the first 120 minutes.

1.6 Von Bertalanffy's Model of Fish Growth

This classical model describes fish length as a function of age, based on the assumption that fish growth is proportional to the difference between the length and a theoretical maximum length. That is, fish grow more rapidly when they are smaller, with growth rate declining as their size approaches the maximum. The differential equation describing this process is

$$\frac{dL}{dt} = k(L_m - L) \tag{1.14}$$

where L is fish length, L_m is the theoretical maximum length, and k is the growth rate constant. This equation integrates to

$$L_t = (L_i - L_m)e^{-kt} + L_m \tag{1.15}$$

where L_t is length at time t and L_i is length measured at $t = 0$.

Exercise 1-8: DeMarais (1985) studied growth of a small flatfish, *Buglossidium luteum*, in a bay of the Mediterranean Sea. During the first year of their life these fish follow the Von Bertalanffy growth model, and may obtain a maximum length of 51.6 mm. Assuming an initial length of 8.2 mm and a growth rate constant of 0.23 month^{-1}, write a program that simulates growth of this species over a period of 12 months.

1.7 Model of Inhibited Growth

A model of population growth that is slightly more realistic than that considered in Section 1.1 can be developed by assuming that a population does not grow beyond some upper limit, L. One form of this model is given in the following differential equation:

$$\frac{dN}{dt} = kN(L - N) \tag{1.16}$$

where N is population number or density as before. (In this model the rate constant k will have different dimensions than the constant as defined in Section 1.1.) Note that as N approaches the limit L, the term inside the parentheses approaches zero, as does the rate of growth, dN/dt. This equation can be integrated to give the following:

$$N_t = \frac{N_0 L}{N_0 + (L - N_0)e^{-Lkt}} \tag{1.17}$$

where N_t is the population number at time t, and N_0 is the initial population size at $t = 0$. This equation for limited growth produces an S-shaped curve of population size plotted against time. This model has been used for simulating the spread of disease, for growth obtained with a given amount of nutrient, and other processes that are limited by resources. Equation 1.16 will be discussed further in Chapter 7.

Exercise 1-9: Using Equation 1.17, write a program that simulates density of a population growing in a limited environment. Assume a limiting density of 800 individuals per unit area. Set $k = 0.0005$ individual^{-1} week^{-1}. Your graph of the simulation data should depict density over a 52-week period, beginning with an initial density of 5 organisms per unit area.

1.8 Kinetics of Bimolecular Reactions

Assume that two chemical reactants, A and B, interact to form a product P, as described in the following reaction:

$$A + B \xrightarrow{\ k\ } P$$

The rate at which reactant B is used up depends upon the concentrations of both A and B:

$$\frac{d[B]}{dt} = -k[A][B] \tag{1.18}$$

where $[A]$ and $[B]$ indicate the concentrations of reactants A and B respectively, and k is the constant for reaction rate. The reaction between A and B will proceed differently depending upon the relative concentrations of A and B.

If $[A] = [B]$, then Equation 1.18 becomes

$$\frac{d[B]}{dt} = -k[B][B] = -k[B]^2 \tag{1.19}$$

This equation may be integrated to obtain the equation

$$\frac{1}{[B]_t} = kt + \frac{1}{[B]_0} \tag{1.20}$$

where $[B]_t$ is the concentration of B at time t, and $[B]_0$ is the initial concentration at $t = 0$. This equation may be rearranged to solve for $[B]_t$:

$$[B]_t = \frac{[B]_0}{1 + ([B]_0 kt)} \tag{1.21}$$

Equations 1.20 and 1.21 are model equations for the kinetics of "second-order reactions", in which the rate is proportional to the product of the concentration of two reactants (or the square of either one, because $[A] = [B]$). When data are collected from such reactions and are plotted with $1/[B]$ on the y-axis and t on the x-axis, the result will be a straight line with slope equal to k and a y-intercept equal to $1/[B]_0$. This is easily seen from Equation 1.20 which has the form $y = a + bx$.

In contrast to the above, if the concentration of reactant A is much greater than that of B, then the concentration of A will not change significantly as the reaction proceeds. $[A]$ may be considered constant in this case, and can be combined with k to produce a new constant, k'. The differential equation describing this is obtained from Equation 1.91 as follows:

$$\frac{d[B]}{dt} = -k[A][B] = -k'[B] \tag{1.22}$$

This equation will integrate to:

$$\ln[B]_t = -k't + \ln[B]_0 \tag{1.23}$$

and can be solved for $[B]_t$:

$$[B]_t = [B]_0 e^{-k't} \tag{1.24}$$

You should recognize this last equation as that of exponential decay (Equation 1.6). Equations 1.23 and 1.24 are the model equations for the kinetics of "first-order reactions", where the rate depends upon the concentration of only one reactant. When data obtained from first-order reactions are plotted with $\ln[B]$ on the y-axis and t on the x-axis, the result will be a straight line with a slope of $-k'$ and a y-intercept of $\ln[B]_0$.

The terms zero-order, first-order and second-order were first employed to describe the kinetics of chemical reactions. However, they are now generally used to describe any rate process which is constant (zero-order), or is dependent on the concentration of a single variable (first-order), or is dependent on the product of two variables or the square of one variable (second-order).

Equations 1.22 and 1.24 are important because they show that reaction order is affected when one reactant is held constant, whether from a high relative initial concentration or from being maintained at constant level by other processes. First-order reactions will be encountered frequently in later chapters.

Exercise 1-10: Write a program that uses Equation 1.21 to simulate a second-order reaction. Start with B having a concentration of 5M, and set $k = 0.20$. Your program should find $[B]$ at one-second intervals from 0 to 20 seconds. Have your computer produce graphs showing both $[B]$ and $1/[B]$ over the 20-second period. (If your graphical capabilities permit, it is instructive to show both $[B]$ and $1/[B]$ on the same graph. In this case, the vertical axis will have to be labeled as arbitrary "units".)

Conclusion

This chapter has briefly introduced some fundamental analytical models that have been developed from differential equations. One objective has been to show that this important technique is useful in describing biological phenomena. Another objective has been to provide an opportunity for some elementary programming of biological simulations. In this chapter the techniques of calculus were used to convert differential equations into usable models. In subsequent chapters different methods of working with differential equations will be introduced.

CHAPTER 2

ANALYTICAL MODELS
BASED ON STABLE STATES

In the last chapter we studied rate equations that were converted to simple biological models through mathematical integration. In this chapter we will continue to use differential equations, but we will use an assumption of stable state to eliminate the differential by setting one rate process equal to another. Although such models are accurate only under stated conditions, they can still provide useful simulations of system behavior over a range of conditions near stability. Before we obtain any models with this approach, let us first examine the two contrasting conditions of stability.

2.1 Stable Conditions

The terminology of unchanging states of systems is not very standard among different fields of science, so that different meanings are attached to concepts of stability, stationary-state, equilibrium, steady-state, and constancy. In this chapter we will be concerned with two stable conditions: equilibrium and steady-state. The key feature of equilibrium models is the assumption of an isolated or closed system. Inside the system, two or more interacting processes compete with one another in such a way that the net change in their variables approaches zero. As an example, consider two substances, A and B, that may each be converted into the other. At equilibrium, the rate of formation of B and the rate of breakdown of B are equal.

$$A \underset{f_2}{\overset{f_1}{\rightleftarrows}} B$$

These rates are termed fluxes, with f_1 being the rate of breakdown of

A or formation of B, and f_2 the rate of formation of A or breakdown of B. The fluxes may be described as

$$f_1 = k_1[A] \qquad (2.1)$$
$$f_2 = k_2[B] \qquad (2.2)$$

where k_1 and k_2 are rate constants and $[A]$ and $[B]$ are concentrations of A and B. At equilibrium the net flux is

$$f_{\text{net}} = 0 = k_2[B] = k_1[A] \qquad (2.3)$$

and

$$f_1 = f_2 \qquad (2.4)$$

By rearranging Equation 2.3, we obtain the expression for the equilibrium constant:

$$k_2[B] = k_1[A] \qquad (2.5)$$

$$\frac{k_1}{k_2} = \frac{[B]}{[A]} = K \qquad (2.6)$$

One advantage of the assumption of equilibrium is that it allows us to use a constant to relate one concentration to another. In other words, if we know K and $[A]$ we can calculate $[B]$ at equilibrium.

In contrast, steady-state conditions are associated with open systems. Let us again consider substances A and B, but arranged now in an open system:

$$\xrightarrow{\;f_3\;} A \; \underset{f_2}{\overset{f_1}{\rightleftarrows}} \; B \xrightarrow{\;f_4\;}$$

As part of the open system, two other transfers are added. Flux f_3 describes the rate at which A enters the system, and flux f_4 describes the rate at which B leaves the system. At steady state, the net change in $[A]$ and $[B]$ will be zero, but f_1 no longer will have to equal f_2. Instead, the following three equations will hold:

$$f_1 = f_2 + f_3 \qquad (2.7)$$
$$f_1 = f_2 + f_4 \qquad (2.8)$$
$$f_3 = f_4 \qquad (2.9)$$

By making assumptions of equilibrium or steady-state conditions, we are able to write equations which relate one variable to another. Even though these conditions are rarely obtained in biological systems, assumptions of stability are the basis of an important technique for formulating certain types of models, as the following examples will show.

2.2 Michaelis-Menten Model of Enzyme Saturation

The concept of saturation is fundamental to biological thinking, and is essentially an equilibrium phenomenon. Models based on saturation are as important as the exponential growth and decay models of Chapter 1, and will appear frequently in subsequent chapters. The basic theory of enzyme reactions is founded on a consideration of saturation, and provides a convenient starting point for studying steady-state models.

The general theory of enzyme action was developed by Michaelis and Menten in 1913. The following representation was given by Briggs and Haldane (1925), with further development by Lehninger (1975). In the simplest case, a single enzyme reacts initially with a substrate to form an enzyme-substrate compound. This compound then breaks down to form a product and free enzyme. These reactions can be shown schematically:

$$E + S \; \underset{k_2}{\overset{k_1}{\rightleftharpoons}} \; ES \; \underset{k_4}{\overset{k_3}{\rightleftharpoons}} \; E + P$$

Here, E and S refer to enzyme and substrate, ES is the enzyme-substrate compound, and P is product. The four rate constants involved are k_1, k_2, k_3 and k_4. Initial velocity of the reaction is assumed to be limited by the breakdown of ES:

$$v = k_3[ES] \tag{2.10}$$

The rate at which the compound ES is formed is given by

$$\frac{d[ES]}{dt} = k_1[E][S] + k_4[E][P] \tag{2.11}$$

Initially, when $[P] = 0$, this becomes

$$\frac{d[ES]}{dt} = k_1[E][S] = k_1([E_t] - [ES])[S] \tag{2.12}$$

where $[E_t]$ = total enzyme in the system = $[E] + [ES]$. The rate of breakdown of ES is

$$\frac{d[ES]}{dt} = k_2[ES] + k_3[ES] = (k_2 + k_3)[ES] \tag{2.13}$$

At steady state, rate of formation of ES will equal rate of breakdown

$$k_1([E_t] - [ES])[S] = (k_2 + k_3)[ES] \qquad (2.14)$$

This can be rearranged to give

$$\frac{([E_t] - [ES])[S]}{[ES]} = \frac{k_2 + k_3}{k_1} = K_m \qquad (2.15)$$

where K_m is the Michaelis-Menten constant. If this is solved for $[ES]$ we get

$$[ES] = \frac{[E_t][S]}{K_m + [S]} \qquad (2.16)$$

Because initial velocity v is proportional to $[ES]$ (Equation 2.10), v will approach a maximum velocity V_{max} as the enzyme becomes totally saturated as ES. As a result,

$$\frac{v}{V_{max}} = \frac{[ES]}{[E_t]} = \frac{[S]}{K_m + [S]} \qquad (2.17)$$

and

$$v = \frac{V_{max}[S]}{K_m + [S]} \qquad (2.18)$$

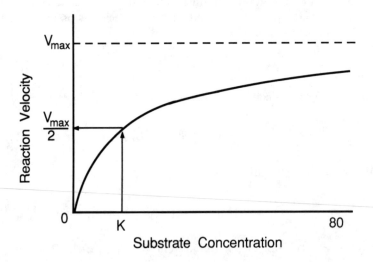

Figure 2.1. Typical hyperbolic curve of enzyme reaction rate as a function of substrate concentration.

Equation 2.18 is the Michaelis-Menten equation described from consideration of saturation, and defines the hyperbolic relationship seen in Figure 2.1. Substrate concentration $[S]$ is usually expressed as moles liter^{-1}, and reaction velocity, v, is usually expressed as moles of product minute^{-1}. Temperature, pH, and enzyme concentration are assumed to be constant for a system described with a particular model equation. If enzyme concentration is increased, both v and V_{max} will be found to increase in direct proportion. K_m is defined as being equal to the substrate concentration when $v = V_{max}/2$, and thus has units of moles liter^{-1}. A key feature of this model is that $V_{max}/2$ is always reached at the same substrate concentration regardless of the value of V_{max}. K_m is independent of enzyme concentration.

Equation 2.18 may be transformed to an equation for a straight line by taking the reciprocal of both sides,

$$\frac{1}{v} = \frac{K_m + [S]}{V_{max}[S]} \tag{2.19}$$

and then rearranging to obtain

$$\frac{1}{v} = \frac{K_m}{V_{max}} \cdot \frac{1}{[S]} + \frac{1}{V_{max}}$$

With this form of the equation, it is a little easier to see that a straight line should result from plotting the reciprocal of velocity (y-axis) as a function of the reciprocal of substrate concentration (x-axis). The y-intercept will be $1/V_{max}$, and the slope will be K_m/V_{max}. In actual laboratory practice, $1/v$ is plotted against $1/[S]$ as one method of estimating K_m and V_{max}.

Exercise 2-1: Write a BASIC program using the Michaelis-Menten model (Equation 2.18) to simulate the initial reaction velocity of an enzyme-activated reaction for different substrate concentrations. Your simulation should produce a plot showing initial velocity for substrate concentrations from 0 to 80 mM. Use values of K_m =10 mM, and V_{max} =0.10 mM min^{-1}.

2.3 Generalized Saturation Models

The basic concept encountered in the enzyme catalytic process described above is not unique. Fundamentally it involves a processing machine (enzyme molecule) which converts a resource material (substrate) into another form (product). The enzyme is effective only during the time it has contact with the substrate. As the concentration of substrate is increased, the enzyme becomes more effective until its maximum effectiveness is reached at saturation, when it has contact with the resource 100% of the time. A reasonable analogy to this process is a machine tool at a factory. The tool can turn out a product only as rapidly as it is supplied with raw stock. As the supply of raw stock is increased, the machine will become increasingly busy. At its maximum efficiency, it will be working 100% of the time. The mathematical description of this process would produce an equation similar to the Michaelis-Menten equation.

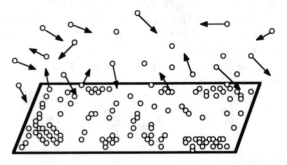

Figure 2.2. Diagram of the Langmuir model of the kinetics of molecular absorption on a surface.

Irving Langmuir (1916) approached the general saturation process similarly. Langmuir was concerned with a description of condensation or absorption of gas molecules on an absorbing surface (Figure 2.2). He assumed that gas molecules would be absorbed in a layer only one molecule thick, so that the number of absorption sites was limited. He next assumed that the rate of absorption depended on the fraction of the surface which was free and available. The rate also was assumed to depend on the concentration of gas (= gas pressure). The following equation then describes the absorption process:

$$\text{rate of absorption} = k_1(1 - \phi)P \qquad (2.21)$$

where k_1 is a rate constant, ϕ is the fraction of surface covered by absorbing molecules with $(1 - \phi)$ the fraction of free surface, and P is the gas pressure. Langmuir assumed that the gas molecules could evaporate or be desorbed from the surface, and that the rate of desorption depended on the fraction of surface that was covered:

$$\text{rate of desorption} = k_2\phi \qquad (2.22)$$

with k_2 the rate constant for evaporation. Then Langmuir made the key assumption that at equilibrium, these two rates would be equal, so that the system might be described with

$$k_1(1 - \phi)P = k_2\phi \tag{2.23}$$

which can be solved for ϕ:

$$\phi = \frac{k_1 P}{k_2 + k_1 P} = \frac{P}{(k_2/k_1) + P} \tag{2.24}$$

Because k_1 and k_2 are constants, they may be replaced by a single constant K. ϕ is given by a/A_T, where a is the area of surface covered by gas molecules, and A_T is the total area. Thus, Equation 2.24 may be written as

$$a = \frac{A_T P}{K + P} \tag{2.25}$$

This equation is strikingly similar to the Michaelis-Menten model (Equation 2.18). The concept of limiting sites and saturation is involved in both models, and the assumption of a stable state is the basis of both formulas.

2.4 A Saturation Model of Predation

A useful model of predation can be developed in the same manner as the molecular models of Langmuir and Michaelis-Menten. Assume a limited number of predators, equivalent to absorption sites or enzyme molecules. In this model the predators exist in two states: satiated and hungry. Satiated predators have captured and consumed a prey, and do not hunt. Hungry predators have digested and assimilated their recent prey and so are unsatiated and hunting for prey. The overall process may be diagrammed:

$$N + P_H \underset{u}{\overset{s}{\rightleftarrows}} P_s$$

where N is prey density, P_H is density of hungry predators, P_S is density of satiated predators. Assuming that the rate of capture and consumption is a function of density of prey and of hungry predators, then

$$\text{rate of satiation} = sNP_H \tag{2.26}$$

Note that this is a second-order process (see Section 1.8), where s is a rate constant for capture and consumption. The rate of return of satiated predators to the pool of hungry, hunting predators is given with

$$\text{rate of unsatiation} = uP_S \tag{2.27}$$

This is a first-order process, where u is the rate constant for unsatiation. For a given constant number of prey, this system should reach a steady-state condition in which rate of unsatiation equals rate of satiation:

$$uP_S = sNP_H \tag{2.28}$$

The total predator population is $P = P_H + P_S$, then

$$uP_S = sN(P - P_S) \tag{2.29}$$

Collecting terms and solving for P_S gives

$$P_S = \frac{NP}{(u/s) + N} \tag{2.30}$$

If the constants for satiation and unsatiation are combined into a single constant k, then we obtain

$$P_S = \frac{PN}{k + N} \tag{2.31}$$

This equation should be familiar to you since it has the form of Equations 2.18 and 2.25. Because of this, the hyperbolic relationship of Figure 2.1 will apply.

The relationship of Equation 2.31 can be used to find predation rates assuming steady-state conditions. For a finite, given time interval, change caused by predation on a prey population can be described with a restatement of Equation 2.26:

$$\frac{\Delta N_p}{\Delta t} = sNP_H \tag{2.32}$$

where ΔN_p is the change in prey numbers due to predation that occurs over a period of time Δt

Since $P_H = P - P_S$, and P_S is defined in Equation 2.31, we can combine the equations and write the following:

$$\frac{\Delta N_p}{\Delta t} = sN\left(P - \frac{PN}{k + N}\right)$$

$$= sNP\left(1 - \frac{N}{k + N}\right)$$

$$= sP\frac{kN}{k + N} = P\frac{uN}{k + N} \tag{2.33}$$

This hyperbolic relationship was recognized by Holling (1959), and was further investigated by Tanner (1975). The model is a useful one which allows us to find loss of prey due to predation as a function of total predator numbers and prey numbers:

$$\Delta N_p = P \frac{uN}{k + N} \Delta t \qquad (2.34)$$

This equation will be discussed further when predation is examined as a component of population dynamics in Chapter 7.

Exercise 2-2: As an example of a predator-prey interaction, suppose a wolfpack with a constant number of 15 individuals is preying exclusively on a population of moose in a large, confined area, such as an island. The moose population fluctuates in size from year to year, and may be as large as 1000 individuals, and as small as 50 individuals. For a particular year when the average number of moose was 500, the pack of 15 wolves is known to have killed 125 moose. For wolf-moose interactions, independent estimates have shown that the value of $k = 200$ moose. With this information and Equation 2.34, write a program that produces a graph showing the relationship between moose population numbers and the number of moose lost to wolf predation. (Note that you will have to solve for the value of the constant u before performing the simulation.)

2.5 Equilibrium Model of Weak Acid Ionization

Ionization of weak acids may be described by the following chemical equilibrium:

$$HA \; \rightleftarrows \; H^+ + A^-$$

where HA is any ionizable weak acid, H^+ refers to hydrogen ions and A^- refers to the conjugate base of the weak acid. This expression produces the following equation for equilibrium:

$$K = \frac{[H^+][A^-]}{[HA]} \qquad (2.35)$$

where K is the equilibrium constant. $[H^+]$, $[A^-]$, and $[HA]$ represent the concentrations of H^+, A^-, and HA at equilibrium. If we solve for $[H^+]$ we obtain

$$[H^+] = K \frac{[HA]}{[A^-]} \qquad (2.36)$$

By taking common logarithms of both sides and multiplying by -1, we get

$$-\log[H^+] = -\log K + -\log\frac{[HA]}{[A^-]} \qquad (2.37)$$

Using common definitions of chemistry,

$$-\log[H^+] = \mathrm{pH} = \mathrm{pK} + \log\frac{[A^-]}{[HA]} \qquad (2.38)$$

The general form of this equation is called the Henderson-Hasselbalch equation:

$$\mathrm{pH} = \mathrm{pK} + \log\frac{[\mathrm{Base}]}{[\mathrm{Acid}]} \qquad (2.39)$$

For the purpose of simulating this equilibrium equation, it is useful to express it in relation to a single variable F_B, the fraction of the acid which is in the base form:

$$\mathrm{pH} = \mathrm{pK} + \log\frac{F_B}{1 - F_B} \qquad (2.40)$$

Exercise 2-3: Write a BASIC program that uses Equation 2.40 to simulate the titration of a weak mono-basic acid with a strong base. For formic acid, which has a pK $= 3.75$, find pH as the fraction of base is increased from 0.01 to 0.99 in increments of 0.01. Your graph of the resulting output will be a simulation of the titration curve for this acid, with pH as a function of the fraction of added base.

A programming note: It is possible to write a BASIC program for this exercise using a FOR...NEXT loop with STEP 0.01. However, with many dialects of BASIC on many computers, it will be somewhat more accurate to use a FOR...NEXT loop with integer steps and obtain the needed variable as a fraction of the counting integer. The reason for doing this is discussed in Chapter 4.

2.6 Effect of pH on Activity of Enzymes

A model of the typical pH optimum curve for enzyme action may be based on the following assumptions. The interaction between an enzyme and its substrate requires that the groups of the enzyme active site are in particular state of ionization. For example, if a carboxylic acid group of the enzyme were involved, only one form of the group would make an effective bond with the substrate (R-COO- but not R-COOH, or vice versa). Once the enzyme-substrate compound is formed, only one particular ionized form undergoes a reaction to produce the product. Finally, the relative rate of formation of product depends on the fraction of total enzyme which is present as the enzyme-substrate compound in the proper ionization state.

These concepts are embodied in the following overall equilibrium reaction discussed in Mahler and Cordes (1966):

$$
\begin{array}{ccccc}
 & & A & & \\
 & & + & & \\
EH_2 & \underset{\longleftarrow}{\overset{K_1}{\longrightarrow}} & EH^- + H^+ & \underset{\longleftarrow}{\overset{K_2}{\longrightarrow}} & E^= + 2H^+ \\
 & & \updownarrow K_a & & \\
AEH_2 & \underset{\longleftarrow}{\overset{K_3}{\longrightarrow}} & AEH^- + H^+ & \underset{\longleftarrow}{\overset{K_4}{\longrightarrow}} & AE^= + 2H^+ \\
 & & \downarrow k & & \\
 & & P & &
\end{array}
$$

In this system, EH_2, EH^- and $E^=$ are variously ionized forms of the enzyme, of which only EH^- is effective in binding with the substrate A to form the enzyme-substrate compound. AEH_2, AEH^- and $AE^=$ are the variously ionized forms of the enzyme-substrate compound; of these, only AEH^- breaks down to yield the product P. K_1, K_2, K_3, and K_4 are ionization constants, K_a the dissociation constant for the enzyme-substrate compound, and k is the rate constant for product formation.

Because the rate of formation of product is a function of AEH^- concentration, the maximum rate of formation would be achieved if the enzyme were all in the form of AEH^-. The fraction of maximum rate can be calculated by finding how much enzyme is in this form relative to the total amount of enzyme. This value is called the saturation fraction:

$$
Y = \frac{[AEH^-]}{[E^=] + [EH^-] + [EH_2] + [AE^=] + [AEH^-] + [AEH_2]} \tag{2.41}
$$

where Y is the fraction of enzyme saturated as AEH^-, $[AEH^-]$ is the concentration of AEH^-, and the denominator of Equation 2.41 represents the sum of the possible states of the enzyme.

To show the relationships among these forms, we first define

$$\alpha = \frac{[A]}{K_a} \qquad (2.42)$$

Then each of the concentrations in Equation 2.41 is related to EH^- in the following manner:

$$K_a = \frac{[A][EH^-]}{[AEH^-]} \qquad [AEH^-] = \frac{[A][EH^-]}{K_a} = \alpha[EH^-]$$

$$K_1 = \frac{[EH^-][H^+]}{[EH_2]} \qquad [EH_2] = \frac{[EH^-][H^+]}{K_1}$$

$$K_2 = \frac{[E^=][H^+]}{[EH^-]} \qquad [E^=] = \frac{K_2[EH^-]}{[H^+]}$$

$$K_3 = \frac{[AEH^-][H^+]}{[AEH_2]} \qquad [AEH_2] = \frac{[AEH^-][H^+]}{K_3} = \frac{[A][EH^-][H^+]}{K_a K_3}$$

$$= \frac{\alpha[EH^-][H^+]}{K_3}$$

$$K_4 = \frac{[AE^=][H^+]}{[AE^-]} \qquad [AE^=] = \frac{K_4[AEH^-]}{[H^+]} = \frac{K_4[A][EH^-]}{K_a[H^+]}$$

$$= \frac{\alpha K_4[EH^-]}{[H^+]}$$

By substitution, we can find

$$Y = \frac{\alpha[EH^-]}{[EH^-]\left(1 + \frac{K_2}{[H^+]} + \frac{[H^+]}{K_1} + \frac{\alpha K_4}{[H^+]} + \frac{\alpha[H^+]}{K_3} + \alpha\right)} \qquad (2.43)$$

$$= \frac{\alpha}{1 + \frac{K_2}{[H^+]} + \frac{[H^+]}{K_1} + \alpha\left(1 + \frac{K_4}{[H^+]} + \frac{[H^+]}{K_3}\right)} \qquad (2.44)$$

Note that K_a, the dissociation constant of the enzyme-substrate compound, is a function of the two rate constants. As such, it is related to the Michaelis-Menten constant discussed in Section 2.2:

$$K_a = \frac{[A][EH^-]}{[AEH^-]} = \frac{k_2}{k_1}$$

From Equation 2.15, we know that $K_m = (k_2 + k_3)/k_1$. Thus, $K_m \cong K_a$ when $k_3 \ll k_2$. The ratio represented by α (Equation 2.42) is therefore very close to being the ratio of substrate concentration $[A]$ to K_m. When $[A] = K_m$ (so that $\alpha \cong 1$), then $v \cong V_{max}/2$.

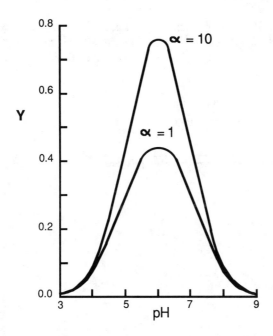

Figure 2.3. Graph of data from simulation of an optimum pH value for enzyme kinetics.

Exercise 2-4: The velocity of an enzymatic reaction is strongly dependent on pH. Write a program using the model of enzyme-pH interaction of Equation 2.44 to determine the saturation fraction (essentially the fraction of V_{max}) at different values of pH. Run your program with two different concentrations of substrate relative to the dissociation constant of the enzyme-substrate compound (i.e. with $\alpha = 1$ and $\alpha = 10$). Use these ionization constants in your simulation:

$$K_1 = 0.0002 \quad K_2 = 0.000001$$
$$K_3 = 0.0004 \quad K_4 = 0.00002$$

The graph of the output from your program should show a plot of Y as a function of pH from 0.1 to 14.0 in increments of 0.1

pH units. (See note to Exercise 2-3.) Your program will have to convert pH values to $[H^+]$ and then find the corresponding Y. It will be instructive for you to show the curves for the simulations with both values of α on the same graph (see Figure 2.3).

Exercise 2-5: The model of the effect of pH on enzyme activity presented above does not consider that the substrate may also be a weak acid, which must be in a particular ionized form to interact with the enzyme:

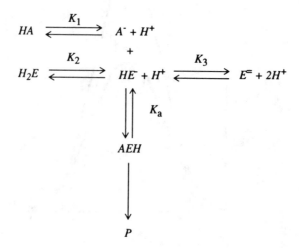

Using the equations provided below, derive the equation for saturation fraction, Y, in terms of α, K_1, K_2, K_3, and $[H^+]$.

$$K_1 = \frac{[A][H^+]}{[HA]} \qquad K_2 = \frac{[HE^-][H^+]}{[H_2E]} \qquad K_3 = \frac{[E^=][H^+]}{[HE^-]}$$

$$K_a = \frac{[HE^-][A]}{[AEH]} \qquad \left[\alpha = \frac{[\Sigma A]}{K_a} = \frac{[A]+[HA]}{K_a}\right]$$

Exercise 2-6: Following the format of Exercise 2-4, write a program for the model you derived in Exercise 2-5. Use these ionization constants: $K_1 = 10^{-6}$, $K_2 = 10^{-7}$, $K_3 = 10^{-8}$. Your output should include a graph of Y as a function of pH. Use values for α of 1 and 10.

2.7 A Concerted Model for Allosteric Enzymes

One of the most important concerns of modern biochemical study is the control of reaction pathways. Some of this control is exerted on the synthesis of enzymes, and some of it is exerted on existing enzymes in such a way that their activity as catalysts is modulated by other metabolites. This latter form of control results when certain metabolites interact with the enzyme molecule at sites other than the catalytic site. The process is therefore called allosteric (other site) control, and the enzymes which respond to this type of control are called allosteric enzymes. A model describing allosteric enzymes was proposed by Monod et al. (1965). This model and others have been reviewed by Newsholme and Start (1976) and by Savageau (1976).

Most allosteric enzymes have been found to consist of multiple subunits each having a catalytic site and one or more allosteric control sites. The typical curve relating a substrate concentration to allosteric enzyme activity or reaction velocity has a sigmoid shape rather than the hyperbolic shape exhibited by other enzymes. This observation supports the assumption that there is strong cooperation between enzyme subunits. The Monod model assumes that the subunits are identical, and are held together in a symmetrical fashion. It also assumes that the enzyme exists in two configurations, one catalytically active, and the other inactive. An equilibrium is assumed to exist between the active and inactive forms. Because of the symmetry of the molecule, all subunits are assumed to shift from one configuration to the other in a concerted fashion. Hence this is called the concerted or symmetrical model.

Any ligands that bind with the active configuration tend to shift the equilibrium in a direction which increases the proportion of enzyme that is catalytically active. Substrate molecules and positive modulators both cause this to occur. On the other hand, negative modulators bind with the inactive enzyme and have the effect of reducing the proportion of enzyme in the active form, thus reducing its overall catalytic activity. The following model is derived from the simple case of an allosteric enzyme composed of two subunits. Effects of modulators other than substrate are ignored.

We will designate the active form of the allosteric enzyme as E and the inactive form as F. These exist in a state of equilibrium:

$$E \; \rightleftharpoons \; F$$

The equilibrium constant L is defined as

$$L = \frac{[F]}{[E]} \tag{2.45}$$

The enzyme interacts with the substrate S to form two species of ES, one for each of the two possible sites that may be occupied by the substrate molecule:

$$ES \; \rightleftharpoons \; E + S$$

$$ES' \; \rightleftharpoons \; E + S$$

$$ES_2 \; \rightleftharpoons \; E + S$$

The free enzyme may be defined by the following expressions where K is the dissociation constant, assumed to be equal for all species:

$$K = \frac{[E][S]}{[ES]} = \frac{[E][S]}{[ES']} = \frac{[ES][S]}{[ES_2]} \qquad (2.46)$$

To simplify some of the following expressions, substrate concentration $[S]$ may be expressed in terms of the dissociation constant K in this fashion:

$$\alpha = \frac{[S]}{K} \qquad (2.47)$$

The saturation fraction Y is defined in the following equation in terms of the concentration of bound substrate and total concentration of binding sites available:

$$Y = \frac{[ES] + [ES'] + 2[ES_2]}{2([E] + [ES] + [ES'] + [ES_2] + [F])} \qquad (2.48)$$

This equation is then substituted so that all terms are expressed relative to the concentration of free enzyme, $[E]$. This results in

$$Y = \frac{[E]\alpha + [E]\alpha + 2[E]\alpha^2}{2([E] + [E]\alpha + [E]\alpha + [E]\alpha^2 + [E]L)}$$

$$= \frac{2[E](\alpha + \alpha^2)}{2[E](1 + 2\alpha + 2a^2 + L)} \qquad (2.49)$$

By cancelling terms and factoring, we obtain:

$$Y = \frac{\alpha(1 + \alpha)}{L + (1 + \alpha)^2} \qquad (2.50)$$

The general form describing this relationship for any number of subunits, n, is the following:

$$Y = \frac{\alpha(1+\alpha)^{n-1}}{L + (1+\alpha)^n} \quad (2.51)$$

Positive modulators (activators), A, and negative modulators (inhibitors), I, affect the equilibrium between the active and inactive forms of the enzyme and therefore affect the L term in the equation as follows:

$$Y = \frac{\alpha(1+\alpha)^{n-1}}{\frac{L(1+\beta)^n}{(1+\gamma)^n} + (1+\alpha)^n} \quad (2.52)$$

For this equation, β gives the concentration of the negative modulator or inhibitor, $[I]$, expressed in terms of the dissociation constant for the inhibitor-enzyme complex, K_i. The equation is similar to Equation 2.47:

$$\beta = \frac{[I]}{K_i} \quad (2.53)$$

Similarly, γ gives concentration of positive modulator or activator, $[A]$, in terms of the dissociation constant for the activator-enzyme complex, K_a:

$$\gamma = \frac{[A]}{K_a} \quad (2.54)$$

Inhibitor is assumed to bind to the enzyme only in the inactive form, and activator is assumed to bind to the enzyme only in the active form. Equation 2.52 may be used to model control of most allosteric enzymes, and is also useful for modeling control of multi-enzyme systems.

Exercise 2-7: Write a program for simulating activity of an allosteric enzyme with four subunits and four active sites. Use Equation 2.52 to determine values of Y as α is varied from 0 to 25 in increments of 0.5. (See note in Exercise 2-3.) Run your simulation for the following three conditions:

(a)	$L = 1000$	$\beta = 0$	$\gamma = 0$
(b)	$L = 1000$	$\beta = 1.7$	$\gamma = 0$
(c)	$L = 1000$	$\beta = 0$	$\gamma = 100$

The curve for condition (a) should show a sigmoid shape characteristic of homotrophic effects in the absence of activation or inhibition. The curves for conditions (b) and (c) should show marked heterotrophic effects due to the positive and negative modulation. It will be instructive for you to have all three curves plotted on the same graph.

2.8 Predatory Maintenance of Polymorphism at Equilibrium

In some predator-prey systems, a prey species may exist in several genetically determined forms having different appearances or behaviors. Some forms will be more open to attack than other forms by being more conspicuous, slower, less alert, tastier, etc. These attractive forms will be removed from the prey population more rapidly as a result. However, predation rate also depends on the frequency of prey (see Equation 2.32), and a scarce form of prey will not be taken very often.

In a simple predator-prey system with two forms of prey and a single predator, the more attractive form will be taken rapidly until it becomes relatively scarce. Then the less attractive prey will be more abundant, and will be taken more often. An equilibrium of predation on the two forms should be established, based on the relative attractiveness of the two forms of prey, and on their relative abundance. The following formulation of this equilibrium is based on Emlen (1973).

The "fitness" or ability of a form of the prey to avoid predation can be given by

$$W_i = c_i(1 - f_i) \tag{2.55}$$

where W_i is fitness, c_i is a measure of non-attraction, and f_i is the relative frequency of form i. As one form of prey becomes more abundant, its frequency will increase and it will be more subject to predation; hence its fitness will decrease. For a prey population with two forms, at equilibrium $W_1 = W_2$. Because $f_1 = 1 - f_2$,

$$\frac{f_1}{f_2} = \frac{c_1}{c_2} \tag{2.56}$$

and at equilibrium, the frequency of the two forms depends on their relative attraction. For convenience, we can define C as a measure of relative attraction:

$$C = \frac{c_1}{c_2} \tag{2.57}$$

C is also a predictive measure of the ratio of the two forms at equilibrium.

In the simple system, suppose the two forms of prey are produced by a simple dominant-recessive genetic system, with the AA and Aa genotypes producing prey form 1 and the aa genotype producing form 2. Elementary population genetics (see Chapter 10) describes the frequency in the prey population of the A gene as p and the frequency of the a gene as $q = 1 - p$. The frequency of the AA genotype is p^2, of the Aa is $2pq$, and of the aa is q^2. Because there are no other A/a-genotypes in the population,

$$p^2 + 2pq + q^2 = 1 \tag{2.58}$$

Because f_1 is the frequency of $(AA + Aa)$ and f_2 is the frequency of aa, at the equilibrium point for predation,

$$C = \frac{f_1}{f_2} = \frac{p^2 + 2pq}{q^2} \tag{2.59}$$

To find the frequency of the A gene (p) in the prey population at some point of equilibrium determined by attractiveness to predators, Equations 2.58 and 2.59 may be combined and solved to give

$$p = 1 - \frac{(C+1)^{1/2}}{(C+1)} \tag{2.60}$$

This expression provides a measure of the ability of predation to modify the genetic makeup of a population, and specifically to maintain morphological variation in a population.

Exercise 2-8: Use Equation 2.60 in a program to find p, the frequency of a simple dominant gene A, in a population at predatory equilibrium between the two dominant and recessive forms. Allow the susceptibility ratio C to vary from 0.01 to 100. Plot your results with C on the x-axis and p on the y-axis. Given the wide range of values for C, they should be plotted as common logarithms (from -2 to $+2$). To provide adequate resolution over the interval 0.01 to 1.0, values of p should be found for increments of 0.01 of C (see note in Exercise 2-3). For values of C from 1 to 100, increments of 1.0 provide satisfactory resolution.

Conclusion

Models involving assumptions of equilibrium or steady-state can be used to simulate directly those biological phenomena known to exist in these stable conditions. The models become less useful the further or longer the systems stray from stability. Because of this, the models are not as broadly useful as those involving the solution of differential equations, and they typically are used where stable states are obtained relatively rapidly. In later chapters equilibrium models will be combined with the use of differential equations to form other simulation models.

Many of the derivations in this chapter depended on the use of rate equations. With the stability approach, the differentials are eliminated when rates of formation are set equal to rates of breakdown. The use of rate equations is fundamental to the development of many models useful in biological simulation, and examples will be found in subsequent chapters.

CHAPTER 3

ESTIMATING MODEL COEFFICIENTS FROM EXPERIMENTAL DATA

In the two previous chapters you learned how model equations may be derived analytically from theory. You may have noticed that the discussion covered only the form of the equation. When you actually worked with the models in the exercises, you used not only the equations, but also some coefficients that were simply given to you without explanation. In this chapter we describe how to find these coefficients.

At the outset of this discussion of models and data, you should understand clearly the differences between theoretical models and empirical models. The equation for a theoretical model is obtained from ideas about how some biological process works. These ideas may be based only on a loose analogy to known processes, but they still lead to an analytical derivation of an equation. The models in the first two chapters were derived from theory.

Empirical equations, on the other hand, do not have a theoretical origin. They are used when the modeler is not interested in the cause-effect relationship of variables used in the equation, but wishes to use the output as part of a larger model. For example, suppose you are making a rather complicated model of plankton populations in a lake, and you need an equation that describes the density of water at different temperatures. There are, of course, many theoretical models for the effect of temperature on water density, but in constructing the plankton model it may not be essential to know this theoretical basis. In such a case, a usable equation and coefficients may be obtained directly from experimental data. The most commonly used empirical equations are of the polynomial type, and their method of derivation is discussed below. Like theoretical models, empirical models require accurate coefficients if they are to be useful.

From a mathematical viewpoint, an empirical equation and a theory-based equation are equally useful for modeling a particular system, assuming they are equally accurate. However, most modelers would prefer to use equations that have some basis in theory, because this will contribute more to an understanding how the system functions.

In general, coefficients for both theoretical and empirical models are obtained by the process of "curve fitting". In the case of theoretical models, we first formulate the model from our conception of how the system might work, next we plot data from the system, and then we try to find coefficients for our equation so that the line drawn by our equation best "fits" the actual data. With empirical equations, we plot the data and then look simultaneously for an equation and coefficients that produce a best-fitting line. Notice the difference in sequence. Thus the sequence for theoretical models is equation, then data, then coefficients, while for empirical models it is data, then equation and coefficients. Because these differences are fundamental, this chapter will be divided into two parts, the first considering theoretical models (Sections 3.1 through 3.6), and the second empirical models (Sections 3.7 and 3.8).

3.1 Selecting a Model Equation

In practice, the sequence for developing theoretical models is sometimes scrambled. We may not have available the knowledge or insight required to formulate an analytical model directly. Unlike our model for cell growth, built in Chapter 1 from ideas about how cells grow, in some cases we may need to collect data to get an idea about how the system works. There is no real "Start Here" point in the overall research process of Figure I.1. Thus, data may suggest an equation which can give us a conceptual notion of the workings of the process.

A first step in working with a theoretical model, even if you are not sure of its final mathematical form, is to plot the data to produce a scatter plot, also referred to as an approximating curve. The shape of this curve will indicate whether you developed your initial model correctly, or it may suggest some equations that describe the phenomenon of interest. Some characteristics of the approximating curve will help in selecting an equation or determining whether an analytical equation is likely to be correct. These are: linearity, presence of minima or maxima, direction of slope, and presence or absence of inflection points. A very large number of equations can be formulated to describe any particular line. In practice, we usually select an equation from several simple equations that have proven useful in biological explanation in other cases that are more or less similar.

For example, suppose we wished to model the increase of biomass in a cell culture based on the following hypothetical data:

Time	Biomass
0	9.9
1	32
2	89
3	271
4	808

When these data are plotted, they appear to follow an exponential growth curve. As you know from Chapter 1, the curve is described by the equation

$$y = Ae^{Bx} \tag{3.1}$$

with x representing elapsed time, A the biomass at $x = 0$, and B the rate of growth. The general problem of this chapter is to find the best values for the coefficient A and the exponent B. The two prevalent approaches to this problem are the use of linear transformation, and nonlinear regression. The techniques for these procedures are discussed in the following sections.

3.2 Linear Transformation Procedures

The technique of linear transformation takes advantage of the fact that many useful equations may be rearranged to a linear form. Experimental data that conform to the equation may then be transformed and fitted to straight line of the transformed equation by the usual linear least-squares methods. The slope and intercept of the resulting straight line equation may be used to find estimates of the values of the coefficients and exponents of the original equations.

In the example case above, the exponential growth equation may be converted to linear form by taking the natural log of both sides to give

$$\ln(y) = \ln(A) + Bx \tag{3.2}$$

Now, if values of y from the data are transformed to $\ln(y)$ and are plotted against time x, the points should fall on a straight line that has a slope of B and an intercept of $\ln(A)$. (You will recall from elementary algebra the expression for a straight line, $y = i + sx$, where i and s are the intercept and slope.) This transformation from the exponential is frequently used to find the growth rate of microorganisms. In this example, the transformed data become

Time	Biomass	log(Biomass)
0	9.9	2.3
1	32	3.4
2	89	4.5
3	271	5.6
4	808	6.7

When the log(Biomass) data are plotted against time we obtain the straight line shown in Figure 3.1.

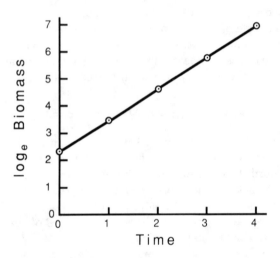

Figure 3.1. Graph of hypothetical data for growth.

The estimates of intercept and slope (A and B) may be derived from the linear plot, and Equation 3.1 may be used to estimate growth as in Exercise 1-1.

It is helpful to look at the values for biomass (y) that were observed and at values that are calculated using the estimates of A and B:

	observed	calculated	
Time	y	y	difference
0	9.9	10	-0.1
1	32	30	-2
2	89	90	+1
3	271	270	-1
4	808	810	+2

The differences, also called residuals or deviations, appear to be relatively small and randomly distributed; the exponential equation with the fitted values of A and B appears to be a reasonable predictor of growth for the organisms. In linear regression, the residual values are important in assessing the fit of the equation to the data.

3.3 Notes on Regression with Linear Transformation

Linear regression should be familiar to all biologists, and is discussed thoroughly in most elementary texts of biological statistics (e.g. Zar 1984). The line obtained with the linear regression procedure minimizes the sum of the squares of the vertical distances (residuals) between the data points and the regression line. In addition to values for slope and intercept, several other statistics based on the residuals are usually obtained from a regression analysis to determine how well the line fits the data.

In the usual linear regression procedure, the fit of the line is assessed from the residuals. While this is also the case for transformed data, the residuals are calculated after back-transformation, using the nonlinear form of the equation. Because you will use the procedure in several exercises in this chapter, it will be mentioned here briefly. Consult your favorite biometry textbook for further discussion.

The residual sum of squares is found from the differences between the observed and predicted values of y, and is given by

$$SS_{\text{residual}} = \sum (y_i - \hat{y}_i)^2 \tag{3.3}$$

with the predicted and observed values derived from the actual data and the theoretical equation. The variance, or residual mean square, is given with

$$MS_{\text{residual}} = \frac{SS_{\text{residual}}}{n - P} \tag{3.4}$$

where n is the number of $x - y$ pairs and P is the number of parameters (coefficients and exponents) obtained by the fitting procedure. In most cases of linear transformation, P has the value of two, for A and B. The standard error of the regression estimate is the square root of the residual mean square.

An F-statistic, used with an F-table to assess whether equations explain significant amounts of variation between the y-values and the overall mean y, is found with

$$F = \frac{SS_{\text{regression}}}{SS_{\text{residual}}} \cdot \frac{n - P}{P - 1} \tag{3.5}$$

Note again that these statistics are found using the residuals from values predicted with the original, untransformed equation. They will differ from statistics calculated from the residuals of the transformed data and the linear equation.

3.4 Useful Equations for Linear Transformation

A variety of equations that produce a diversity of curve shapes may be linearized with transformation procedures; see Hoerl (1954) for examples. A number of these equations that have been useful in biological research are given in Table 3.1. Each equation is presented in both the nonlinear standard form, and in the linear form.

Some forms of the curves produced by these equations are shown in Figure 3.2.

The values of the parameters producing the curves are given to show the role they play in determining the shape of the curve.

You will notice that the same general shape of curve may be produced by equations with different forms. For example, a negatively accelerated rise to a plateau may be produced by the hyperbolic (Eq. 2), reciprocal exponential (Eq. 5), and exponential saturation (Eq. 7) equations. Thus, before undertaking a curve-fitting analysis you should have a good idea about the mechanism that may produce your results.

Several commercial programs for large and small computers are available that can transform sets of x-y data and perform a linear regression for several of the equations in Table 3.1. We have incorporated these curvilinear equations, along with the straight-line equation into a BASIC program, CURFIT. The program can assess the fit of sets of x-y data to these equations, and determine the best-fit coefficients. The procedures used in running this program are described in Appendix 3. The program is designed to make it easy to perform the different transformations of data, and to assess the goodness of fit of the back-transformed data, as described above.

If the specific equation is not known from theoretical equations, the CURFIT program permits a quick comparison among several that might describe the set of data. (If an equation is selected in this manner, you should not attach a great deal of weight to the biological mechanism implied by the equation.) The shape of the curves in Figure 3.2 may help you to decide which equations to select in fitting data with the CURFIT program.

Different statistics are used to compare and assess the relative goodness of fit among several equations. It is generally held that the best fitting equation will have the minimum residual mean square. This statistic is better than the residual sum of squares because it takes into account the different degrees of freedom associated with different model equations. Other statistics which may be employed to evaluate goodness of fit include the F-statistic and the standard error of the regression estimate.

The program finds the coefficient of determination, R^2, for the linear equation. This value, which has a maximum of 1, compares variance of the predicted y-values with the variance of the actual y-values. This coefficient is not a direct function of fit when nonlinear equations are involved. In fact, for nonlinear equations, higher R^2 values may be obtained from equations that are obviously poorly fit. Also, values of R^2 that exceed 1.0 may be obtained when the variance of data predicted by the nonlinear model actually exceeds that of the original data.

3.5 Notes on Some of the Equations

Several of the equations listed in Table 3.1 and Figure 3.2 and used in the CURFIT program deserve some comment. The hyperbolic equation, for example, should be familiar from our discussion of Michaelis-Menten kinetics in Chapter 2. The equation may be linearized in a variety of ways. The usual Lineweaver-Burk plot (see Figure 3.3) is the poorest of the possibilities (Dowd and Riggs 1965). It is based on the following equation:

$$\frac{1}{y} = \frac{1}{A} + \frac{(B/A)}{x} \tag{3.6}$$

The Woolf transformation (Table 3.1) used in CURFIT is a better estimator of the parameter values. Raaijmakers (1987) provides a recent review of the problems in estimating enzyme kinetics with the equations.

Several of the curves used in CURFIT require an estimate for a constant used in the equation. These are usually an asymptote or an intercept of the approximating curve. A trial value to enter to begin the analysis can be obtained by a visual approximation of the value from a rough curve. From the initial value, the curve-fitting process may be repeated to find the value for the constant that maximizes the value of the F-statistic or produces the minimal residual mean square.

The exponential saturation equation (Eq. 7 in Table 3.1) appeared in Chapter 1 as the Von Bertalanffy growth model. The equation requires an estimated value for A in order to calculate the slope ($\ln A$) and intercept of the linear form of the equation. Note that A-value appears twice in the equation. If you select this equation to fit to your data, you may have to repeat the curve-fitting procedure several times so that your estimate of A equals the computed value.

Equation 10, labelled "sigmoid", can in fact produce a variety of curve shapes other than sigmoidal. This is shown in Figure 3.2. This equation can have rather high values of F for a variety of data sets. (This provides another argument for knowing which equation is likely to explain the underlying mechanism of your system.)

No.	Name	Standard Function	Linear Transformation	Definition of Parameters by Slope (S) and Intercept (I)		
				A	B	K
1.	straight line	$y = A + Bx$	$[y] = A + B[x]$	S	I	
2.	hyperbolic	$y = Ax/(B + x)$	$[x/y] = (B/A) + (1/A)[x]$	1/S	I/S	
3.	modified inverse	$y = A/(B + x)$	$[1/y] = (B/A) + (1/A)[x]$	1/S	I/S	
4.	exponential	$y = A \cdot \exp(Bx)$	$[\log_e y] = \log_e A + B[x]$	exp(I)	S	
5.	exp. reciprocal	$y = A \cdot \exp(B/x)$	$[\log_e y] = \log_e A + B[1/x]$	exp(I)	S	

6.	maxima function	$y = Ax \cdot \exp(Bx)$	$[\log_e(y/x)] = \log_e A + B[x]$	exp(I)	S
7.	exp. saturation	$y = A[1 - \exp(Bx)]$	$[\log_e(A - y)] = \log_e A + B[x]$	exp(I)	S
8.	logistic	$y = K/[1 + A \cdot \exp(Bx)]$	$[\log_e(K/y - 1)] = \log_e A + B[x]$	exp(I)	S	estimate
9.	logarithmic	$y = K + (Ax^B)$	$[\log_e y] = \log_e A + B[\log_e x]$	exp(I)	S	estimate
10.	"sigmoid"	$y = K/(1 + Ax^B)$	$[\log_e(K/y - 1)] = \log_e A + B[\log_e x]$	exp(I)	S	estimate

Table 3.1. Listing of some equations that occur often in biological and biomedical research, and their linearized forms. The transformed values of x and y are given in brackets in the linear form. The definitions of parameters show how they are derived from the linear form. Equation numbers refer to the CURFIT program.

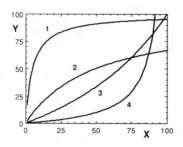

Eq. 2 Hyperbolic   y = Ax/(B + x)

1) A = 100   B = 5
2) A = 100   B = 50
3) A = -100   B = -200
4) A = -10   B = -101

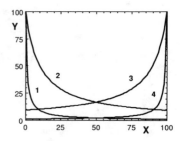

Eq. 3 Modified Inverse   y = A/(B + x)

1) A = 100   B = 1
2) A = 1000   B = 10
3) A = -1000 B = -110
4) A = -100   B = -101

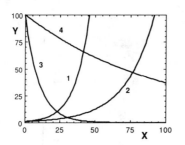

Eq. 4 Exponential y = A exp(Bx)

1) A = 1   B = 0.1
2) A = 1   B = 0.05
3) A = 100   B = -0.1
4) A = 100   B = -0.01

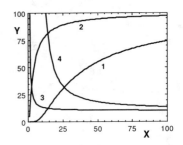

Eq. 5 Reciprocal Exp.   y = A exp(B/x)

1) A = 100   B = -30
2) A = 100   B = -3
3) A = 10   B = 3
4) A = 10   B = 30

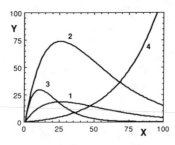

Eq. 6 Maxima Function   y = Ax exp(Bx)

1) A = 2   B = -0.04
2) A = 8   B = -0.04
3) A = 8   B = -0.1
4) A = 0.1   B = 0.025

Figure 3.2. Curves from equations of Table 3.1.

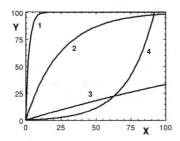

Eq. 7 Exp. Saturation y = A[1-exp(Bx)]

1) A = 100   B = -0.4
2) A = 100   B = -0.04
3) A = 100   B = -0.004
4) A = -1   B = 0.05

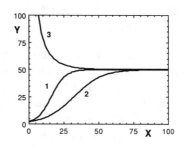

Eq. 8 Logistic y = K/[1+A exp(Bx)]

1) K = 50   A = 25   B = -0.2
2) K = 50   A = 25   B = -0.1
3) K = 50   A = -1   B = -0.1

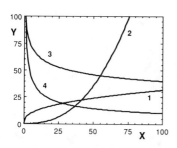

Eq. 9A Log Function   y = K + Ax^B

1) K = 0   A = 5   B = 0.4
2) K = 0   A = 0.002   B = 2.5
3) K = 0   A = 100   B = -0.2
4) K = 0   A = 100   B = -0.5

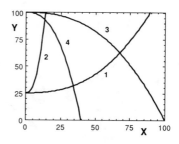

Eq. 9B Log Function   y = K + Ax^B

1) K = 25   A = 0.001   B = 2.5
2) K = 25   A = 0.1   B = 2.5
3) K = 100   A = -0.001   B = 2.5
4) K = 100   A = -0.01   B = 2.5

Eq. 10 'Sigmoid'   y = K/(1 + Ax^B)

1) K = 50   A = 5000   B = -3
2) K = 50   A = 1   B = 0.5
3) K = 50   A = -0.0001   B = 2
4) K = 50   A = -0.04   B = 0.5

Figure 3.2. (continued).

3.6 Practicalities and Problems of Regression with Transformation

Linear regression with transformed data has several attractive features. The procedures are easy to understand, they have a long history in the biological literature, and they can be carried out with pencil and paper with smaller data sets. However, there are some more or less severe statistical difficulties associated with the procedure.

Linear regression is properly performed when the variability in values of y are not related to x or to y. Unfortunately, this is often not the case in biological research, and a transformation may aggravate the problem. In addition, after transformation the data points may have different weights than before transformation.

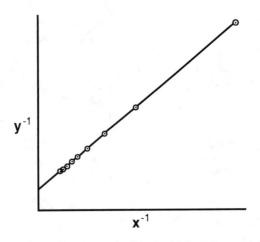

Figure 3.3. Plot of reciprocal of hypothetical data, with uniform spacing of intervals of x. The reciprocal plot results in a non-uniform clustering of data points.

For example, in Figure 3.3, uniformly distributed values of x have been transformed to $1/x$, so that most are clustered at low values of $1/x$. The points with high $1/x$ values have a great influence on the slope of the line, which is almost certain to pass through the extreme point. In the hyperbolic curve, this point has the lowest value of both x and y. This point probably has the greatest percentage error, because it is located on the steepest part of the hyperbolic curve.

Other problems arise from the appearance of negative or zero values in the data. Log transformation of such values is impossible, of course, as is reciprocal transformation of zero values. Approximating zero with a very small number (e.g. 10^{-6}) will distort greatly the impact of the value on the linear regression. Such data may be normalized to positive real numbers by adding a constant to all the affected values before linear analysis. Depending on the equation, this may introduce further bias into the estimates of the coefficients. Alternatively, the data pair may simply be dropped from the analysis.

Problems with the analysis of linearly transformed data may be discovered sometimes by inspecting a plot of the residuals vs. the x-value. (CURFIT provides such a plot.) If the residuals are not distributed uniformly, nonlinearity is indicated. If the residual plot tends to look wedge-shaped, it may indicated the overly strong impact of an extreme point. If the residual plot is U-shaped, it may indicate that the transformed data are not well-described by the particular equation.

Exercises: The following sets of data have been synthesized from standard equations. Fit at least two theoretical equations to each using linear transformation techniques. Decide on the best-fitting equation based on either the residual mean square or the F-statistic.

Exercise 3.1		Exercise 3.2		Exercise 3.3		Exercise 3.4	
X	Y	X	Y	X	Y	X	Y
1	5	1	1	0	100	0	110
2	9	2	4	1	91	1	100
5	20	5	20	2	83	2	92
10	33	10	50	5	67	5	71
15	43	15	69	10	50	10	47
20	50	20	80	15	40	15	32
25	56	25	86	20	33	20	24
30	60	30	90	25	29	25	18
35	64	35	92	30	25	30	15
40	67	40	94	35	22	35	13
45	69	45	95	40	20	40	12
50	71	50	96	45	18	45	11
				50	17	50	10.7

3.7 Nonlinear Regression

The use of linear transformation and regression is limited by the problems outlined above. In addition, many useful equations cannot be made linear with a transformation. Nonlinear regression is a useful alternative to linear transformation, and may provide the only method for working with some equations. The subject will be given a brief introduction here, to permit some elementary work with CURNLFIT, a nonlinear regression program in Appendix 5.

The objective of nonlinear regression is the same as linear regression: to select values of coefficients that minimize the sum of the squared values of residuals. Thus, nonlinear regression is also a least-squares procedure. However, unlike linear and polynomial regression methods, nonlinear regression requires the computer to make several steps to solve for least-squares coefficients. These steps or iterations begin with a first guess about the values of the coefficients. These values are then adjusted to improve the fit of the curve to the data, and then adjusted again. These iterations will proceed until little improvement is noticed in the values. Several methods of iteration to a solution are commonly available in different computer programs for nonlinear regression; Motulsky and Ransnas (1987) provide a clear and helpful discussion of these from a biologist's perspective.

In using any nonlinear regression program, several decisions about the type of data have to be made, because different calculating procedures are used based on the decisions. Obviously, the initial decision will involve the selection of one or perhaps two equations that might describe the relationship between x and y. The data should be scaled so that the units of the data are neither very large nor very small. The regression program will work more accurately with values of 0.1 to 10 micrograms, rather than 0.000001 grams. The user will have to specify accurate estimates for initial values. A poor selection of values will slow up the computer's search for the best estimates, and may send the nonlinear regression program on a hunt for values in the wrong direction, so that it never converges, or perhaps converges on incorrect values. In some cases, reasonable estimates of values can be made by inspecting the data. If possible, values from a regression based on linear transformation (e.g. CURFIT) can be used for initial estimates.

The weighting scheme used with the data will be important. Linear regression assumes that the variability in y is unrelated to the magnitude of x and y. However, in biological research it is often the case that variability increases with increasing y; in many biological procedures, the experimental uncertainty associated with a measured value is related to the size of the value. In this case, data points with large values of y are scattered about the regression line more than points with small y values. In estimating regression coefficients, a least-squares program would emphasize points with larger values of y. To avoid this problem with nonlinear regression, different weights may be given very high and very low values of y, depending upon the nature of the data. The CURNLFIT program offers the option of selecting for an assumption of no weighting, or weighting that assumes a variation proportional to values of y, or an "in between" weighting scheme. If the variability of measurement is assumed to be proportional to y, the program will minimize the sum of squares of the relative distances of the points from the curve, rather than the measured distances. In general, it is probably best to select "no weighting" if you do not have information about the variability of the data.

The problem of how to treat "outlying values" occurs in both linear and nonlinear regression. The point(s) furthest from the line will be the most important in increasing the value of the residual sum of squares. It is tempting simply to discard such values as being mistakes in data collection, and thus to increase the goodness of fit considerably. The problem then is to determine what is a mistake, and what is real variation in the system. A technique that is less biased is to modify the residuals so that less weight is given to outlying values. A variety of such systems have been used. CURNLFIT allows the selection of a bisquare weighting system for outlying values (Duggleby 1981).

In addition to having the equations from CURFIT built into the program, CURNLFIT also permits the user to write into the program any two- or three-parameter nonlinear equation for a least-squares estimate. The details of the procedure are in Appendix 5.

In assessing goodness of fit with nonlinear regression, the plot of residuals is particularly important as a way of comparing more than one equation. Values of R^2, the coefficient of determination, are not particularly useful, nor are values of F. An F-test can be used to compare the fit of two equations; Motulsky and Ransnas (1987) describe the procedure.

Exercises: Use the data and results from the linear regression Exercises 3-1 to 3-4 above to explore the procedures of nonlinear regression. For any of the data sets, pick the one or two best-fitting equations and perform a nonlinear regression using the same equation, and the linear estimates of coefficients as initial values for the iterations.

Exercise 3-5: Use data and results from Exercise 3-1.

Exercise 3-6: Use data and results from Exercise 3-2.

Exercise 3-7: Use data and results from Exercise 3-3.

Exercise 3-8: Use data and results from Exercise 3-4.

3.8 Polynomial Regression

Fitting empirical equations to experimental data is generally easier to accomplish than fitting theoretical equations because there are fewer decisions to be made. However, from the viewpoint of discovering or explaining the function of a system, polynomials and other empirical equations have little value; they should be used sparingly. If you immediately jump to the use of an empirical equation without trying to use one based on insight and theory, you may be passing up an opportunity to understand some components of the system.

Assuming you are satisfied that an empirical equation is needed, you may find the best fit to the following equation:

$$y = A_0 + A_1 x + A_2 x^2 + A_3 x^3 + A_4 x^4 + \ldots \qquad (3.7)$$

The aim of polynomial regression is to find the values for A_0, A_1, etc., that make the best fitting curve. When using polynomial regression you must specify the order of the regression, which amounts to the number of coefficients to be fit. When only A_0 and A_1 are used, the equation will define a straight line or a first-order equation. The remaining equations formed by adding successive coefficients are quadratic (second-order), cubic (third-order), quartic (fourth-order), and quintic (fifth-order). Ideally, you would fit these equations with the usual procedure of polynomial regression analysis, starting with the lowest order and stopping when the addition of more terms does not improve the fit.

In Appendix 4 of this text you are provided with a curve-fitting program called POLYFIT, which allows you to fit up to five different polynomial equations to a set of x-y data.

Exercises: The following data sets were obtained from various sources in the scientific literature. For each exercise, select at least two theoretical equations from Table 3.1 that appear to match the form of the curve. Then use the CURFIT program or a similar program to analyze the data. Based on your CURFIT results, fit the best equation with nonlinear regression using CURNLFIT or other nonlinear regression program. For those data sets which do not appear to fit any available equation, use the POLYFIT program.

Exercise 3-9: Data from Machado et al. (1989)

X: Minutes after injection of radiolabeled glucose

Y: Specific radioactivity of blood glucose in a Brazilian carnivorous fish, *Hoplias malabaricus*, starved for 10 months

Exercise 3-10: Data from Sirko et al (1989)

X: minutes after injection of endotoxin

Y: body temperature of cats after injection of endotoxin in hypothalamic region of brain

Exercise 3-11: Data from Kooloos and Zweers (1989)

X: depth of water in drinking vessel, mm

Y: ml of water swallowed in a single drinking cycle (bill dipped and raised) by mallard ducks (*Anas platyrhynchos*)

Exercise 3-12: Data from Fargo and Bonjour (1988)

X: Constant incubation temperature in $°C$

Y: Developmental rate (time^{-1}) of third instar nymphs of a squash bug, *Anasa tristis*

Exercise 3.9		Exercise 3.10		Exercise 3.11		Exercise 3.12	
X	Y	X	Y	X	Y	X	Y
0	60.0	0	38.4	2.5	0.7	20.0	0.0885
15	52.1	30	38.5	5.0	1.6	21.7	0.1162
30	49.0	60	38.7	10.0	2.0	23.3	0.1724
45	45.7	90	39.4	12.5	2.3	26.7	0.2778
60	39.3	120	39.8	17.5	2.5	28.9	0.3333
90	35.3	150	40.0	22.5	2.7	31.1	0.3571
120	29.2	180	40.2	27.5	2.8	33.3	0.4167
180	22.5	210	40.4	30.0	3.2	35.6	0.4348
240	18.3	240	40.4	35.0	3.1	36.7	0.4000
300	15.4			40.0	3.2	37.8	0.4000
360	12.0					38.9	0.4167
						40.0	0.3030

Exercise 3-13: Data from Mukhtar et al. (1989)

X: Concentration (μM) of LTB$_4$ (leukotriene B$_4$) , a pro-inflammatory agent

Y: Activity of rat hepatic microsomal LTB$_4$ ω-hydroxylase (pmol min^{-1} mg^{-1} protein), measured after 20 min of aerobic incubation of microsomes at 37°C.

Exercise 3-14: Data from Miyazono and Heldin (1989)

X: Log concentration (nM) of transforming growth factor–$\beta1$ (TGF–$\beta1$) in normal rat kidney cell cultures

Y: Percentage of radiolabeled TGF–$\beta1$ remaining bound to cells after incubation with unlabeled TGF–$\beta1$.

Exercise 3-15: Data from Tulasi and Ramana Rao (1989)

X: Environmental oxygen tension in mm Hg

Y: Equilibrium percentage oxygen saturation of blood of a freshwater crab, *Barytelphusa guerini*

Exercise 3-16: Data from Inagaki and Yamashita (1989)

X: Age (days) of fifth-instar larvae of silkworm (*Bombyx mori*)

Y: Radiolabeled acetate uptake by lipids of larval fat body during 1 hr incubation after surgical removal of fat body of female silkworm larvae

Exercise 3.13		Exercise 3.14		Exercise 3.15		Exercise 3.16	
X	Y	X	Y	X	Y	X	Y
0.5	2.7	-3.3	98.5	2.7	12.4	1	0.13
1.5	6.9	-2.7	96.1	3.3	15.1	2	3.50
3.3	10.6	-2.1	93.0	7.0	32.5	3	3.60
4.5	12.8	-1.5	78.3	12.4	52.1	4	3.29
8.5	17.8	-0.9	56.6	15.7	63.3	5	2.35
14.0	18.6	-0.3	24.8	20.6	75.7	6	1.48
20.0	17.9	+0.3	9.3	24.9	86.4	8	0.34
		+0.9	7.0	28.9	92.3	12	0.03
		+1.5	2.3	33.3	96.2		
				41.4	100.0		

Exercise 3-17: Data from Fukuhara and Takao (1988)

X: Days after hatching

Y: Time (seconds) spent swimming during a 1-minute period by larval anchovy, *Engraulis japonica*

Exercise 3-18: Data from Lynch (1989)

X: Concentration (μg ml^{-1}) of food in culture water

Y: Mean lifespan (days) of a small freshwater crustacean, *Daphnia pulex*

Exercise 3.17		Exercise 3.18	
X	Y	X	Y
0	0.5	0.04	14.9
1	1.5	0.12	19.8
2	4.1	0.16	36.4
3	15.9	0.31	32.4
4	49.5	0.55	51.3
5	58.5	0.78	53.9
6	59.5	1.55	44.4
		2.32	39.3
		3.08	33.2

Conclusion

This chapter has shown how to find coefficients of the best fitting theoretical or empirical equations using experimental data. In each case we have assumed that the cause and effect relationship between the dependent (y) variable and the independent (x) variable was not in question. Regression techniques are used here to find coefficients, not to assess the significance of the relationship. The resulting best-fit equation is used only for predicting values of y for a given x. In several exercises throughout this book, you will use the techniques of curve-fitting to find coefficients from data.

The statistics such as F-values that are used by the curve-fitting programs should be used only for establishing the condition of best fit. You should consult any good reference on statistics before you test hypotheses about the relationship of one variable to another using regression.

CHAPTER 4

PLANNING AND PROBLEMS
OF PROGRAMMING

At this point you have written several short programs in BASIC. Some exercises in the following chapters will require writing programs that are slightly more complicated. Before you go on to these it will be helpful to review some aspects of the BASIC language, and some elementary points about planning programs. A well-planned program is easier to enter into the computer, and in particular it is much easier to find and remove programming errors.

BASIC was the language of choice for this introduction to biological simulation because it is simple to learn, relatively powerful, versatile, and widely available. BASIC is not "structured", because it permits the free use of GOTO and similar statements to transfer control within the program. Overuse of such statements can create "spaghetti-code" which can be extremely difficult to decipher and to correct should errors occur. BASIC dialects are available which provide some degree of structure but are more difficult to learn. Writing in structured language is somewhat like writing poetry in sonnet form; after the form is learned, great beauty and creativity are possible. The unstructured form of BASIC is like the unstructured nature of blank verse: great beauty and creativity are likewise possible, but some extremely ugly results are also possible, and perhaps more likely. More self-discipline is needed for working successfully in a free environment than in a regimented one, and good BASIC programmers must provide their own structure. The purpose of this chapter is to show some possible problems with BASIC, and to demonstrate how to avoid and correct them.

4.1 Planning Programs

The short programs you wrote in the first two chapters did not require much planning. However, some of the programs that are to be written in the upcoming chapters are more complex, and a plan for writing them probably cannot be held in detail in your mind. Often a mental program outline that is quite clear when you begin will be muddled by the practical difficulties of entering the program into the computer. The planning that precedes a program can begin with a few notes and sketches, and become a detailed description of the program before any BASIC code is written for the machine.

Flowcharts are an old and still useful method of representing program operation and flow in diagrammatic form. Such diagrams provide a quickly grasped reference to the overall logic of a program. They are useful at all stages of developing programs for simulation models. A roughly sketched flowchart is helpful in designing the primary input, processing and output steps of a program. The rough flowchart may be expanded and elaborated as the details of the programming sequence are established. With the rough chart, one may run through the sequence to test the correctness of the logic. After the program has been written, the flowchart is also useful in finding and correcting mistakes in the implementation of the program. Finally, after the program is operational, the flowchart is useful as documentation for the program. In this final, carefully prepared form the flowchart may be used to explain the programming logic to other modelers, and as a reference for the programmer should the program need modifications at some later date.

4.2 Flowcharting Guidelines and Symbols

Flowcharts should be designed so that the program can be implemented in almost any programming language. The flowchart is therefore not just a series of boxes filled with BASIC or other code. Unfortunately, there are no standardized forms or techniques for writing flowcharts. Numerous symbols have been developed for flowcharting use; these rarely serve to clarify the logic of a program. A minimal set of five standard symbols (Figure 4.1) will be adequate to represent almost any program.

1. Terminal Blocks are used to indicate where programs start and stop. The starting block may contain the name of the program. The stop block may contain the word "stop", or "end" if it terminates a main program, or "return" if it terminates a subroutine. Only one starting and stopping block will occur in any program, and only one line should be attached to either block.

2. The Process Block indicates calculations, definitions of variables, and other processing that is not indicated by another symbol. There should be a single line into and out of a process block.

3. The Decision Block represents an either/or decision. One line will enter the symbol, and two will exit, labeled to indicate the result

of the decision.

4. The Input/Output (I/O) Block shows all the types of input and output. Again, a single line will enter and leave the block.

5. The Connector circle indicates linkage between parts of the same program.

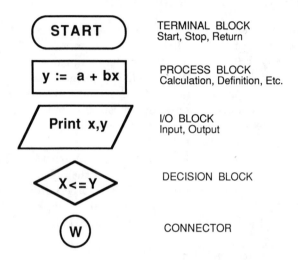

Figure 4.1. The minimal set of five flowcharting symbols.

The following suggestions may help to keep your flowcharts clear and useful:

1. Keep the chart simple and uncluttered. Be concise, but clear.

2. Limit the number of symbols.

3. The chart should flow from top to bottom, and from left to right. Reverse flows should be indicated with connectors or arrowheads.

4. Use connectors to keep flowlines from crossing. Connectors should be labelled and paired; only two connectors should have the same label in any program.

5. Any symbol should have a single entering line. If needed, flowlines should be joined before entry points.

6. Only the decision symbol should have more than one exiting flowline. These lines should be labeled with a 'Yes' or 'No' answer to the question in the decision symbol.

7. Processes may include algebraic expressions. For clarity, the symbol for replacement or definition ($\leftarrow$ or $:=$) may be used instead of the ($=$) symbol used in BASIC programs.

Flowchart representation of the construction of a FOR-NEXT loop is not easy, and a discussion of some specific problems is warranted.

4.3 FOR-NEXT Loops in BASIC

Loops are important in any computer language. The FOR-NEXT loop is a significant part of BASIC. It has been used in most of the programs you have already written from previous chapters and it will be used in most of the remaining simulations of biological phenomena. Because of its usefulness and importance, its functioning should be clearly understood. Equivalent constructions are found in almost all computer languages, so the following notes will apply generally to these languages as well as to different versions or dialects of BASIC.

FOR-NEXT loops are simple in form, but even experienced programmers may have difficulties with them. Some of the difficulties arise from the differences among dialects of BASIC, and others from a misunderstanding of what the loops will accomplish. For example, try to predict the output from the following elementary FOR-NEXT loop:

```
10 REM LISTING 4.1
20 FOR J1 = 1 TO 5 STEP 1
30    PRINT J1
40 NEXT J1
50 PRINT J1
60 END
```

The program should first print out the digits 1 through 5. The variable J1 is used as a counter, but the method of its use varies among different BASICs. This can be seen by trying to predict the result of Line 50. Some versions of BASIC will print 5, others will print 6, and some will even print 4. In effect, the dialects differ on the sequence of incrementing the counting variable, and testing it against the exit-loop value of 5. You can explore the method used by your BASIC version in Exercise 4-1 below.

In general, BASIC can be relied upon to complete a FOR-NEXT loop satisfactorily when the STEP value is an integer. However, some dialects are quite unreliable when the increment is a decimal. The following loop provides an example of unexpected results for some BASICs, notably for Applesoft BASIC and for Microsoft BASICA and GW-BASIC interpreters running on IBM personal computers:

```
210 REM LISTING 4.2
220 FOR J2 = 0.1 TO 1.0 STEP 0.1
230    PRINT J2
240 NEXT J2
250 END
```

Rather than the expected printing of 0.1, 0.2, 0.3, ..., 1.0, the listing may terminate at 0.9, with the loop being executed 9 times. The problem is evidently caused by round-off error. Similar results may occur if the loop has a starting value of 1.1, and an exit value of 2.0. However, many starting values will produce the expected printing of 10 values, with the loop executing 10 times. Unless you are very knowledgeable about the performance of your particular BASIC, it is probably not a good idea to use a fractional increment. If the value of a looping variable involves a fraction, you should be wary of the values it might supply for use in a simulation.

Exercise 4-1: To find how the FOR-NEXT loop functions with your computer and version of BASIC, go through the short BASIC programs listed as 4.1 (above), and below as 4.3, 4.4, and 4.5. Write out what you think will be the output from these programs. Then type them into your computer and run them, and compare the result with your prediction. You should be aware that other versions of BASIC may produce different results than you obtained.

```
REM LISTING 4.3          410 REM LISTING 4.4
320 FOR J3=10 TO 9 STEP 1  420 FOR J4 = 1 TO 5
330    PRINT J3            430    PRINT J4
340 NEXT J3               440    J4 = J4 - 1
350 PRINT J3              450 NEXT J4
360 END                  460 END
```

```
510 REM LISTING 4.5
520 P = 6
530 Q = 1
540 FOR J5 = 1 TO P STEP Q
550    P = 10
560    Q = 2
570    PRINT J5
580 NEXT J5
590 END
```

Also write the program in Listing 4.2 to find the output for the FOR-NEXT loop executed with fractional increments for the interval 0.1-1.0. Then alter Line 220 to find the output for the interval 1.1 to 2.0, then 2.1 to 3.0, and so on through 10.1 to 11.0.

The FOR statement combines initializing and incrementing. In some BASICs it also includes the condition test; in others, the test is made at the NEXT statement. Figure 4.2 shows alternate and unambiguous methods of programming the FOR-NEXT loop based on the dialect-specific construction. From your results of Exercise 4-1, you should know which to use with your BASIC dialect. To indicate the FOR-NEXT loop, a comment may be attached to the process symbol used to initialize the loop counter.

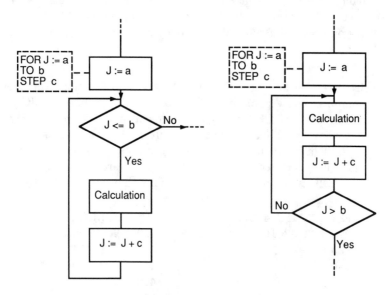

Figure 4.2. Two ways of representing BASIC-language FOR-NEXT loops in flowcharts. The proper method will depend upon the logic of the program, and upon the action of a FOR-NEXT loop in a given dialect of BASIC.

A sample flowchart for the population growth exercises of Chapter 1 is shown in Figure 4.3.

Exercise 4-2: Draw up a flowchart for Exercise 1-3, the simulation for exponential decay.

Exercise 4-3: Draw up a flowchart for Exercise 1-5, the description of deepwater crab distribution around hydrothermal vents.

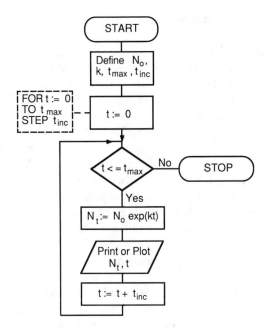

Figure 4.3. Flowchart for BASIC program of population growth (Exercise 1-1).

4.4 Notes on Debugging Programs

Except for the most elementary programs, you will spend more time removing the mistakes (bugs) in your programs than in designing them and entering them into the computer. Programmers develop their own methods of finding and fixing problems that appear in execution of programs. However, it is possible to suggest several general approaches that most experienced programmers use to detect and remove errors in programs. They are for the most part intuitive, and are reinvented by every self-taught programmer. The following discussion applies to most languages, not only BASIC.

Perhaps the easiest mistakes to correct are those that the BASIC interpreter or compiler detects as "syntax errors" or something more specific, such as a "NEXT without FOR error in Line 2310". A variety of errors may occur to interrupt the execution of a program, such as a division by zero, or a request to take the log of a variable which has been set to zero. These errors may be difficult to trace. With most BASIC interpreters, it is possible to use the PRINT command in immediate mode to discover which variables are causing the problems. That is, entering something like "PRINT X,Y,Z" can show which variables have taken on unexpected values.

If the program executes without halting, then other errors may remain. Some obvious mistakes may appear in the formatting of output. In fact, a large proportion of programming time is often spent getting the output to "look right". More difficult to detect may be problems with the accuracy of output. The two principal methods for detecting errors in accuracy involve desk checking and isolation-and-correction.

Desk checking is a method of testing the program by "playing computer". The programmer uses pencil, paper, and perhaps a hand calculator to work through the program with example data. Variables are changed on paper as they are encountered in the code, with a running tabulation for each variable. This procedure can be used to verify both the correctness of the overall program logic and the particular computer code.

The isolation-and-correction procedure can use the information obtained from the procedures above to find where the problems occur in a program. Most isolation procedures call for inserting temporary lines of code into a malfunctioning program to learn what the program is actually doing. If the code produces an obscure flow, it may be useful to insert lines that print out some identifying message when they are executed. Thus, you can put in a temporary Line 155 PRINT "AT LINE 155"; whether and how often the line is printed can let you decide the number of times a FOR-NEXT loop is executed, or whether an expected branch is taken out of a decision. Similarly, it is possible to insert temporarily a STOP or END statement that will halt the program execution of a program at a particular location. Such a modification can serve the dual purpose of letting you know that the program has reached the location, and of allowing you to examine the values of various program variables in the immediate or direct mode. (Some BASIC dialects will let you restart your program after it has been halted with STOP or END by entering CONT in the immediate mode.) Instead of using the STOP statement, the same dual function can be elaborated and automated by inserting temporarily a program line that will print out values of variables when the line is executed. This procedure allows you to obtain "snapshots" of the program as it executes. The values can be compared with values calculated by hand.

Conclusion

This chapter has explained some problems that may arise with programs written in BASIC, and has suggested some ways to minimize or avoid the problems. As you work through exercises in the following chapters, you will have opportunities to put some of these suggestions into practice. If you carefully plan the programs, they will be more likely to run properly the first time. You will find it also much easier to discover and correct errors in well-planned programs.

CHAPTER 5

NUMERICAL SOLUTION
OF RATE EQUATIONS

Chapter 1 showed briefly that rate equations (differential equations) could be solved analytically using integral calculus. However, many differential equations of biological interest are not easily solved analytically because of their nonlinear nature. These more complex equations may be solved with numerical integration, which requires a lot of repeated arithmetical calculation. This is the principal reason that biological simulation depends heavily on computers. Some elementary techniques of numerical integration will be introduced below, using simple equations.

This chapter is the key to understanding much of the whole field of biological simulation. Many exercises in following chapters require you to use numerical integration. Try to learn the techniques thoroughly before going on to other material.

5.1 The Fundamental Ideas of Numerical Integration

In its simplest form, numerical integration is easily understood. It employs the straightforward idea of finding an incremental change in a variable, and adding the change to obtain a new value for the variable. A bank does this, for example, when it calculates interest and adds it to your savings account. Likewise, to find the size of a population of organisms at time $t + \Delta t$, you would find the change in population that occurred in the interval Δt, and add it to the size of the population at time t. The size of the population (or of your savings account) may be found after any length of time by repeating the calculation and addition.

The computer is ideally designed to perform repetitive operations of this sort. The results of numerical integration are approximations of the analytical solution, but they can be highly accurate. The difference (error) between the analytical solution and the approximation is produced by various factors. Several exercises in this chapter are concerned with estimating error, and you will find that it can be minimized by selecting appropriate numerical techniques.

79

In most cases, the errors caused by using numerical integration are not serious in biological simulation, because we are more interested in the behavior of models than in precisely duplicating real data. We will discuss in this chapter a simplified version of only one elementary technique and some of its variations. More complete discussions of these and other methods may be found in Bronson (1973), Carnahan et al. (1969), Macon (1963), and Davies (1971).

5.2 The Finite Difference Approach

You will remember that in Chapter 1 we worked with the derivative of the differential equation, dy/dx. All numerical integration techniques are based on the fact that for small incremental changes in x (the independent variable), the difference quotient $\Delta y/\Delta x$ approximates dy/dx. You may recall from elementary calculus that dy/dx is actually defined as:

$$\frac{dy}{dx} = \lim_{\Delta x \to 0} \frac{\Delta y}{\Delta x}$$

Therefore, if

$$\frac{dy}{dx} = f(y) \tag{5.1}$$

then

$$\frac{\Delta y}{\Delta x} \cong f(y)$$

and with rearrangement

$$\Delta y \cong f(y) \cdot \Delta x \tag{5.2}$$

This leads to the general equation upon which the techniques of numerical integration are based:

$$y_{x+\Delta x} = y_x + \Delta y_x \tag{5.3}$$

Equations 5.2 and 5.3 are used together, with 5.2 defining the change in y for a finite change in x, and 5.3 defining the new value of y after the change.

The mathematically knowledgeable reader will notice that Equation 5.1 is a limited case of the customary expression for the first-order differential equation $dy/dx = f(x, y)$. The simpler form of Equation 5.1 is used here because in most biological simulation models, y is not a function of x.

5.3 The Euler Technique

Equations 5.2 and 5.3 actually describe a technique for solving rate equations called the Euler method, or Euler-Cauchy method. To show

how this method would work with a specific example, we will examine the differential equation for exponential growth:

$$\frac{\Delta N}{\Delta t} \cong f(N) = kN \tag{5.4}$$

You will recall from Section 1.1 that N is population density and k is the constant for growth rate. Like many differential equations in biological models, the independent variable is time. This differential equation is approximated by the following difference equation:

$$\frac{\Delta N}{\Delta t} \cong kN \tag{5.5}$$

or

$$\Delta N_t \cong kN_t \Delta t \tag{5.6}$$

and the new value of N is obtained with the equation

$$N_{t+\Delta t} = N_t + \Delta N_t \tag{5.7}$$

To carry out an Euler integration, increment time by units of size Δt, and at each time interval, calculate the change in N, and then the new value of the population density, $N_{t+\Delta t}$. This new value of N then becomes the N_t for the following time interval.

This integration is simple enough to be calculated with a single equation combining Equations 5.6 and 5.7. However, it is better to develop the habit of first calculating the change with an equation like 5.6, and then updating the variable with an equation like 5.7. The computer may be programmed to carry out these calculations with two steps:

$$\Delta N \leftarrow k * N * \Delta t$$
$$N \leftarrow N + \Delta N$$

This "two-stage" approach will help to keep your programs clear when the models become more complex in the following chapters. The left-arrow symbol ($\leftarrow$) here indicates the computer operation of performing the calculation on the right-hand side and storing the result under the variable label on the left-hand side. In BASIC this operation is indicated by the equal sign ($=$).

Exercise 5-1: Examine the model for inhibited growth from Chapter 1. For Equation 1.16 find the difference equations analogous to 5.6 and 5.7. Write and implement a computer program using your equations to find and plot population size from $t = 0$ to 52 weeks.

Use the appropriate constants from Chapter 1, and a time increment Δt of 1. For comparison, have your program also plot values from Equation 1.17, the analytical solution. (Try to plot the Euler values as circles and the analytical values as a continuous line on the same graph.)

5.4 Time Increment and Error in Numerical Integration

The size of Δt is important in determining the error between the analytical solution of a differential equation and the numerical solution. The farther Δt departs from the limit of 0, the farther the results of the simple Euler method will depart from the true value of the integral. This is shown graphically in Figure 5.1, which compares the analytical solution of the exponential growth model (Equation 1.1) with Euler integration, using two different values for Δt.

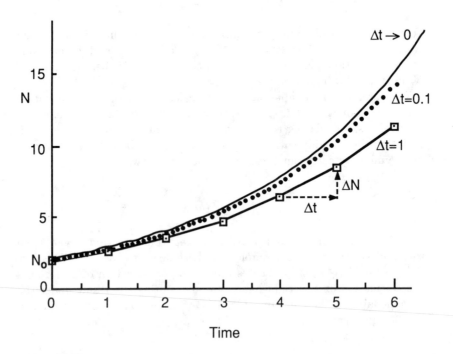

Figure 5.1. The effect of the time increment value on a simple Euler integration.

Error for a single step is directly proportional to the size of Δt. In addition, error is cumulative, because each new value of N depends on the previous value. Selection of smaller values of Δt will decrease both kinds of error. The error in a single step is termed local error or truncation error, while the cumulative error due to previous steps is termed propagated error. Both can be minimized by selecting more accurate methods, as well as by decreasing Δt. A third type of error, termed rounding or round-off, occurs because computers can hold only a finite number of digits in their memories. The magnitude of this error will vary among types of computers.

Exercise 5-2: To estimate the magnitude of the errors due to Δt, write a computer program to compare three solutions for the exponential growth model (Equation 1.3):

(A) the analytical solution (Equation 1.6);

(B) Euler integration with $\Delta t = 1.0$;

(C) Euler integration with $\Delta t = 0.1$.

Use the two-stage approach with Euler integration. Set $N_0 = 10$, $k = 0.10$, and let t run from 0 to 30. Your output should be in the form of a table, with the following information printed out at each whole-integer time interval:

(1) time t;

(2) the true value of N_t calculated with the analytical solution;

(3) N_t from the Euler integration with $\Delta t = 1$;

(4) N_t from the Euler integration with $\Delta t = 0.1$;

(5) the difference between (2) and (3);

(6) the difference between (2) and (4).

Note that your program will not require the graphing subroutines of previous exercises. An easy method of programming this problem is to use nested FOR...NEXT loops, one loop for the integer time interval, and an inner loop for the smaller time increments.

Exercise 5-3: Modify and run the program from Exercise 5-2, using Δt values of 1.0 and 0.01.

5.5 The Improved Euler Method

As you discovered in Exercise 5-3, reducing the time increment Δt decreases the error, but also causes a proportional increase in computer time required to perform a given integration. Complex biological simulations may require highly accurate integration to work properly and will cause great increases in computer times that become inconvenient, expensive, or both. Hence, there has been a lot of mathematical research on methods which reduce the error other than by decreasing Δt.

Most of the error in the simple Euler method applied to the growth model occurs because it proceeds as if the slope in the interval Δt has a constant value which is a function of N_t. In fact, the slope changes continuously between N_t and $N_{t+\Delta t}$.

The Improved Euler is based on finding a better estimate of slope within the interval Δt. (The description here is based on the general notation of Equations 5.1-5.3 above.) First one finds an estimate of slope at y_t and then an estimate of slope at $y_{t+\Delta t}$. The improved slope is the average of these two estimates. The equations which accomplish this estimation are as follows:

$$\Delta y'_t = f(y_t)\,\Delta t \tag{5.8}$$

$$y'_{t+\Delta t} = y_t + \Delta y'_t \tag{5.9}$$

$$\Delta y''_t = f\left(y'_{t+\Delta t}\right)\Delta t \tag{5.10}$$

$$\Delta y_t = \frac{\Delta y'_t + \Delta y''_t}{2} \tag{5.11}$$

$$y_{t+\Delta t} = y_t + \Delta y_t \tag{5.12}$$

In this set of equations, $\Delta y'_t$ is the first estimate of the slope and is used to find the second estimate, $\Delta y''_t$. These two estimates are averaged in Equation 5.11, and the average is used to update y_t in Equation 5.12. The technique is illustrated graphically in Figure 5.2. The analogous equations for exponential growth (Equation 1.3) would be:

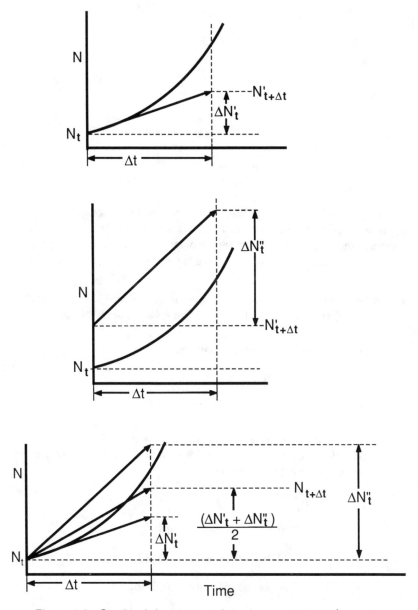

Figure 5.2. Graphical description of the Improved Euler (second-order Runge-Kutta) method. Top: first estimate of slope, $\Delta N'$. Middle: second estimate of slope, $\Delta N''$. Bottom: estimate of slope ΔN from the average of $\Delta N'$ and $\Delta N''$.

$$\Delta N'_t = k N_t \, \Delta t$$

$$N'_{t+\Delta t} = N_t + \Delta N'_t$$

$$\Delta N''_t = k N'_{t+\Delta t} \, \Delta t$$

$$\Delta N_t = \frac{\Delta N'_t + \Delta N''_t}{2}$$

$$N_{t+\Delta t} = N_t + \Delta N_t$$

The Improved Euler method has been identified by various terms, including Heun's method, second-order Runge-Kutta, and a simplified predictor-corrector method. In the latter terminology, Equation 5.9 is the predictor equation, and Equations 5.10-5.12 are the corrector step.

Exercise 5-4: Write a computer program to compare three methods for integrating Equation 1.3, the exponential growth model:

(A) The analytical (true) solution;

(B) the simple Euler integration;

(C) Improved Euler integration.

Use the same approach and parameters as in Exercise 5-2, with $\Delta t = 0.1$. Your output should also follow the format of Exercise 5-2, showing a tabulation of:

(1) time (t);

(2) Values of N_t based on method (A) above;

(3) Values of N_t based on method (B) above;

(4) Values of N_t based on method (C) above;

(5) the difference between (A) and (B);

(6) the difference between (A) and (C).

5.6 A Fourth-Order Runge-Kutta Method

A great variety of techniques are available for numerical solutions to differential equations, of which Runge-Kutta methods are a particular set. In fact, the Euler method is a first-order Runge-Kutta method and the Improved Euler is a second-order Runge-Kutta. While Runge-Kutta methods of almost any order are possible, the fourth-order is convenient and most commonly used. The estimates it provides can be quite accurate, even with the simplest form of the method given here. The derivation of Runge-Kutta methods may be found in most textbooks of numerical analysis; the practical discussion of Carnahan et al. (1969) is particularly clear. For the following example, we will use the diffusion model (Equation 1.12):

$$\frac{dC}{dt} = -k\,(C - C_x)$$

The simple Euler (first-order Runge-Kutta) two-stage approach with this equation is

$$\Delta C \leftarrow -k(C - C_x)$$
$$C \leftarrow C + \Delta C$$

A BASIC program to accomplish a fourth-order Runge-Kutta integration is given here:

```
100 REM DIFFUSION MODEL
110 REM 4TH-ORDER RUNGE-KUTTA
120 K=0.3 :  C=100 :  CX = 50
130 DT = 0.1
140 FOR T=0 TO 20
150     IF T=0 THEN GOTO 330
160     FOR I=1 TO 10
170:
180        D1 = -K * (C - CX) * DT
190        C1 = C + D1/2
200:
210        D2 = -K * (C1 - CX) * DT
220        C2 = C + D2/2
230:
240        D3 = -K * (C2 - CX) * DT
250        C3 = C + D3
260:
270        D4 = -K * (C3 - CX) * DT
280:
```

```
290        DC = (D1 + (2*D2) + (2*D3) + D4) / 6
300        C = C + DC
310:
320     NEXT I
330     PRINT T,C
340 NEXT T
350 END
```

In essence, this method provides four different estimates of the slope (ΔC) over the increment Δt (lines 180, 210, 240, 270) and then finds a weighted average of these estimates (line 290). The variable is updated with the weighted average in line 300. As part of the updating process, ΔC_1 and ΔC_2 are divided by 2 (lines 190 and 220). Note that this listing is designed for clarity, not computational efficiency.

Exercise 5-5: Write a program to compare four methods for integrating Equation 1.3 (exponential growth):

(A) the analytical (true) solution;

(B) the simple Euler integration;

(C) the Improved Euler integration;

(D) the 4th-order Runge-Kutta method.

Implement your program using the approach and parameters in Exercise 5-2, with $\Delta t = 0.1$. Your output should show a tabulation of time, and the values from methods (A), (B), (C), and (D).

Exercise 5-6: Using the model for exponential growth (Equation 1.3), write a program that will allow you to vary Δt values for numerical integration with the simple Euler method. Find the largest Δt value that will give the same accuracy (rounded to 5 decimal places) as a 4th-order Runge-Kutta solution with $\Delta t = 0.1$, evaluated after 20 time units.

5.7 Systems of Equations

Many of the models to be investigated in subsequent chapters are developed using several equations that contain two or more common variables.

In a simple, two-variable form, these will make up a system of equations:

$$\frac{dy}{dx} = f_1(y, z) \qquad (5.13a)$$

$$\frac{dz}{dx} = f_2(y, z) \qquad (5.13b)$$

with values available for y and z at $x = 0$. An elementary example of a system like this has already been examined in Section 1.8. Such systems of equations are easily solved with the numerical integration methods you have looked at in the sections above. The only difficulty is to keep in mind the proper sequence.

The fundamental rule for solving systems of equations is that they are taken through the steps of numerical integration in parallel. If you are using a two-equation system with the simple Euler method, both equations will be taken through the first stage (calculating the changes), and then through the update stage (finding the new value).

Listed below is another example, a program that uses the Improved Euler method with the bimolecular reaction model (Equation 1.18):

$$\frac{d[B]}{dt} = f([A], [B]) = -k[A][B]$$

Note the second equation is

$$\frac{d[A]}{dt} = f([A], [B]) = -k[A][B]$$

In this case the function equations are the same, although this will rarely be the case in the chapters that follow. (For this example, we will assume that the starting concentrations of A and B are similar but not equal, so that neither Equations 1.19 nor 1.22 apply.)

```
100 REM BIMOLECULAR REACTIONS
110 REM IMPROVED EULER METHOD
120 A=100:  B=110:  K=0.1
130 DT = 0.1
140 FOR T = 0 TO 25
150    IF T=0 THEN GOTO 330
160    FOR I = 1 TO 10
170:
180       D1A = -K * A * B * DT
190       D1B = -K * A * B * DT
200:
```

```
210      AX = A + D1A
220      BX = B + D1B
230:
240      D2A = -K * AX * BX * DT
250      D2B = -K * AX * BX * DT
260:
270      DA = (D1A + D2A) / 2
280      DB = (D1B + D2B) / 2
290:
300      A = A + DA
310      B = B + DB
320    NEXT I
330    PRINT T,A,B
340 NEXT T
350 END
```

The key point of this listing is that the two variables $[A]$ and $[B]$ are taken through the updating process in parallel. (As before, this program listing is designed for clarity, rather than efficiency in computation.)

Exercise 5-7: Following the format of the listing of the Improved Euler method above, write and implement a program for the bimolecular reaction model using a fourth-order Runge-Kutta method, with $k = 0.025$, $[A]_0 = 95$, and $[B]_0 = 115$. Your output should be a table of $[A]$ and $[B]$ for values of t from 0 to 20.

5.8 Fifth-Order Runge-Kutta

In most cases that are encountered in biological simulation, the Improved Euler or 4th-order Runge-Kutta methods will be adequate. Where extreme accuracy is required, a 5th-order Runge-Kutta method will usually provide it with minimal increases in computing time. The following listing uses a set of equations developed by Butcher (1964), which Waters (1966) found to have the best balance of accuracy and computing time (James et al. 1977).

```
100 REM EXPONENTIAL GROWTH
110 REM 5TH-ORDER RUNGE-KUTTA
120 N = 10
130 K = 0.10
140 DT = 0.1
150 FOR T=0 TO 30
```

```
160    IF T=0 THEN GOTO 390
170    FOR I = 1 TO 10
180:
190       D1 = K * N * DT
200       N1 = N + D1/4
210:
220       D2 = K * N1 * DT
230       N2 = N + D1/8 + D2/8
240:
250       D3 = K * N2 * DT
260       N3 = N - D2/2 + D3
270:
280       D4 = K * N3 * DT
290       N4 = N + D1*3/16 + D4*9/16
300:
310       D5 = K * N4 * DT
320       N5 = N - D1*3/7 + D2*2/7 + D3*12/7 + D4*12/7 +
330:
340       D6 = K * N5 * DT
350:
360       DN = ( 7*D1 + 32*D3 + 12*D4 + 32*D5 + 7*D6 ) / 90
370       N = N + DN
380    NEXT I
390    PRINT T, N
400 NEXT T
410 END
```

Here again, this listing is designed for clarity, not ease of computation. (The omission of a term containing D2 from line 360 is intentional.) An experienced BASIC programmer can shorten the listing and decrease computational time considerably with algebraic manipulation of the equations and use of the DEF FN statement.

Conclusion

This chapter has described simple numerical methods for solving differential equations. The exercises have been designed to give you some idea of the accuracy you might expect with each method. Most of our simulations in the following chapters can be performed with the simple Euler or Improved Euler techniques. We will be primarily concerned with the behavior of the models rather than precise duplication of complex analytical solutions.

In this chapter, we have used simple equations that have analytical solutions. In the following chapters, we will be concerned with differential equations that are nonlinear, and the value of the methods you have practiced here will become obvious.

PART TWO
MODELS WITH
MULTIPLE COMPONENTS

You have been introduced to some basic methods of modeling using some simple models in Part One of this book. In this second part we will be concerned with models that involve the concerted interaction of several equations. Such models are used to simulate biological systems with multiple components, and are based on the summation principle of modeling.

The summation principle may be stated as follows: If a single equation gives verifiable output when implemented independently, then it may be combined with similar equations to model multicomponent systems. Stated this way, the summation principle is a concept about scientific procedure. Some biologists argue that behavior of biological systems is not necessarily the sum of behaviors of the parts. Supporters of the principle reply that in instances where the principle does not hold, there must exist unrecognized nonlinearities or feedback interactions. The principle can never be falsified therefore. However, acting as if the principle is correct provides a structured framework for investigating biological systems. Simulation plays a key role in the procedure. The general concept of the replacement principle (see the Introduction to Part One) acting in conjunction with the summation principle is illustrated in Figure P2.1.

The models described in this section are all of the deterministic type. That is, random elements are assumed to be absent. In such models, the state of a given variable at any time is entirely determined by previous states of that variable, as well as other variables influencing it. You should understand that deterministic models can depart considerably from the real system, depending on the importance of the random components. This fact must be accepted as one of the fundamental assumptions of deterministic model-building. Stochastic or probabilistic models which include random effects are considered in Part 3 of this book.

Your major objective in this second part of the book should be to understand the application of the general numerical approach to the simulation of complex systems. More specific goals may include the following:

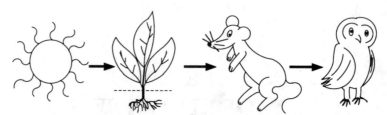

STEP ONE: CONCEPTUAL MODEL OF THE REAL SYSTEM

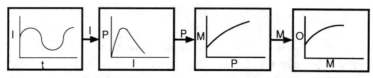

STEP TWO: RESPONSE DATA IN GRAPHICAL FORM

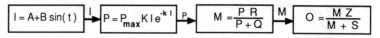

STEP THREE: BLOCK DIAGRAM OF MATH MODELS

Figure P2.1. A simple illustration of stepwise replacement. Conceptual components of a hypothetical system are replaced by equations to form a multicomponent model of the system.

1) Learning how to solve several differential equations simultaneously using techniques of numerical integration.
2) Understanding the place of deterministic simulation in the scheme of biological investigation.
3) Learning to use the concepts of compartmental models.
4) Learning to use matrix algebra in implementing biological simulations.

Each chapter of this section includes several introductory simulations as exercises at the beginning of each chapter. More complex simulations are included in most chapters to provide follow-up practice for modeling systems. The chapters cover a range of biological subject material, from molecular interactions through ecosystems. It is good practice to select an area of particular interest and to try some of the more complex simulations.

CHAPTER 6

KINETICS OF
BIOCHEMICAL REACTIONS

Some fundamental principles of simulation are introduced in this chapter, based on examples taken from the study of biochemical reactions. The kinetics of chemical reactions have been studied for a very long time, and the elementary descriptive mathematics is well established. The concepts are not specialized, and the models derived from them are straightforward. The exercises of this chapter will provide practice in techniques of numerical integration that will be useful in most of the following chapters involving deterministic models.

It would be a mistake to skip this chapter even if you usually prefer to avoid thinking about molecules. Models in all fields of biology are based on rate processes and the law of mass action. Many current models in physiology, population ecology, and ecosystem analysis, for example, are founded in some of the reaction models discussed below.

6.1 Kinetics of Bimolecular Reactions

In Chapters 1 and 2 you had opportunity to look at reactions with rather restricted models based on integration or on assumptions of stable states. The direct approach of numerical integration that you studied in Chapter 5 will let you perform simulations without these restrictions.

We will deal first with a simple reaction between two reactants forming a product:

$$A + B \quad \xrightarrow{k} \quad P$$

The rate of formation of the reaction's product, P, will depend upon the concentrations of the reactants, $[A]$ and $[B]$, and may be expressed as:

$$\frac{d[P]}{dt} = k[A][B] \tag{6.1}$$

As before, the rate at which A and B are used up will depend on their concentrations:

$$\frac{d[A]}{dt} = \frac{d[B]}{dt} = -k[A][B] \qquad (6.2)$$

Whenever several differential equations like these are used simultaneously they form a system. It becomes essential to use the two-stage approach to numerical integration described in Chapter 5. To model the reaction with the simple Euler method, the changes in concentration would be found using these equations as the first stage:

$$\Delta[B] = \Delta[A] = -k[A][B]\Delta t \qquad (6.3)$$
$$\Delta[P] = +k[A][B]\Delta t \qquad (6.4)$$

The second stage involves updating the variables with these equations:

$$[A] \leftarrow [A] + \Delta[A] \qquad (6.5)$$
$$[B] \leftarrow [B] + \Delta[B] \qquad (6.6)$$
$$[P] \leftarrow [P] + \Delta[P] \qquad (6.7)$$

It is important to put all the variables through the first stage before going on to the second stage.

Implementing the bimolecular reaction model on the computer would typically involve entering initial concentrations of A and B, with output consisting of concentrations of A and B over subsequent time intervals. As discussed in Section 1.8, the order of the reactions depends on the relative concentrations of A and B. If reactant A greatly exceeds reactant B, the reaction will follow first-order kinetics, and a straight line will be observed when $\ln[B]$ is plotted against time. If the two reactants are equal in concentration, they will follow second-order kinetics, and $1/[B]$ will plot as a straight line vs. time (Figure 6.1). Where concentrations are unequal but not extremely different, both $\ln[B]$ and $1/[B]$ will be curved.

Exercise 6-1: Write a program using simple Euler integration (Equations 6.3-6.7) to simulate bimolecular reactions. Set $\Delta t = 0.1$ and test your simulation with two sets of values:

Set 1 :	$[A] = 100$	$[B] = 100$	$k = 0.0003$
Set 2 :	$[A] = 1000$	$[B] = 10$	$k = 0.0003$

As the output for each set of values, obtain graphs of $1/[B]$ and $\ln[B]$ plotted against time to determine whether the data sets produce first- or second-order kinetics. You may write your program to produce two lines on the same graph. (Be aware that all the exercises in this chapter employ arbitrary units of concentration and time. They are models of types of reactions, and are not based on real examples.)

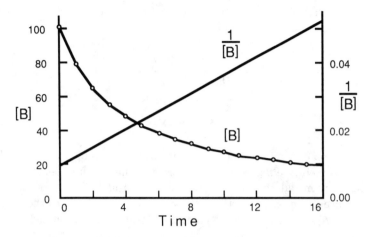

Figure 6.1. Simulation results from a bimolecular reaction showing second-order kinetics. In this example, $[A]_o = [B]_o = 100$ and $k = 0.002$.

6.2 Chemical Equilibrium Model

The model of the approach to chemical equilibrium is described here to demonstrate the use of flux rates in simulation. We will work with a simple model of this type:

$$A + B \underset{k_2}{\overset{k_1}{\rightleftharpoons}} C + D$$

We could use the same two-stage Euler technique as in Section 6.1 above to find changes in concentration of A, B, C, and D. (The changes in the concentrations would be calculated as in Equations 6.3-6.4, and then would be updated as in Equations 6.5-6.7.) An alternate approach uses flux rates to simplify programming by dividing the first stage of the two-stage Euler process into two parts. In Part 1, the instantaneous flux rates for the forward and reverse reactions are defined:

$$F_1 = k_1[A][B] \tag{6.8}$$
$$F_2 = k_2[C][D] \tag{6.9}$$

In Part 2 of the first stage, the estimates of change in concentration are calculated, using the flux rates:

$$\Delta[A] = \Delta[B] = (-F_1 + F_2)\Delta t \tag{6.10}$$
$$\Delta[C] = \Delta[D] = (+F_1 - F_2)\Delta t \tag{6.11}$$

The second stage of the Euler process is not changed. Concentrations are updated with:

$$[A] \leftarrow A + \Delta[A] \tag{6.12}$$
$$[B] \leftarrow B + \Delta[B] \tag{6.13}$$
$$[C] \leftarrow C + \Delta[C] \tag{6.14}$$
$$[D] \leftarrow D + \Delta[D] \tag{6.15}$$

This extension of the two-stage procedure will be valuable in later models also, and is readily adapted to the Improved Euler technique. As in the previous section, all the variables are taken through Part 1 of the first stage, before going on to Part 2 of the first stage, and then to the second stage. The equilibrium constant for this general model is defined by

$$K_{eq} = \frac{[C][D]}{[A][B]}$$

where the concentrations of A, B, C, and D are measured at equilibrium.

Exercise 6-2: Write and implement a computer program to simulate the approach to chemical equilibrium using the model above. Use simple Euler methods with $\Delta t = 0.1$. Set $k_1 = 0.0025$ and $k_2 = 0.0015$. Test your simulation with these values for initial concentrations:

$$[A] = [B] = 100 \qquad [C] = [D] = 0$$

Output should be in tabular form showing concentrations of each of the four reactants for times 0, 1, 2, 3, ..., 30, by which time your simulation should have approached a reasonable equilibrium. Have your program find and print the equilibrium constant after 30 time units. Compare this value with the calculated value of k_1/k_2.

Exercise 6-3: Expand your program from Exercise 6-2 to print columns for each of eight variables: elapsed time, $[A]$, $[B]$, $[C]$, $[D]$, the Net Forward Reaction Rate $(F_1 - F_2)$, the ratio $[C][D]/[A][B]$, and $\ln([C][D]/[A][B])$. Again, print the results for times 0 to 30.

Also set up your program to produce graphs of $[C][D]/[A][B]$ and $\ln([C][D]/[A][B])$ as functions of Net Forward Reaction Rate (NFRR). These graphs are useful in estimating the equilibrium constant when NFRR = 0, by extrapolating to zero net rate; see Figure 6.2.

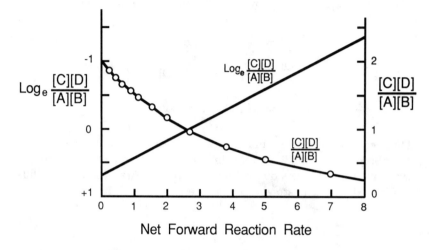

Figure 6.2. Simulation results showing the approach to chemical equilibrium. $K_{eq} = 2.0$.

When you think your program is working successfully, alter the initial concentration of reactants and products, and produce another set of output data. You can confirm the success of the simulation by checking your output against the following properties of chemical equilibria: 254050

1) Computed and extrapolated K_{eq} values should equal k_1/k_2.

2) NFRR should approach zero as concentrations of reactants and products approach equilibrium.

3) Output should show the same K_{eq} regardless of initial concentrations of reactants and products.

6.3 Kinetics of a Sequential Reaction

A sequential reaction is one in which the product of one reaction is a reactant of a second reaction. We will assume a simple reaction in which B reacts with C to yield an intermediate D which then breaks down to yield a product E, as follows:

$$B + C \underset{k_2}{\overset{k_1}{\rightleftharpoons}} D \xrightarrow{k_3} E$$

The kinetics are described by the following differential equations:

$$\frac{d[B]}{dt} = \frac{d[C]}{dt} = -k_1[B][C] + k_2[D] \qquad (6.16)$$

$$\frac{d[D]}{dt} = +k_1[B][C] - k_2[D] - k_3[D] \qquad (6.17)$$

$$\frac{d[E]}{dt} = +k_3[D] \qquad (6.18)$$

These equations can be numerically integrated following the procedure described in Section 6.2 above, using flux rates with the first stage of the two-stage Euler process divided into two parts. First, the reaction fluxes are defined for each rate-governed process:

$$F_1 = k_1[B][C] \qquad (6.19)$$
$$F_2 = k_2[D] \qquad (6.20)$$
$$F_3 = k_3[D] \qquad (6.21)$$

Then, the changes in concentration are defined using the flux rates:

$$\Delta[B] = (-F_1 + F_2)\Delta t \qquad (6.22)$$
$$\Delta[C] = (-F_1 + F_2)\Delta t \qquad (6.23)$$
$$\Delta[D] = (+F_1 - F_2 - F_3)\Delta t \qquad (6.24)$$
$$\Delta[E] = (+F_3)\Delta t \qquad (6.25)$$

The updating process then proceeds as usual: $[B] \leftarrow [B] + \Delta[B]$, etc.

Exercise 6-4: Write a program to simulate the kinetics of sequential reactions. Test your simulation with these coefficients:

$$k_1 = 0.002 \qquad k_2 = 0.04 \qquad k_3 = 0.15$$

For initial conditions, set concentrations as follows:

$$[B] = 100 \qquad [C] = 100 \qquad [D] = 0 \qquad [E] = 0$$

Use simple Euler methods, with $\Delta t = 0.1$. Your output should be in the form of a graph showing $[B]$, $[C]$, $[D]$, and $[E]$ plotted against time, for at least 50 time units.

6.4 The Chance-Cleland Model for Enzyme-Substrate Interaction

The numerical solution of the classical Michaelis-Menten enzyme model (see Section 2.2) was first reported by Chance (1960), and has also been attributed to Cleland by several authors including Mahler and Cordes (1966). The Chance-Cleland model uses techniques similar to those of the previous sections to simulate the interaction of enzyme and substrate to yield a product. The reaction assumed in this model is:

$$E + S \underset{k_2}{\overset{k_1}{\rightleftharpoons}} C \xrightarrow{k_3} E + P$$

where S is the substrate, E is the enzyme, C is the enzyme-substrate compound, and P is the product. The rate constants are k_1, k_2, and k_3. The differential equations for the kinetics of the reactions are:

$$\frac{d[S]}{dt} = -k_1[S][E] + k_2[C] \tag{6.26}$$

$$\frac{d[C]}{dt} = \frac{-d[E]}{dt} = k_1[S][E] - (k_2 + k_3)[C] \tag{6.27}$$

$$\frac{d[P]}{dt} = k_3[C] \tag{6.28}$$

Exercise 6-5: Program the Chance-Cleland model to simulate enzyme-substrate interactions. Use the two-stage approach described in Section 6.3, based on the simple Euler method with $\Delta t = 1$. Use the following coefficients:

$$k_1 = 0.005 \qquad k_2 = 0.005 \qquad k_3 = 0.1$$

For initial conditions, set

$$[S] = 100 \qquad [E] = 10 \qquad [C] = 0 \qquad [P] = 0$$

Your output should consist of a graph showing $[S]$, $[E]$, $[C]$, and $[P]$ vs. time, for 200 time units.

6.5 Autocatalytic Reactions

Autocatalytic reactions are those which catalyze the formation of the catalyst itself. This property causes autocatalytic reactions to show positive feedback characteristics that resemble the growth of organisms or populations of organisms. Autocatalytic reactions have been used frequently as models for growth of living organisms.

Classically, these reactions are exemplified by the activation of trypsin in the gastrointestinal tract. Trypsin is formed in the intestine from trypsinogen, a product of the pancreatic cells. This proenzyme or zymogen is converted to the active form by enterokinase of the small intestine. Trypsin, however, is itself capable of activating trypsinogen in the same way. The activation of the enzyme is therefore autocatalytic.

The process may be described by the following:

$$G + E \; \underset{\longleftarrow}{\overset{\longrightarrow}{\rule{3em}{0pt}}} \; C \; \xrightarrow{\rule{3em}{0pt}} \; T + E$$

where G represents trypsinogen, E is enterokinase, the activating enzyme, T is trypsin, and C is the enzyme-substrate complex.

Assuming steady-state conditions, this will simplify to

$$G \; \xrightarrow{E + T} \; T$$

Assuming further that the reaction is catalyzed equally by E and T, we can adapt the Michaelis-Menten model (Equation 2.18) to find the rate of formation of trypsin:

$$F = V \frac{[G]([E] + [T])}{K + [G]} \tag{6.29}$$

where V is the maximum rate of formation for a given unit of enzyme, K is the Michaelis-Menten constant, and $[E] + [T]$ is the total concentration of activating enzyme.

In this system, defining the flux F is Part 1 of the first stage of the simple Euler integration process. Part 2 is a definition of changes in concentration with

$$\Delta[G] = (-F)\Delta t \tag{6.30}$$
$$\Delta[T] = (+F)\Delta t \tag{6.31}$$

The second stage of numerical integration follows as usual:

$$[G] \leftarrow [G] + \Delta[G] \tag{6.32}$$
$$[T] \leftarrow [T] + \Delta[T] \tag{6.33}$$

This is an example of a simulation in which an analytical model (Equation 6.29) is used in conjunction with numerical integration.

Exercise 6-6: Write and implement a program which simulates the effect of an autocatalytic reaction. Use $K = 10$, $V = 0.1$, and employ simple Euler methods with $\Delta t = 0.1$. For initial conditions, use the following:

$$[G] = 100 \qquad [E] = 1.0 \qquad [T] = 0$$

As output, plot $[G]$ and $[T]$ simultaneously over a period of at least 60 time units.

Exercise 6-7: Modify the Chance-Cleland model (see Exercise 6-5) to include rate constants k_4, k_5, and k_6. These involve the interaction of product with enzyme to produce an enzyme-product compound which then converts to an enzyme-substrate compound as follows:

$$S + E \underset{k_2}{\overset{k_1}{\rightleftharpoons}} C \underset{k_6}{\overset{k_5}{\rightleftharpoons}} D \underset{k_4}{\overset{k_3}{\rightleftharpoons}} P + E$$

where C is the enzyme-substrate compound and D is the enzyme-product compound. Assume the following coefficients in addition to those listed in Exercise 6-5:

$$k_4 = 0.0013 \qquad k_5 = 0.11 \qquad k_6 = 0.1$$

Begin with $[D] = 0$ at $t = 0$. Graphical output for your simulation should show $[S]$, $[E]$, $[C]$, $[D]$, and $[P]$ over 40 units of time.

Exercise 6-8: The Chance-Cleland model may be modified further to simulate catalysis of bimolecular reactions. Consider a possible sequence of reactions in which $S + E$ produces an intermediate C_1, which reacts in an ordered fashion with a second substrate B, to produce intermediate C_2. This intermediate breaks down sequentially to product F and an intermediate C_3, which breaks down further to product D. The reactions are:

$$S + E \underset{k_2}{\overset{k_1}{\rightleftharpoons}} C_1$$

$$C_1 + B \underset{k_4}{\overset{k_3}{\rightleftharpoons}} C_2 \underset{k_6}{\overset{k_5}{\rightleftharpoons}} C_3 + F$$

$$C_3 \underset{k_8}{\overset{k_7}{\rightleftarrows}} D + E$$

Set up your simulation with these initial concentrations of reactants:

$$[S] = 100 \qquad [B] = 90 \qquad [E] = 10$$

Let other starting concentrations $= 0$. For the rate constants k_i:

$$k_1 = 0.005 \qquad k_2 = 0.005 \qquad k_3 = 0.01 \qquad k_4 = 0.001$$
$$k_5 = 0.1 \qquad k_6 = 0.1 \qquad k_7 = 0.01 \qquad k_8 = 0.1$$

Graphical output should show the concentrations of the eight reactants over at least 40 units of time. Use simple Euler methods with $\Delta t = 0.1$.

Conclusion

This chapter has presented examples of systems which lend themselves to deterministic simulations involving numerical integration. Only Exercise 6-1 could have been solved analytically. The other exercises involved sets of equations that would have been very difficult, if not impossible to solve by this method. They provided good examples of the power and simplicity of numerical integration.

CHAPTER 7

MODELS OF HOMOGENEOUS POPULATIONS OF ORGANISMS

In Chapter 6 we simulated chemical reactions among populations of molecules using deterministic models. We assumed that all the molecules of a given type were the same (homogeneous), and ignored the differences that exist among individual molecules. Our reaction rates were really average rates, and would not necessarily apply to any given molecule. Average rate is a reasonable predictor of overall reaction rate, given the large numbers of molecules involved in most chemical reactions.

When the first attempts were made to construct models of homogenous populations of organisms, it was logical to use some of the same principles as in models of homogeneous molecules. Particularly important was the "law of mass action" which states that the rate of interaction depends directly on the product of the concentrations of the interacting types of molecules. The pioneering work of Lotka (1925) frequently made use of this law. (This law is probably impossible to prove in a formal sense; its validity is based upon a very large number of observations, made without finding an exception.)

Homogeneous populations of organisms are assumed to be composed of a single type and described by a single variable, density. Differences in age, sex, genotype, phenotype, etc., are ignored or assumed to be irrelevant for the model. The density variable for organisms is identical to the concentration variable for molecules. Density is typically expressed as number per unit area (e.g. wolves ha^{-1}) or number per unit volume (e.g. *Paramecium* ml^{-1}). The models for homogeneous populations assume that interactions (predation, parasitism, competition) between populations will proceed at rates directly proportional to the product of the densities of the populations.

Most populations of most organisms are not homogeneous, and the unreality of the models in this chapter is admitted at the outset. However, they are constructed from principles that make elementary biological

sense. These models are important in the scientific literature of populations, even if most current references are critical. More importantly for us,
they provide simple models for simulation, and the understanding gained
in working with them will prepare you for the more complex models in
following chapters.

7.1 The Verhulst-Pearl Equation

The first model in Chapter 1 was an equation for simulating growth of
biological populations in an unlimited environment, with constant growth
rate:

$$\frac{dN}{dt} = kN \tag{1.3}$$

You used this equation extensively in the exercises of Chapter 5. Ev-
eryday observation suggests that this equation is not a good model of
biological growth, because we are not up to our ears in bacteria or ele-
phants. The simplest modification for limiting growth is to assume that
k in Equation 1.3 is not a constant, but decreases as N increases. The
simplest assumption is that k decreases linearly with N, so that

$$k = a - bN \tag{7.1}$$

If the organisms are scarce so that resources are assumed to be plentiful,
growth rate will approach a. There is also a greater, limiting density of
the population where growth rate is zero, and at densities greater than
this, growth rate is negative. If this expression is substituted for k in
Equation 1.3, the following equation results:

$$\frac{dN}{dt} = (a - bN)N = aN - bN^2 \tag{7.2}$$

This equation is one form of the Verhulst-Pearl logistic, first proposed
by Verhulst in 1838 and rediscovered by Pearl and Reed in 1920 (see
Hutchinson 1978). Another form was given in Chapter 1:

$$\frac{dN}{dt} = cN(L - N) \tag{1.16}$$

The equation is usually expressed with the following terminology:

$$\frac{dN}{dt} = rN\left(1 - \frac{N}{K}\right) \tag{7.3}$$

All of these forms are formally identical, with $a = r$, $L = K$, and $b = c =
r/K$. Here, r is the term for rate of per capita growth at minimal density,

and K is the term for the population density at which the growth rate is zero.

Equation 7.3 is perhaps the most easily interpreted form of the logistic. When N is small relative to K, $(1 - N/K)$ is close to 1 and growth is almost exponential at rate r. When $N = K$, the growth rate of the population is zero. Hence, a population will grow asymptotically to the value K. If N exceeds K, the growth rate is negative and the population declines toward K.

The form of the logistic in Equation 7.2 shows clearly the relationship of the logistic to the law of mass action. The first term aN is that for unlimited growth, and represents the growth potential of the population. The second term is negative, and is a function of $N \times N$. It may be thought of as a loss of potential population, resulting from the negative effects of the interaction of organisms, one with another. With time, population density will approach a steady-state in which the potential for increase is exactly balanced by the potential loss. The second term provides a good example of the application of the law of mass action to population dynamics.

Exercise 7-1: Pearl (1927) collected these data from a culture of yeast cells:

Hours :

0	1	2	3	4	5	6	7	8	9	10	12	14	18

Yeast :

4	7	12	19	28	48	70	103	140	176	205	238	256	265

Yeast were measured as biomass (mg 100ml^{-1}), and this may be taken as a suitable measure of population density. You should find estimates of values for r and K of yeast based on Pearl's data. This may be done by fitting constants using the techniques of Chapter 3. (Note that the form of the logistic used in Chapter 3 is like that of Equation 1.16.)

After finding values for r and K, simulate the growth of yeast using Equation 7.3 with numerical integration. Plot your simulation data (N) as a continuous line through time. On the same graph, also plot Pearl's data as discrete points, using circles if possible. To obtain a good fit of the simulation data to the actual data, it will be necessary to use the Improved Euler method with $\Delta t = 0.1$, or to use a small value of Δt (e.g. 0.0001) with the simple Euler method. (The latter approach may consume a lot of computing time.)

Exercise 7-2: Write and implement a program to simulate logistic growth using Equation 7.3. Set $K = 100$ and $r = 0.1$. Use a simple Euler method with $\Delta t = 0.1$, and begin the simulation with $N = 2$. Allow the simulation to run sufficiently long for N to approach K very closely. On the same graph, also show the result of a simulation for a population that begins with $N = 120$.

Write a short program to plot population growth rate (dN/dt) for values of N from 0 to 120. Use the values of K and r above. (Your graph will have to show negative rate values where N exceeds K.) Also produce a plot showing per capita growth rate $(dN/N/dt)$ for values of N from 1 to 120. This latter plot should produce the straight line of Equation 7.1. The x-intercept for both of these plots should be K. The first graph should show a maximum population growth rate of $rK/4$, at a density of $K/2$.

7.2 Time Lags and Oscillations in Population Growth

Population growth following the Verhulst-Pearl logistic results in a stable population size near K, the "carrying capacity". This stability is the result of a balance between the two elements of Equation 7.2, the positive feedback of the aN term, and the negative feedback of the $-bN^2$ term. A basic principle of control theory (see Chapter 17) is that oscillations around a stable limit may be caused by delay in the time that information about the current state is fed into the negative feedback loop.

This idea is interesting here because logistic growth to a stable upper limit is rarely displayed by organisms more complex than yeast and bacteria. Instead, their populations oscillate above and below an apparent upper limit. The classic observations of Slobodkin (1954) on *Daphnia* seem to be good examples of this. Evidently the fluctuations result from the passage of time between egg formation and egg hatching. The number of eggs a daphnid produces is not determined by resources available at the instant of hatching, but instead at the time the eggs are formed. Several days may elapse between egg formation and hatching, and the population may reach the stable limit during this interval. The result is a population that overshoots the limit, and must decline in density. There are time lags associated also with mortality and with recovery of reproductive potential. The result is an oscillating population.

Delays and lags are best simulated with age-class models like those discussed in Chapter 9. However, some approximations may be obtained by building a time lag into the logistic. For example, Equation 7.3 may be modified to be

$$\frac{dN_t}{dt} = rN_t\left(1 - \frac{N_{t-f}}{K}\right) \tag{7.4}$$

where N_{t-f} refers to the population density at f intervals of time prior to t:

$$N_{t-f} \ldots \to N_{t-2} \to N_{t-1} \to N_t$$

Writing a computer simulation of the time lag model requires that several previous population densities be retained so that N_{t-f} is available when needed.

Another type of time lag is the delay between birth and reaching reproductive age. This lag would appear in the rN term of the logistic (Equation 7.3). This lag may be simulated by using rN_{t-f} instead of rN_t. Here, f would represent the number of time intervals required to reach reproductive age. Both of these time lags are described by Wangersky and Cunningham (1957).

Exercise 7-3: Write a computer program to simulate populations growth in which there is a time lag in resource use, based on Equation 7.4. Set $K = 1000$, $N_0 = 2$, and $r = 0.5$. Allow your simulation to proceed for about 50 time intervals, plotting N_t against time. You should produce 5 different graphical outputs, for time lags of 0 to 4 time units. Use $\Delta t = 1$; otherwise, saving and updating of the previous N_t values will be cumbersome. The flowchart in Figure 7.1 and the program listing in Figure 7.2 may provide some assistance in writing your program.

7.3 Variable Carrying Capacity and the Logistic

The value of the carrying capacity K is constant in the above discussions of the logistic. Because organisms rarely live in environments with a constant K, it is instructive to consider simulations in which K is a variable. K might vary in an orderly fashion if the food supply of the population were tied to an annual cycle, for example. Consider an insect in the tropics, consuming vegetation whose amount is determined by wet and dry seasons. In Chapter 11 we shall work with several models for such weather variation. Here, we shall consider a very simple annual cycle based on a modified sawtooth or ramp function, as shown in Figure 7.3.

Such a line can be obtained with:

$$K_t = C + k\tau \tag{7.5}$$

where

$$\tau = \begin{cases} t & 0 \le t \le 26 \text{ weeks} \\ 52 - t & 26 < t \le 52 \text{ weeks} \end{cases}$$

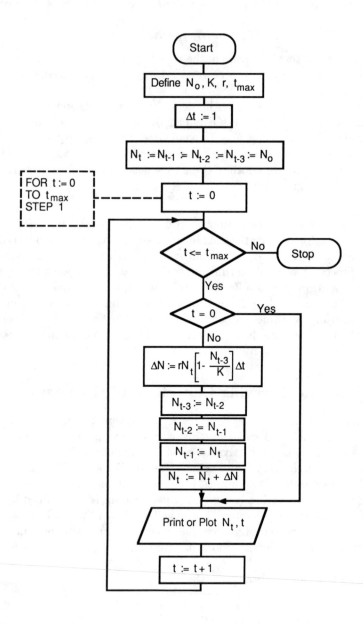

Figure 7.1. Flowchart for program to simulate population growth with the Verhulst-Pearl model with a time lag of three time units.

```
10 REM    THIS PROGRAM IS BASED IN THE
20 REM    GRAPH PROGRAM (APPENDIX 2)
30 REM ***********************************
80 REM        NT = POP SIZE AT TIME T
              N1 = POP SIZE AT TIME T-1
              N2 = POP SIZE AT TIME T-2
              N3 = POP SIZE AT TIME T-3
              N4 = POP SIZE AT TIME T-4
              N5 = POP SIZE AT TIME T-5
90 REM         K = CARRYING CAPACITY
               R = GROWTH RATE CONSTANT
              DN = DELTA N
100 :
110 XM = 50 :  YM = 4000
120 X$ = "TIME" :  Y$ = "DENSITY"
130 GOSUB 3000 :  REM DRAW GRAPH AXES
140 :
150 K = 1000:  R = .5
160 NT = 2
170 DT = 1
180 N1 = NT : N2 = NT : N3 = NT
190 N4 = NT : N5 = NT
200 XA = 1 :  XD = 0 :  REM LINE PLOT
210 :
220 FOR T = 0 TO 50
230    IF T = 0 GOTO 320
240    DN = R * NT * (1 - N5 / K) * DT
250    N5 = N4
260    N4 = N3
270    N3 = N2
280    N2 = N1
290    N1 = NT
300    NT = NT + DN
310    IF NT < 0 THEN NT = 0
320    Y = NT : X = T : GOSUB 4000
330 NEXT T
340 END
```

Figure 7.2. A sample BASIC program for simulating population growth based on the Verhulst-Pearl model, with a time lag of five time units.

In this equation C is the annual minimum of carrying capacity K, and k is the slope relating its increase to time expressed as a week of the year. The equation produces an increase in K for weeks 0 through 26 of a calendar year, and a decline for the remainder of the year. The upper population limit thus fluctuates on a roughly seasonal basis.

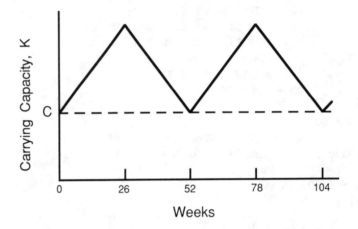

Figure 7.3. Diagram of a modified sawtooth pattern of annual variation in carrying capacity based on seasonal fluctuations with a peak in mid-year.

Exercise 7-4: Implement .a simulation for population growth in an environment in which K varies in an annual sawtooth pattern (Equation 7.5). Set $C = 30$ individuals, $N_0 = 2$ individuals, $r = 0.3$ week^{-1}, and $k = 6$ individuals week^{-1}. Run your simulation for 2 years (104 weeks), using a time unit of 0.1 week. Plot N_t and K_t vs. time on the same graph. Simple two stage Euler integration will be adequate for this simulation with $\Delta t = 1$.

7.4 A Simple Model for Harvesting Populations

Most biological populations are harvested or exploited by other biological populations. We can imagine a cat exploiting a mouse population in a barn. Left to itself, we assume the mouse population would follow a logistic growth pattern. However, the cat removes a constant number of mice per day, so that the logistic is modified with another term:

$$\frac{dN}{dt} = rN \left(1 - \frac{N}{K}\right) - C \qquad (7.6)$$

where C represents the removal rate of mice by the cat. (This term is unrealistic in the case where N is zero or less, because it indicates that the cat can catch mice when none exist. We will increase the realism of this term in the next section of this chapter.) Even this simpleminded model of harvest demonstrates that the cat has to plan her harvesting strategy with some care. The value of C has a realistic impact on the stable upper limit of population size, and it also imposes a critical lower limit. If the cat begins to harvest a small population of mice at rate C, the population may decline, but the same harvest rate applied to a larger population may permit the mouse population to grow.

Exercise 7-5: Using Equation 7.6, simulate harvest of a population growing logistically. Use the approach and values of Exercise 7-2 to set up the simulation. Use three values of C: 0, 1, 2. For each value of C, produce a graph showing N vs. time starting with populations of size 2, 15 and 30. To gain some insight into your results, again follow Exercise 7-2 in producing a graph of dN/dt vs. N, using the three values of C.

7.5 The Lotka-Volterra Model of Predation

Lotka in 1925 and Volterra in 1926 independently developed a model for predator-prey interaction (Hutchinson 1978). Although this model is as primitive as the logistic and the harvesting models above, it has served as a basis for many predation models and it is a useful starting point for an investigation of predation.

The model consists of two equations. The first describes the changes in prey population density:

$$\frac{dN}{dt} = rN - gNP \qquad (7.7)$$

In this equation, N is the density of prey and r is rate of prey population growth as in Equation 7.3. P is density of predators and g is a rate constant representing the efficiency of predation. Note that the prey population may grow without limits, except for whatever loss occurs by predation. If predators are absent ($P = 0$), then Equation 7.7 becomes the equation for exponential growth (Equation 1.3).

The equation for change in predator population density is

$$\frac{dP}{dt} = hNP - mP \tag{7.8}$$

In this equation the growth of the predator population is seen to depend solely on the existence of prey (hNP), so that when N is zero, the predator population declines exponentially with rate m, the predator's death-rate constant. The constant h includes the capture rate (g in Equation 7.7) multiplied by a factor for the efficiency of the conversion of captured prey into predators. Both equations include the cross-product term, (NP), that is typical of models based in the law of mass action.

To help in visualizing the behavior of models for two interacting species, it is useful to show the population densities graphed against each other, making a phase plot. The time dimension does not appear directly on these plots. It is also valuable for phase plots to include lines, called isoclines, that indicate equilibrium values for each species. An isocline effectively divides the phase plot into two regions, one in which the population will increase, and another in which it will decrease. With predator-prey models, prey density is conventionally given on the x-axis, and predator density on the y-axis. The prey isocline is a line comprising all the points for which $dN/dt = 0$. The equation for the line is found by setting Equation 7.7 to zero and solving to find

$$P = \frac{r}{g} \tag{7.9}$$

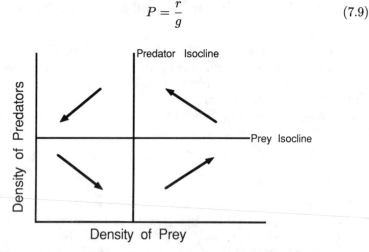

Figure 7.4. A phase plot of predator-prey population interactions, based on the simple Lotka-Volterra model. Construction of isoclines is described in the text.

This describes a line running parallel to the prey axis. Hence, for combinations of predator and prey densities above the line, the prey population will decrease; below it they will increase. An isocline for predator numbers may be constructed similarly by setting Equation 7.8 equal to zero and solving:

$$N = \frac{m}{h} \tag{7.10}$$

The predator isocline is a vertical line, to the right of which predators will increase, and to the left, decrease. A phase plot for predator-prey interactions following the Lotka-Volterra model is shown in Figure 7.4.

Exercise 7-6: Write and implement a computer simulation using numerical integration of the Lotka-Volterra predation model (Equations 7.7 and 7.8). Be sure to use the two-stage approach that was described in Chapter 6. Set the following values for the constants of the model:

$$r = 0.1 \qquad g = 0.002 \qquad m = 0.2 \qquad h = 0.0002$$

Start your simulation with $N = 1500$ and $P = 50$. Your output should consist of two graphs. The first should show density of prey and density of predators vs. time. (It may be necessary to divide or multiply predator or prey density by some constant to show meaningful fluctuations on the graph.) The second graph should be a phase plot of predator density vs. prey density. For clarity, your program should draw the predator and prey isoclines (use Equations 7.9 and 7.10), and plot the densities as a continuous line. For both graphs, allow the simulation to proceed for at least two complete predator-prey cycles.

The Lotka-Volterra model is very sensitive to small error in the numerical integration process. This simulation requires the accuracy of the Improved Euler method with a $\Delta t = 0.1$ or less. The simple Euler may be used with $\Delta t = 0.001$ or less, but this will result in a relatively slow output of results.

Exercise 7-7: Your simulation in Exercise 7-6 should display the stable oscillations characteristic of the Lotka-Volterra model. Ginzburg and Golenberg (1985) used the model to show that use of pesticides can be complex, even with this simple model. Assume the prey to be an agricultural "pest" insect that harms a crop. It is preyed upon by another insect which is therefore "beneficial". The population cycles of predator and prey follow the Lotka-Volterra model. Set up your program from Exercise 7-6 so that it will run for at least 4 complete cycles. Then modify the program to simulate the one-time application of a pesticide near the peak of the prey (pest) density of cycle number 2, so that the density of prey is suddenly lowered to 55 percent of its pre-pesticide density. After you have obtained the simulation results for this pesticide application, make another simulation run, but with the program modified so that the pesticide is applied at a time near the minimal prey density, instead of the maximum. The simple change in timing of pesticide application can have great impacts on dynamics of predator-prey systems.

7.6 Modifications of the Lotka-Volterra Predation Model

The Lotka-Volterra model is attractive because it can produce the oscillations that sometimes have been observed in some predator-prey interactions. As a result, the model has been studied, criticized and improved by dozens of modelers, who have published enough modifications of the basic model to fill a small book (Wangersky 1978). We will discuss briefly two of these modifications, chosen because they produce interesting simulations that mimic the fascinating graphical analyses of Rosenzweig (see Rosenzweig and MacArthur 1963, Rosenzweig 1969, Ricklefs 1979).

A rather obvious criticism of the basic model (Equations 7.7 and 7.8) is that the cross-product expression is unrealistic. The model follows the law of mass action, and is based on the assumption that predation losses always are proportional to prey density. Thus, doubling the prey population will result in a doubling of predation rate, regardless of the predator population. This provokes some mental images of slow-growing predators becoming exhausted with capturing and consuming rapidly-growing prey. Several models have incorporated a term to model satiation of predators, thus limiting predation rate. One of the simpler models uses the hyperbolic approach to an upper limit:

$$\frac{dN}{dt} = rN - \frac{uNP}{k + N} \qquad (7.11)$$

In this equation, u is a constant representing the maximum predation rate under satiating conditions, and k describes how rapidly the asymptote is approached. Asymptotic terms of this form have been used by Holling (1959) and Tanner (1975). One method of deriving this equation was shown in Section 2.4.

Another obvious problem with the Lotka-Volterra models is that there are no limits to the growth of either predator or prey population other than that imposed by each other. Leslie and Gower (1960) added some logistic-like terms to the predation model:

$$\frac{dN}{dt} = rN\left(1 - \frac{N}{K} - gP\right) \tag{7.12}$$

$$\frac{dP}{dt} = sP\left(1 - \frac{P}{jN}\right) \tag{7.13}$$

In Equation 7.12, growth of the prey population is limited both by the carrying capacity K, and by predation. The growth of predators (Equation 7.13) is a function of a growth rate constant s, and a constant j that relates maximum predator density to prey density. When predators are scarce and prey are abundant, predator growth rate is high and approaches sP as an upper limit. As prey becomes rare or predators more abundant, the parenthetical term for predator growth becomes small or negative, thus controlling predator growth rate.

Exercise 7-8: The following equation incorporates both the Holling-Tanner and the Leslie-Gower modifications to the Lotka-Volterra model for prey growth:

$$\frac{dN}{dt} = rN\left(1 - \frac{N}{K} - \frac{uP}{k+N}\right) \tag{7.14}$$

Implement a simulation of predator-prey dynamics with this equation and Equation 7.13, using the two-stage procedures of Euler numerical integration. The following are suggested values for the constants of the equations:

$r = 0.4 \quad K = 900 \quad u = 2.5 \quad k = 200 \quad s = 0.1 \quad j = 0.5$

Simple Euler integration is adequate with $\Delta t = 0.01$, or use the Improved Euler method with $\Delta t = 0.1$. Begin your simulation with initial values of $N = 25$ and $P = 5$. Plot the densities of predators and prey vs. time on a single graph. Allow sufficient time for the simulation to proceed to a steady-state.

Exercise 7-9: Solve Equations 7.13 and 7.14 for the values of the predator and prey isoclines. Modify the program from Exercise 7-8 so that it will draw these isoclines on a phase plot of prey density (x-axis) vs. predator density (y-axis). Run the simulation using the parameters and starting values of Exercise 7-8. Then repeat the simulation with these constants:

$r = 0.3 \quad K = 1000 \quad u = 1 \quad k = 200 \quad s = 0.03 \quad j = 0.45$

Begin the simulation with $N = 60$ and $P = 10$.

7.7 Volterra's Model for Two-Species Competition

Models for competing species are attempts to describe dynamics of populations of species that inhabit the same environment and use the same resources. These models are generally based on a pair of modified logistic equations that were first proposed by Volterra (Hutchinson 1978). The equations were developed subsequently in the classical publications of Gause and Witt (1935) and Gause (1934). The assumptions for this competition model are expanded slightly from those of the logistic growth model. The two populations involved are assumed to follow a pattern of logistic growth, and to be in an environment where both use and compete for some resource that limits their population densities. Thus the growth of each species will be controlled not only by its own density, but also by the density of the competitor.

The differential equations involved are

$$\frac{dN_1}{dt} = r_1 N_1 \left(1 - \frac{N_1 + \alpha N_2}{K_1}\right) \tag{7.15}$$

$$\frac{dN_2}{dt} = r_2 N_2 \left(1 - \frac{N_2 + \beta N_1}{K_2}\right) \tag{7.16}$$

In these equations, N_1 and N_2 are the densities of species 1 and 2, r_1 and r_2 are the growth rate constants for the two species, K_1 and K_2 are the carrying capacities for each species when growing alone, and α and β are competition coefficients. These equations are like the logistic (Equation 7.3) with the addition of another term describing the effect of the other species on population growth. This extra term is a function of the product $N_1 N_2$ that represents another application of the law of mass action.

The coefficient α is a factor that "converts" individuals of species 2 into individuals of species 1; α has units of N_1/N_2. For example, if α is 1.5, then from the view of species 1, an individual of species 2 uses as much of the limiting resource as 1.5 individuals of species 1. Likewise, the coefficient β converts individuals of species 1 into individuals of species 2, from the viewpoint of species 2; β has units of N_2/N_1. Rarely will $\alpha = 1/\beta$ because species generally do not have identical competitive abilities.

Like the Lotka-Volterra predation equations discussed above, it is helpful to plot data from simulations of competition as phase plots, graphing densities of N_1 vs. N_2. As in the case of predation, drawing isoclines divides the phase plots into regions where the populations increase and decrease. These lines are similarly solved by setting Equations 7.15 and

7.16 to zero. The isoclines for species 1 and 2 are, respectively,

$$N_1 = K_1 - \alpha N_2 \qquad (7.17)$$

$$N_2 = K_2 - \beta N_1 \qquad (7.18)$$

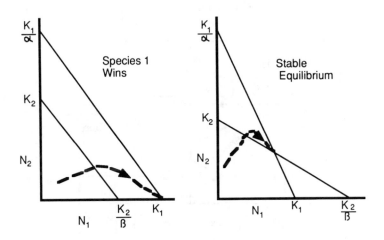

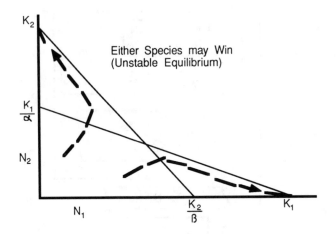

Figure 7.5. Sample phase plots for competitive interactions between two populations, based on the model of Volterra.

When drawn on a phase plot, Equation 7.17 has an intercept of K_1 on the N_1-axis and K_1/α on the N_2-axis. Similarly, Equation 7.18 has

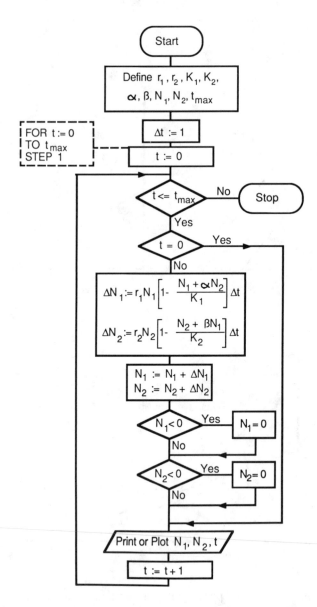

Figure 7.6. Flowchart for program to simulate population growth for two competing populations following the Volterra model.

an intercept of K_2 on the N_2-axis and K_2/β on the N_1-axis. Because Equations 7.17 and 7.18 describe straight lines on a phase plot, the iso-clines are easily drawn, defining the regions of the graph where the species grow and decline. The result of some competitive interactions may be pre-dicted easily from phase plots with isoclines. Figure 7.5 illustrates some examples. In general there are four possible graphical arrangements of the isoclines, representing the different possible combinations of K_1, K_2, α, and β in Equations 7.17 and 7.18.

Exercise 7-10: Write and implement a computer simulation for the competition model of Volterra. The flowchart of Figure 7.6 may be of assistance. Your output should be in the form of a phase plot diagram, as illustrated in Figure 7.5. If you are using the GRAPH program to produce your output, write your program so that it first draws the isoclines for the two species, and then plots the points for N_1 vs. N_2 through time as circles. Test your simulation first with these values for the constants of Equations 7.17 and 7.18:

$$r_1 = 0.8 \quad r_2 = 1.0 \quad K_1 = 300 \quad K_2 = 300 \quad \alpha = 0.5 \quad \beta = 0.6$$

Begin your simulation with $N_1 = N_2 = 10$. Use two-stage simple Euler integration, with $\Delta t = 1$. Allow the simulation to run for enough time to reach an equilibrium, about 25 to 30 time units. De-cide which of the possible four types of interaction this simulation demonstrates: stable equilibrium, unstable equilibrium, species 1 always wins, or species 2 always wins. Then, modify the constant values and rerun the simulation to produce three other graphical results that demonstrate the other three competitive interactions.

Conclusion

This chapter has been concerned with the classical techniques for mod-eling the dynamics of homogeneous populations. The models we have considered are based in the law of mass action. The simple cross-product approach lacks realism for most population interactions. The unrealistic assumptions of some of these models have prompted some biologists to question the validity of modeling and simulation generally. These models, however, have provided a starting point for descriptions of more complex systems, and were never really intended to be much more than this.

CHAPTER 8

SIMPLE MODELS
OF MICROBIAL GROWTH

In laboratory cultures bacteria and yeast follow the population models of the previous chapter more precisely than more complex organisms. This is not surprising, because microbes are more likely to behave as homogenous populations. Their life cycles are simpler and large numbers are usually involved in growth experiments. However, microbiologists seldom use the models of Chapter 7. The logistic model has been a starting point for biologists interested in more complex organisms and processes (e.g. predation and competition). Microbiologists have developed a number of models which are based on different assumptions.

Different models have been chosen probably because laboratory culture procedures differ. Water fleas, protozoans, flour beetles, fruit flies, and guppies were used in classic experiments describing logistic growth. Usually these animals were transferred regularly to containers of fresh medium to resupply food and remove waste products. When food was the limiting factor in the experiment, reproduction in populations grown to carrying capacity would halt until death released enough food for reproduction to resume.

The resupply of nutrients is difficult with microbial populations, which are generally grown in batch cultures or continuous cultures. So, while microbiologists have studied population growth intensively for decades, they have used a number of interesting models other than those we studied in the previous chapter. These models are the subject of this chapter.

8.1 The Monod Model of Microbial Growth

For a long time, microbiologists have recognized that microbes respond to their nutrient sources much like enzymes act on substrates. The most striking resemblance is the way growth rate of microbes responds to an increase in concentration of a limiting nutrient. Monod (1942) noticed that this response was hyperbolic, like the saturation behavior of enzymes (Figure 8.1).

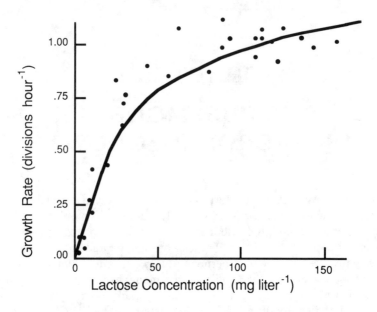

Figure 8.1. Graph of growth rate of the bacterial species *Es-cherichia coli* at different concentrations of lactose. Based on data from Monod (1942).

This behavior suggested that equations for enzymes could serve as models for some types of microbial growth, for example:

$$\frac{dB}{dt} = \mu B = \left(\mu_m \frac{[S]}{K_s + [S]} \right) B \qquad (8.1)$$

where B is cell density, μ is the specific growth rate constant, S is the concentration of nutrient, and μ_m is the maximal growth rate when nutrient concentration is completely unlimited. K_s is the half-saturation constant, which represents the substrate concentration that permits growth at half the maximum rate; thus, $[S] = K_s$ when $\mu = \mu_m/2$. In effect, K_s determines how rapidly a hyperbolic curve such as Figure 8.1 approaches the asymptote. B may be measured as biomass (e.g. mg liter^{-1}), or as density of cells (e.g. number ml^{-1}) assuming uniform cell size. μ has units of biomass (unit biomass)$^{-1}$ time^{-1}.

A formal derivation of this equation is probably not possible, but it may be obtained by analogy to the Michaelis-Menten equation. An enzyme-substrate interaction may be diagrammed with:

$$E + S \underset{k_2}{\overset{k_1}{\rightleftarrows}} ES \underset{k_4}{\overset{k_3}{\rightleftarrows}} P + E$$

where E is free enzyme, S is the substrate, ES is the enzyme-substrate compound, P is the product, and k_1, k_2, k_3, and k_4 are rate constants for the reactions. As you saw in Section 2.2, these reactions can be modeled with the following equation:

$$v = V_{\max} \frac{[S]}{K_m + [S]} \tag{8.2}$$

You may recall that this equation was derived by assuming a steady-state concentration for ES and assuming that v approached $V_{\max}$ as ES approached E_{total}.

In a similar way, the growth of cells in a culture may be described by a "reaction" like the following:

$$S + B \underset{k_2}{\overset{k_1}{\rightleftarrows}} SB \underset{k_4}{\overset{k_3}{\rightleftarrows}} B + P$$

In this hypothetical and not very precise description, B refers to living cells, S is the limiting nutrient, SB is the combination of cells plus absorbed but unassimilated nutrient, and P is the new cell biomass derived from the metabolized nutrient. k_1 is rate of nutrient intake, k_2 is the rate of nutrient loss through excretion, and k_3 is rate of formation of new cellular material that includes the nutrient. In this scheme, k_4 is zero because P is new cell biomass and is not distinguishable from B. (k_4 might take on a positive value in the case of cannibalism.) Because of metabolic requirements, more S must be used by the cells than appears in the amount of new biomass, P. The process may exhibit the positive feedback characteristics of the autocatalytic enzyme reaction (see Section 6.5).

As with enzymes, new cell material is formed fastest when cell biomass and associated systems are saturated with nutrient in the form of SB. Considerable evidence for the validity of the Monod model has accumulated since it was first proposed (Dugdale 1967, Powers and Canale 1975). In practice, values of μ are found for different nutrient concentrations S in experiments. The curve-fitting techniques of Chapter 3 are then used with the hyperbolic equation to find values for K_s and μ_m, the constants in Equation 8.1.

In many experiments, the occurrence of respiration, mortality and autolysis have been found to decrease growth rates independently of nutrient

concentration (Herbert 1958). For these cases, Equation 8.1 is modified
to include an additional term to account for these losses:

$$\frac{dB}{dt} = \mu B = \left(\mu_m \frac{[S]}{K_s + [S]} - R \right) B \qquad (8.3)$$

where R is a rate constant describing loss of biomass.

8.2 Batch Culture of Microorganisms

Batch cultures are started with a few microorganisms introduced into
a container of sterile nutrient solution. The population grows until the
nutrient supply is depleted; then it begins to decline as organisms die or
become dormant. The typical S-shaped portion of the growth curve may
be logistic. However, when a microorganism in a batch culture dies, it
may not release nutrients that other microbes can use for reproduction
to replace the dead microbe. Hence, batch cultures may not conform to
logistic assumptions. A population that does meet these assumptions will
neither decline nor increase when it is at carrying capacity.

Exercise 8-1: Use Equation 8.3 in a simulation of bacterial growth in a
batch culture with a limiting nutrient. Assume a fixed amount of
nutrient, $[S]_{total}$. The nutrient found in cells, free nutrient concentration $[S]$ in the culture medium, and total nutrient concentration
follow this simple relationship:

$$[S]_{total} = [S] + aB \qquad (8.4)$$

where a is the proportion of biomass concentration B that is made
up of absorbed nutrient. Use the following values for constants in
your simulation:

$$R = 0.03 \, \text{hr}^{-1} \qquad a = 0.03 \qquad \mu_m = 0.3 \, \text{hr}^{-1}$$
$$[S]_{total} = 100 \, \text{mg liter}^{-1} \qquad K_s = 25 \, \text{mg liter}^{-1}$$

Assume an initial biomass of 1 mg liter^{-1} and allow the simulation
to proceed to steady-state. Be sure to adjust your initial value
of $[S]$ to include the amount of nutrient contained in the initial
inoculum. Use the usual two-stage Euler method to perform the
integration. Because of the potential for instability with numerical
integration, use an Improved Euler integration with $\Delta t = 0.1$, or
a simple Euler with $\Delta t = 0.01$. Output for your simulation should
be B and $[S]$ plotted on the same graph through time.

8.3 Continuous Cultures in Chemostats

Chemostatic cultures are used both in laboratory research and in industrial production of the useful results of microbial growth (e.g. antibiotics). Some sewage treatment plants are designed as chemostats.

A chemostat is set up with the microorganisms in a container (reactor) of liquid nutrient medium. The medium in the container is usually stirred or agitated continuously to prevent development of nutrient gradients and settling of the microorganisms. At a carefully controlled rate, fresh nutrient medium is put continuously into the reactor from a reservoir. Medium is removed from the container at the same rate as it is added; this effluent contains unused medium, microbial cells, and metabolic products. The approach used here follows the description of Novick and Szilard (1950a,b).

A chemostat may be diagrammed in this way:

$$V$$

$$F,[S_i] \longrightarrow [S_i] \longrightarrow D$$

$$B$$

V is the volume of the chemostat culture vessel, F is the flow rate through the chemostat, and B is cell biomass density. $[S_i]$ is the concentration of nutrient entering the chemostat, and $[S]$ is concentration of nutrients in the container and in the outflow. D is the dilution rate, found by F/V, with a dimension of time^{-1}. The rate of growth (or decline) of cells in the chemostat is described by

$$\frac{dB}{dt} = \mu B - DB \qquad (8.5)$$

where μ is the growth rate constant, and can be obtained from Equation 8.1 for any given $[S]$. When the chemostat population is at steady state, $dB/dt = 0$ and $\mu = D$.

The change in substrate concentration in the chemostat is defined with the equation:

$$\frac{d[S]}{dt} = D[S_i] - D[S] - \mu Ba \qquad (8.6)$$

Here, $D[S_i]$ defines the rate of nutrient input to the chemostat and $D[S]$ is the rate at which unused nutrients are washed out. The constant a is the fraction of cell biomass that is comprised of nutrient (see Equation 8.4), so that μBa represents the rate of uptake of nutrients by the cells.

Equations 8.5 and 8.6, coupled with 8.1, allow a complete simulation of chemostat behavior. A variety of objectives are possible for different combinations of input data. For example, it is possible to determine

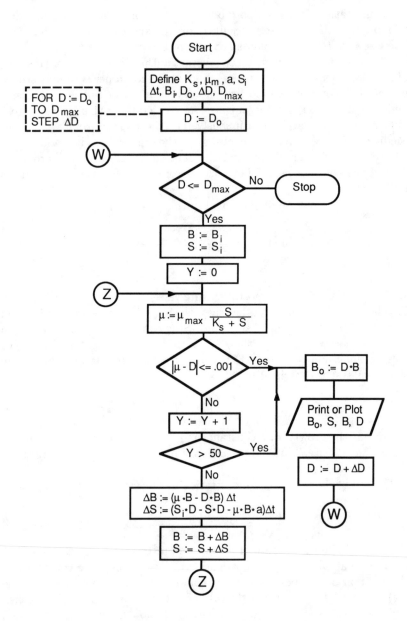

Figure 8.2. Flowchart for program for steady-state simulation of chemostat (Exercise 8-3).

dilution rates that will maximize cell biomass output, B_o, from the system for a given $[S_i]$.

Exercise 8-2: Use Equations 8.5, 8.6 and 8.1 to simulate growth of a microorganism in a chemostat. Assume an input nutrient concentration, $[S_i] = 100$ mg liter^{-1}. Set initial biomass $B = 10$, and initial nutrient concentration $[S] = [S_i]$ in the chemostat. Set the constants of the system as follows:

$$K_s = 75 \qquad \mu_m = 1.5 \qquad a = 0.013$$

Examine the effects on B and $[S]$ of different dilution rates: 0.05, 0.25, 0.50, and 0.70. Your output should consist of two graphs, the first showing B for the 4 levels of D, and the second showing $[S]$ for the 4 levels. Show these plotted over about 50 units of time. Numerical integration can be unstable at low rates of dilution, so use the simple two-stage Euler procedure with $\Delta t = 0.1$.

Exercise 8-3: Considerable insight into the behavior of a chemostat can be obtained from a simulation of steady-state conditions. The basic simulation of Exercise 8-2 can be inserted into a program which varies D, and allows the simulation to iterate to a steady state. The program can show the steady-state values for different characteristics of the chemostat. Write a program that will vary

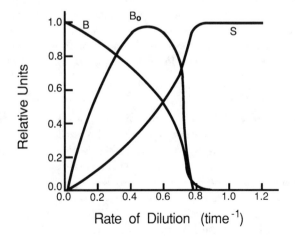

Figure 8.3. Graph of data from simulation of steady-state conditions in a chemostat at various rates of dilution. See Exercise 8-3.

dilution rate from 0.02 to 1.2 in increments of 0.02. Use $\Delta t = 1.0$. Plot out the results as in Figure 8.3, which shows typical steady-state results. A flowchart for the simulation is given in Figure 8.2.

8.4 Multiple Limiting Nutrients

The Monod equation for microbial growth has been used in a variety of simulation studies. The Monod equation may not be precisely correct, but it provides a convenient starting point for modeling populations of cells and even rather complex animals, e.g. rotifers (Boraas 1983) and daphnids (Vyhnalek 1987). Because it is an analogue of the Michaelis-Menten equation, modelers are able to draw upon a large body of literature that deals with enzyme kinetics. Chen (1970) used the Monod equation as a starting point for describing the effect of multiple nutrients:

$$\mu = \mu_m \left(\frac{[S]}{K_s + [S]} \right) \left(\frac{[N]}{K_n + [N]} \right) \left(\frac{[P]}{K_p + [P]} \right) \tag{8.7}$$

This equation could be employed as a model for the growth of diatoms, for example, where S, N, and P might represent the three most likely limiting nutrients: silicon, nitrogen, and phosphorus. The half-saturation constants for each of these elements would then be given with K_s, K_n, and K_p. The growth rate μ_m is presumably determined with all three elements in saturating concentrations. The origin of this and similar equations is evidently Cleland's (1970) equation for enzyme kinetics with multiple substrates.

The free nutrients in this system at any time may be determined from a set of three equations, each following Equation 8.4. For example:

$$[S]_{\text{free}} = [S]_{\text{total}} - a_s B$$

where $a_s B$ represents the fractional biomass of the organism that is derived from S. B is the biomass density, and a is the coefficient relating a nutrient to biomass. We will assume that a is constant, although this is probably not the case in many circumstances (Boraas 1983). The use of Equation 8.4 implies a batch culture procedure, rather than the continuous chemostatic culture of the preceding section.

Equation 8.7 may be used to simulate an ecological or agricultural principle, Liebig's Law of the Minimum. Briefly, this law states that growth of a population will be limited by the factor which exists in least supply relative to the need of the population. Liebig's idea was that the factors limiting the yield of crops could be identified and removed in turn (Hutchinson 1973).

Exercise 8-4: Write and implement a program to simulate multiple nutrient limitation of diatom growth based on Equation 8.7. (See also Equations 8.3 and 8.4.) For the first part of the batch culture simulation, set up your simulation with the following constants and initial values:

Total nutrient concentrations (mg liter^{-1}):

$$S_t = 10.0 \qquad N_t = 1.00 \qquad P_t = 0.10$$

Fractional composition of diatom biomass:

$$a_s = 0.15 \qquad a_n = 0.10 \qquad a_p = 0.005$$

Half-saturation constants (mg liter^{-1}):

$$K_s = 0.10 \qquad K_n = 0.15 \qquad K_p = 0.04$$

Initial diatom biomass (mg liter^{-1}):

$$B = 0.10$$

Rate of loss for mortality, etc. (hr^{-1}):

$$R = 0.005$$

Maximum growth rate at saturation (hr^{-1}):

$$\mu_m = 0.120$$

Use a two-stage simple Euler integration in this simulation, with $\Delta t = 1$. Let the simulation proceed for about 120 hours, and plot 4 lines on the same graph, showing B, $[S]$, $[N]$ and $[P]$. Based on these simulation results, identify the nutrient that is "limiting" in this system.

Exercise 8-5: Modify the simulation of Exercise 8-4 to test your identification of the limiting nutrient. First plot B over time using the values for S_t, N_t and P_t given in Exercise 8-4. Then rerun the simulation using 100 times each of the baseline nutrient concentrations in turn, to discover which nutrient is limiting to this hypothetical diatom population.

You should discover that one of the nutrients limits the final level of B, but another limits rate of growth. This distinction between different limits for Monod-Herbert models of Leibig's Law is discussed in O'Brien (1972, 1973), Holmes (1973), and Kelly and Hornberger (1973).

8.5 Competition for Limiting Nutrients

In Chapter 7 we discussed competitive interactions among organisms, using equations for interaction based on the logistic limits to population growth. The Monod-Herbert model (Equation 8.3) has been used for some time in equivalent models to describe the interaction of two or more species competing for some nutrients.

Powers and Canale (1975) developed a model of green algae and blue-green algae (cyanobacteria) competing for nitrogen and phosphorus, two important limiting nutrients for plants in lakes. Some blue-green algae differ significantly from green algae because some species can "fix" molecular nitrogen. These blue-green algae are able to use the atmospheric nitrogen gas dissolved in water as a source of nitrogen that is not available to the green algae. The interaction between these two groups is important, because blue-green algae are capable of causing "problems" for humans when they are abundant in water. The following description of this model is simplified from the original and omits many factors known to be significant in determining nutrient dynamics, including runoff, mixing, sinking of algae, and predation by herbivorous zooplankton. However, it still proves to be an interesting and challenging simulation.

We will assume a lake with two types of algae, blue-green and green, and assume that they respond as homogeneous populations. We will further assume that their growth is limited by nitrogen and phosphorus dissolved in the water, and that we can treat the lake as a batch culture, with the initial nutrients either remaining dissolved in the lake water or incorporated in the algae. We will assume that algal growth rates can be described with equations for multiple limiting nutrients (see Equation 8.7):

$$\mu_G = \mu_{mG} \left(\frac{[P]}{K_{pG} + [P]} \right) \left(\frac{[N]}{K_{nG} + [N]} \right) \tag{8.8}$$

$$\mu_B = \mu_{mB} \left(\frac{[P]}{K_{pB} + [P]} \right) \left(\frac{[N] + [n]}{K_{nB} + [N] + [n]} \right) \tag{8.9}$$

The inclusion of $[n]$ as an extra term in Equation 8.9 is to permit a rough simulation of the fixation of dissolved molecular nitrogen gas by the blue-green algae. The terms in these two equations are defined as follows:

μ_G is growth rate of green algae;

μ_B is growth rate of blue-green algae;

μ_{mG} is growth rate of green algae at saturation with N and P;

μ_{mB} is growth rate of blue-green algae at saturation with N and P;

$[P]$ is concentration of available phosphorus;

$[N]$ is concentration of available nitrogen;

$[n]$ is concentration of dissolved molecular nitrogen gas;

K_{pG} is the half-saturation constant for phosphorus and green algae;

K_{pB} is the half-saturation constant for phosphorus and blue-green algae;

K_{nG} is the half-saturation constant for nitrogen and green algae;

K_{nB} is the half-saturation constant for nitrogen and blue-green algae.

In this simplified model, light and temperature variation are assumed to be negligible, although Powers and Canale (1975) considered them to have important impacts on μ_{mG} and μ_{mB}. The equations describing change in biomass will be based on the Monod-Herbert model (Equation 8.3):

$$\frac{dG}{dt} = \mu_G G - R_G G \qquad (8.10)$$

$$\frac{dB}{dt} = \mu_B B - R_B B \qquad (8.11)$$

where G and B refer to the biomass concentrations of the green and blue-green algae, and R_G and R_B are the mortality and respiration loss constants for each algal type. Unlike the description of bacterial growth of Section 8.2 above, algal death and respiration will release the nutrients P and N to the lake water. As a result, the available phosphorus concentrations can be found with

$$[P] = [P]_t - a_{pG}G - a_{pB}B \qquad (8.12)$$

where $[P]_t$ is concentration of total phosphorus, a_{pG} is the fraction of phosphorus in green algal biomass, and a_{pB} is the fraction of phosphorus in blue-green algal biomass.

Because the blue-green algae can fix molecular nitrogen, $[N]_t$ will not be constant, but will increase as the concentration of blue-green biomass increases. In addition, $[N]$ may be lost from the system by bacterial breakdown of nitrogenous compounds to molecular nitrogen, a process termed denitrification. $[N]_t$ must be adjusted to account for these two processes:

$$\frac{d[N]_t}{dt} = -D[N] + a_{nB}B\mu_{mB}\left(\frac{[N] + [n]}{k_{nB} + [N] + [n]} - \frac{[N]}{k_{nB} + [N]}\right) \qquad (8.13)$$

where $[N]_t$ is the total nitrogen concentration, D is the rate constant for denitrification, and the parenthetical expression accounts for the increase of blue-green algal biomass that is attributable to the fixation of molecular nitrogen. Because most lake waters are saturated with molecular nitrogen, $[n]$ may be considered a constant for the lake. After adjusting the total amount of nitrogen $[N]_t$ for changes due to nitrogen fixation and denitrification, the concentration of available nitrogen can then be found following the form of Equation 8.12:

$$[N] = [N]_t - a_{nG}G - a_{nB}B \qquad (8.14)$$

Exercise 8-6: Use the information and equations above to write and implement a computer program to simulate the competition between green and blue-green algae for limiting P and N. Use the following constants:

$$K_{pG} = 0.05 \text{ mg liter}^{-1} \qquad a_{pG} = 0.01 \text{ mg } P \text{ (mg biomass)}^{-1}$$
$$K_{pB} = 0.03 \text{ mg liter}^{-1} \qquad a_{pB} = 0.01 \text{ mg } P \text{ (mg biomass)}^{-1}$$
$$K_{nG} = 0.30 \text{ mg liter}^{-1} \qquad a_{nG} = 0.08 \text{ mg } N \text{ (mg biomass)}^{-1}$$
$$K_{nB} = 0.20 \text{ mg liter}^{-1} \qquad a_{nB} = 0.08 \text{ mg } N \text{ (mg biomass)}^{-1}$$
$$\mu_{mG} = 2.0 \text{ day}^{-1} \qquad R_G = 0.06 \text{ day}^{-1}$$
$$\mu_{mB} = 1.0 \text{ day}^{-1} \qquad R_B = 0.04 \text{ day}^{-1}$$
$$D = 0.05 \text{ day}^{-1}$$

Set the constant concentrations as follows:

$$[n] = 0.02 \text{ mg liter}^{-1} \qquad [P]_t = 0.05 \text{ mg liter}^{-1}$$

Set the variable concentrations with these initial values:

$$G = 0.01 \text{ mg liter}^{-1}$$
$$B = 0.01 \text{ mg liter}^{-1} \qquad [N]_t = 0.1 \text{ mg liter}^{-1}$$

Use the two-stage simple Euler integration for this simulation, with $\Delta t = 1.0$. The output for this simulation could take many different forms. As a minimum, show $[N]_t$ and the concentrations of B, G and available N over a simulated 240 days.

8.6 Toxic Inhibition of Microbial Growth

As concern about the occurrence of toxic materials in the environment has risen in recent years, so has the number of attempts to model the effect of toxic materials on microbial populations (e.g. Bates et al. 1982, Rai et al. 1981, Vaccaro et al. 1977). Modifications of the elementary Monod model for microbial growth can approximate the effect of toxic materials in some circumstances.

Most biochemistry textbooks describe ways of modifying the basic Michaelis-Menten enzyme model (Equation 8.2) to show different types of inhibition or interference with enzyme activity. The Monod model for microbial growth may be modified similarly. For example, the following equation modifies the growth rate constant for the Monod model (Equation 8.1) to describe a reaction to toxicants similar to non-competitive enzyme inhibition:

$$\mu = \left(\frac{1}{1 + [T]/K_T} \right) \mu_m \left(\frac{[S]}{K_s + [S]} \right) \tag{8.15}$$

K_s, $[S]$, μ_m and μ are defined as in the Monod model (Equation 8.1), and $[T]$ is the concentration of toxicant. K_T is a toxicity constant and is equal to the concentration of toxicant that causes a halving of the value of μ_m. The action of the toxicant is to reduce the value of μ_m so that regardless of the concentration of nutrient $[S]$, the population cannot grow as rapidly with the toxicant present as it can without the toxicant.

The Monod growth constant may be modified also to show another possible toxic effect, this one resembling the activity of competitive enzyme inhibition:

$$\mu = \mu_m \left(\frac{[S]}{[S] + K_s \left(1 + [T]/K_T \right)} \right) \tag{8.16}$$

Here the toxicant increases the concentration of limiting nutrient required to achieve a particular level of growth. In effect, K_T is equivalent to the concentration of the toxicant that will produce a doubling of the value of K_s.

In actual practice it is difficult to discriminate between these two modes of toxic effect. In most toxicity experiments, different levels of toxicant are added to replicate cultures of the microorganism, and observations are made of the decline in growth rate that results (e.g., Gillespie and Vaccaro 1978). If you wanted to decide experimentally which of the two modes of toxic activity a particular toxicant follows, you would add a range of toxicant concentrations across a range of substrate concentrations. This can be tricky because of the possibility of interactions between a toxicant and increased levels of nutrients. Molecules of toxic materials often become attached to organic molecules, which may serve to increase or decrease their toxicity in unexpected ways.

Exercise 8-7: Select either Equation 8.15 or 8.16 as the basis of a simulation of the effect of adding a toxicant to a batch culture (Exercise 8-1). Use the following constants:

$$\mu_m = 1 \quad K_s = 25 \quad R = 0.03 \quad a = 0.03 \quad K_T = 0.01$$

Initiate your batch cultures with biomass concentrations of 1 mg liter^{-1}. Make your simulation of two parts. First plot the rate of change of biomass, dB/dt, against toxicant concentration from 0 to 0.04 mg l^{-1}, in increments of 0.001 mg l^{-1}. Do this for at least 4 different values of $[S]$. These values should cover a fairly wide range, including much less than K_s, about equal to K_s, greater than K_s, and very much greater than K_s to simulate saturation. Secondly, plot the growth of the culture through time with at least six different concentrations of toxicant: 0, 0.001, 0.01, 0.1, 1, and 10. This log series of toxicant levels will approximate the first-trial procedure of an experimental determination of the reaction to various toxicant levels.

Conclusion

The Monod model provides an alternative to the logistic for describing how homogeneous populations grow in limited environments. Like the logistic, it is based on simple assumptions which are infrequently met even in carefully controlled experiments. A fair amount of printer's ink has been spilled describing the failures of the Monod model. For example, the concentrations of nutrients are known to differ among cells of different ages; some nutrients will affect growth of biomass and others affect rates of cell division; and, the "luxury consumption" of nutrients like phosphorus is not considered.

Like the Verhulst-Pearl logistic, the Monod model is used not because it is a "good" or a "bad" model, but because it is simple. In the present case, it serves as a tractable model that provides the basic starting point for models of microbial growth.

CHAPTER 9

POPULATION MODELS
BASED ON
AGE-SPECIFIC EVENTS

In the previous two chapters we studied homogeneous populations, with all members of populations assumed to be identical. We essentially ignored birth and death processes, lumping them together as "growth rate". However, age makes a difference in the performance of complex organisms, and this must be considered if we wish to make realistic models of their populations. Ability to reproduce depends on age, with some members of a population more likely to reproduce than others. Death rates change with age of organisms, and age is important in changing the impact of disease, parasitism, predation, etc. on an individual. Age-class models that consider these differences are the subject of this chapter.

9.1 Age-Specific Survival and Reproduction

Members of populations of most organisms do not all die at the same age. (The life insurance industry is based on this observation.) If enough data about age of death can be collected for a population, an overall pattern can be seen for age at death, also called age-specific mortality or survivorship. Three common patterns are diagrammed in Figure 9.1 (Pearl and Miner 1935, Deevey 1947, Slobodkin 1980). Type A is characteristic of species with relatively high death rates for very young individuals, low and constant mortality rates for intermediate ages, and higher mortality rates again for older individuals. This type of curve is found for most human populations and for many mammals. Type B is produced by a constant mortality rate, with a constant percentage of individuals dying for each unit of time. (This is identical with the exponential die-off simulation of Section 1.2.) Adults of some species of birds and bats may follow this curve (Deevey 1947, Keen and Hitchcock 1980). Type C is characteristic of most organisms, with a high rate of mortality early in life, and a relatively low rate for the later periods. This curve is typical of many fish

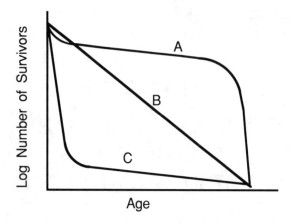

Figure 9.1. Three common patterns of survival in populations. See text for identification and discussion of these types.

and marine invertebrates and of most plants; an organism surviving the hazards of early life has a good chance of living to a relatively old age.

Just as plants and animals in a population usually die at different ages, so do they reproduce at different ages, producing different numbers of offspring. Figure 9.2 shows three common patterns of variability of reproduction with age. Type A describes organisms that reproduce once in their lifetime. This is common in annual plants, and in species of salmon. Other organisms such as trees and many species of fish show type B, in which young animals do not reproduce, and in which newly reproductive individuals do not produce as many offspring as older individuals. Type C is found in species that show increasing reproductive ability after sexual maturity, and then a decline toward zero reproduction at older ages. This type is typical of most populations of humans.

9.2 Life Tables

A life table presents age-specific information about the survival and reproduction of a population. The table is a handy way of summarizing the data shown in Figures 9.1 and 9.2. The following statistics are usually presented in a life table:

x = the age of the organisms at the beginning of the time interval;
N_x = number of individuals alive at age x;
d_x = number of individuals dying between ages x and $x + 1$;
s_x = proportion of individuals alive at x that are also alive at $x+1$;
m_x = number of offspring produced between x and $x + 1$ by the average individual alive at x.

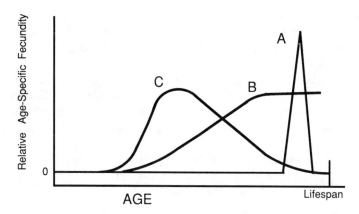

Figure 9.2. Three common patterns of variation in reproductive capacity with age in populations.

Other terms are frequently incorporated in a life table, depending upon its objective. The terms s_x and m_x are the age-specific survival and natality rates. These two are the terms used in modeling age-structured populations.

The data used to construct a life table may be obtained in a variety of ways. One method involves following the performance of a group of individuals from birth, recording their ages at death and the number of offspring at each age. Another method requires a single census over a unit period of time, looking at the reproductive rates of the different ages present, and the rates of death of the age groups in the population. For either method, the population is assumed to be under fairly stable environmental conditions, so that life table values can be applied through time.

The usual life table is set up for females only. The assumption behind this procedure is that males and females make up constant proportions of the population. In such female life tables, age-specific natality will refer only to female births. Some populations may exhibit significant survival differences between the sexes; certain populations of deer, for example, have higher mortality among older males which are hunted intensively. Adequate models of such populations often require use of a different life table for each sex, as in Section 9.5 below.

Life tables are most often based either on a description of the fate of 1000 births, or on a proportional survivorship basis. Where life tables are used to simulate the performance of populations, more realistic numbers may be employed, based on actual population sizes or densities. However, survival and natality will be expressed as rates, usually as number per female per unit of time.

Age-specific survival will always be a decimal fraction less than unity, and can be considered as the probability that an organism of age x will survive to age $x + 1$. The idea of probability will be useful in stochastic simulations later. Age-specific natality can be greater or less than unity, depending upon the organism and the time intervals. The units of time used in life tables will vary among different organisms. Life histories of some small invertebrates may be described in hours, and some longer-lived species in years. Life tables for humans are often given with five-year intervals.

Table 9.1 is a sample life table for a population of a common terrestrial crustacean, and contains the information needed to develop an age-class simulation of a population of this animal.

Age in years x	Number living at age x N_x	Number dying x to $x+1$ d_x	Rate of survival x to $x+1$ s_x	Per capita birth rate x to $x+1$ m_x
0	10000	8896	0.1104	0.00
1	1104	989	0.1042	3.13
2	115	99	0.1391	42.53
3	16	14	0.1250	100.98
4	2	2	0.0000	118.75
5	0			

Table 9.1. Life table for females of a California population of a sowbug, *Armadillidium vulgare*. Adapted from data by Paris and Pitelka (1962).

In this chapter we will use an easily understood stepwise approach to simulating events in a population that follows a particular life table. An approach using matrices is simpler to program for computers, but is more complex in concept and will be deferred to Chapter 16.

The equation used to describe the reproduction that takes place in one time period would be

$$\Delta N = \Sigma N_x m_x \tag{9.1}$$

$$= N_0 m_0 + N_1 m_1 + N_2 m_2 + N_3 m_3 + N_4 m_4 \tag{9.2}$$

The survival of each age class is calculated with the following equations, which also move the survivors into the next age class. The order of their solution is important in writing a life table program:

$$N_5 \quad \leftarrow \quad N_4 s_4 \qquad (9.3)$$

$$N_{x+1} \quad \leftarrow \quad N_x s_x \qquad (9.4)$$

$$N_1 \quad \leftarrow \quad N_0 s_0 \qquad (9.5)$$

$$N_0 \quad \leftarrow \quad \Delta N \qquad (9.6)$$

$$\Sigma N = N_0 + N_1 + N_2 + N_3 + N_4 + N_5 \qquad (9.7)$$

Note that the number of births is calculated first with Equation 9.1, and saved in a temporary variable ΔN. The value of ΔN is assigned to N_0 only after the previous N_0 has been used to find N_1. The total population size ΣN calculated in Equation 9.7 is the total number existing at the end of the time interval, or the start of the next time interval.

Exercise 9-1: Write a program for the simple age-class simulation of a sowbug population that has constant survival and natality characteristics as described in Table 9.1. Set up your simulation to start at time = 0 using the values from the N_x column of Table 9.1 as initial values for the various age classes. Generate 12 years of simulation data. For each year from 0 to 12, your program should print out a table that shows the number of individuals in each age class, as well as the total population size. Figure 9.3 is a flowchart to help you set up your simulation.

Exercise 9-2: Populations that follow a fixed schedule of age-specific natality and survival are known to attain a "stable age distribution" after passing through several generations. With a stable age distribution, the size of each age class is a constant proportion of total population size. Alter the program for Exercise 9-1 to follow a sowbug population for 20 years, printing out the size of each age class at the end of the 20-year period. Also print out the proportion of the total size represented by each age class. Perform 20-year simulations with at least four different sets of initial age-class numbers, to demonstrate that the population does in fact attain a stable age distribution. You should be as inventive as possible in selecting your data.

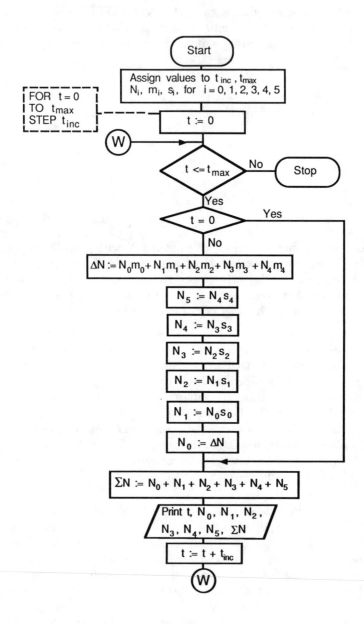

Figure 9.3. Flowchart for program to simulate age-class survival and reproduction in the population of sowbugs discussed in the text.

9.3 A Logistic Modification for Life Table Simulations

Simulations that involve fixed natality and survival schedules assume these rates to be independent of density effects produced by growth or shrinkage of the population. The population of sowbugs following the schedule of reproduction and mortality in Table 9.1 will grow exponentially. An age-class model can be modified readily to simulate a logistic growth pattern. We will make one of the easier changes here, using the sowbug example.

We will assume that survival rates are independent of population density, so that only rates of natality are affected by density. As a first step in the modification, natality rates are adjusted so that the population will just replace itself when the age distribution is stable. (See Exercise 9-2 for a description of such stability.) Ideally these rates would be derived from experiment, but here we will simply reduce all the m_x values by a constant fraction, so that $N_0 = \Sigma N_x m_x$ when the population has a stable age distribution. For the sowbugs, the adjusted set of natality rates are: $m_0 = 0$, $m_1 = 3.06$, $m_2 = 41.66$, $m_3 = 98.27$, $m_4 = 115.392$.

The second step in the modification is to change Equation 9.1 of the life-table model so that reproduction is limited to replacement of dying animals when population abundance is at the carrying capacity K:

$$\Delta N = (N_0 m_0 + N_1 m_1 + N_2 m_2 + N_3 m_3 + N_4 m_4) \cdot \left(2 - \frac{\Sigma N}{K} \right) \quad (9.8)$$

The m_x values in this equation must be the modified values given above. The rate-limiting term in the equation has been altered from the $(1 - \Sigma N/K)$ term of the Verhulst-Pearl model of Equation 7.3. In Equation 7.3 the growth constant r includes both birth and death rates. However, in Equation 9.8 the term $(2 - \Sigma N/K)$ is used to modify only the number of population births, ΔN. Here, when $\Sigma N = K$, the rate-limiting term $(2 - \Sigma N/K)$ equals 1 and reproduction is held to the replacement level. When the population size ΣN is less than K, birth rate increases proportionally, reaching a factor of 2 as ΣN approaches zero. If your simulation is likely to encounter total population sizes ΣN much greater than K, your program should include a statement to limit ΔN to zero whenever Equation 9.8 might produce a negative number of births.

Natality at different densities can be modified by changing the logistic term in Equation 9.8. For example, using $(4 - 3 \cdot \Sigma N/K)$ will produce an almost 4× increase in birth rate at low ΣN, and $(1.1 - 0.1 \cdot \Sigma N/K)$ will provide about a 1.1× increase.

Exercise 9-3: Substitute the modified m_x values and Equation 9.8 into the simulation of Exercise 9-1. Allow the simulation to proceed for 20 years with a carrying capacity of $K = 12000$. At time interval 21, simulate a single year of poor reproduction by letting ΔN be 25% of the value calculated with Equation 9.8. Allow the simulation to proceed for an additional 20 years after the perturbation. Your output for this simulation should be a plot of ΔN and ΣN over the 40-year period.

9.4 Use of Subscripted Variables in Life Table Simulations

A program for a large age-class model using the stepwise approach involves many equations of almost identical form. Whenever this occurs in writing computer programs, it is a good idea to consider using subscripted variables to reduce the number of programming instructions. The age-class model is a good example where this simplification may be used. The following BASIC program employs subscripted variables to accomplish the same calculations as the sequence of Equations 9.1-9.7.

```
100 REM AGE-CLASS MODEL USING SUBSCRIPTED VARIABLES
110 DN = 0
120 FOR J = 0 TO 4
130    DN = DN + N(J) * M(J)
140 NEXT J
150 SN = 0
160 FOR J = 4 TO 0 STEP -1
170    N(J+1) = N(J) * S(J)
180    SN = SN + N(J+1)
190 NEXT J
200 N(0) = DN
210 SN = SN + N(0)
220 END
```

Although this listing is about the same length as a program that uses the direct approach for the sowbug life table, with minor changes this one could also be used for models having very many age classes. For such larger models, using subscripted variables is always more efficient. Look for other opportunities to use this technique.

Age class	Class ages (years)	Number of females (millions)	Age-class survival rate	Age-class natality rate
1	0-4	8.806	0.996	0
2	5-9	8.231	0.999	0
3	10-14	8.340	0.998	0.003
4	15-19	9.107	0.997	0.124
5	20-24	10.479	0.997	0.262
6	25-29	10.865	0.996	0.264
7	30-34	10.171	0.995	0.162
8	35-39	8.967	0.993	0.056
9	40-44	7.116	0 989	0.010
10	45-49	5.969	0.981	0.001
11	50-54	5.660	0.971	0
12	55-59	5.957	0.955	0
13	60-64	5.877	0.932	0
14	65-69	5.151	0.899	0
15	70-74	4.415	0.849	0
16	75-79	3.311	0.777	0
17	80-84	2.293	0.676	0
18	85-89	1.073	0.544	0
19	90-94	0.536	0.385	0
20	95-99	0.268	—	0

Table 9.2. Age-class data for female population of the U.S.A. in 1985. Survival and natality are for an age group for a 5-year period. (For example, the probability of an individual in age-class 7 surviving to age-class 8 is 0.995, and the probability is 0.162 that an individual in age-class 7 will produce a female offspring in 5 years). Data are derived from U.S. Bureau of the Census (1986).

Exercise 9-4: The demographic data in Table 9.2 describe the female population of the U.S.A. in 1985. The tabulated rates of survival and natality are based on five-year intervals. Although age classes exist for greater ages, they are ignored for purposes of this exercise. Assume the given rates of survival and natality will hold for 40 years. Write a simulation to predict the female population for each five-year interval from 1985 through the year 2025. Your output for each interval should be in the form of a table showing age class, age group, and number of individuals (in millions). After you have your simulation working properly, send output to the printer for the years 1985, 2010, and 2025.

9.5 Simulating Sex-differentiated Survival and Reproduction: Deer Hunting

Age-specific survival rates will differ for the sexes of most populations. Males of some animals may be more important economically than females, so that a female-only life table does not adequately describe the population. Males of some game animals, for example, are particularly subject to hunting pressure. We will illustrate this with an age-class model of a population of white-tail deer (*Dama virginiana*) hunted under the "bucks only" regulations common to many areas of North America.

The basic life-history information of age-specific survival and natality is given in Table 9.3. The table also presents a hypothetical population composed of the indicated number of males and females, M_x and F_x, for different age groups x, 0 through 12. From the table, $\Sigma M_x = 3351$ and $\Sigma F_x = 4465$. The area involved is assumed to be 200 square miles. Deer are sexually mature at the age of one year, assuming an adequate diet. For convenience in calculations using the model equations below, we will define the number of adult reproductive females F_a and adult males M_a as

$$M_a = \Sigma M_x - M_0 \tag{9.9}$$

$$F_a = \Sigma F_x - F_0 \tag{9.10}$$

In working with this model, separate sets of simulation data must be maintained for males and females. Both sexes are assumed to follow the given survival rates in the absence of hunting pressure. With hunting pressure, males are subject to additional mortality.

In most deer populations, male-female ratio at birth is not 1.0, but 1.12. The m_x values of Table 9.3 are age-specific birth rates giving the production of male and female offspring combined. The fraction of these births that are male is about 0.528; 0.472 is the fraction that are female. To find F_0 and M_0, use equations that are analogous to Equation 9.6:

$$F_0 = 0.472 \, \Delta N \tag{9.11}$$

$$M_0 = 0.528 \, \Delta N \tag{9.12}$$

x	s_x	m_x	F_x	M_x	P_x
0	0.62	0.00	1000	1000	0.00
1	0.87	0.50	620	694	0.80
2	0.88	0.60	539	543	0.89
3	0.89	0.63	474	406	0.92
4	0.90	0.66	422	289	0.95
5	0.88	0.68	380	195	0.98
6	0.86	0.70	334	120	0.99
7	0.78	0.68	287	62	0.99
8	0.62	0.60	224	29	0.99
9	0.30	0.50	139	11	0.99
10	0.10	0.40	42	2	0.99
11	0.00	0.30	4	0	0.99
12	—	—	0	0	

Table 9.3. Life history information and population sizes for a hypothetical deer herd in an area of 200 square miles. x = age in years; s_x = age-specific rate of survival for unhunted deer of age x; m_x = age-specific rate of production of fawns of both sexes by females of age x, assuming 100 % fertilization; F_x = number of female deer of age x; M_x = number of male deer of age x; P_x = male mortality probabilities for an encounter between a hunter and a male deer of age x. Data are derived from Dahlberg and Guettinger (1956).

The natality rates of Table 9.3 are adjusted to maintain a steady-state population of females, assuming they follow the given survivorship schedule. The hypothetical female population of the table has been set up specifically to demonstrate such a stationary population. If 1000 0-aged females follow the survival schedule given in the table, their numbers at each age will equal those given for each age-group of the hypothetical population. If the hypothetical population reproduces following the given m_x schedule, and ΔN is found as usual with Equation 9.1, then $\Delta N = 2120$. The fraction (0.472) of these births that are females is 1000, which will replace the F_0 age-group and hold the female population at a constant size.

In populations of deer with a small number of males relative to the number of females, some females may not be fertilized during the mating season. Intense hunting pressure may lower sufficiently the density of male deer to cause a decline in female mating success. The deer-hunting model should account for this source of reproductive failure. We can reasonably assume that mating success is a positive function of the number of reproductive males. We assume that mating success is 0 when M_a is 0; from observation it is known that success is 90% when M_a is about 870. An exponential equation like Equation 7 of Chapter 3 can be used to simulate this process:

$$R = 1 - \exp\left(kM_a\right) \qquad (9.13)$$

where R is the fraction of adult females fertilized, M_a is the number of mature breeding males, and the constant $k = -0.002656$. Thus, Equation 9.1 for finding the number of deer born each year is modified to become

$$\Delta N = \left(\Sigma F_x m_x\right) R \qquad (9.14)$$

This population, that is just capable of replacing itself, is assumed to be at "carrying capacity". One may also assume that as population density declines below this level, the fertility of the deer will increase. This effect can be simulated with a variation of Equation 9.8. We will assume that the annual production of young deer increases by a factor of 2.5 when deer populations are low and food is plentiful. It will be necessary therefore to multiply Equation 9.14 by the type of limiting logistic term introduced in Section 9.3:

$$2.5 - \frac{1.5\left(\Sigma M_x + \Sigma F_x\right)}{6000}$$

Although both males and females follow the schedule of survival given in Table 9.3, males are also subject to hunting mortality. To account for hunting mortality, an equation must consider the number of male deer and the number of hunters. The predation equations of Chapter 7 are not suitable because the hunting season is brief, not continuous through the year. An equation used by O'Neill et al. (1972) appears suitable (see also Watt 1975 and DeAngelis et al. 1975). The general form of the equation is:

$$\Delta M_H = \frac{P\, H\, M_a}{H + M_a} \qquad (9.15)$$

Here, ΔM_H is the mortality in the males due to hunting, H is the density of hunters, and P is the probability that an encounter between a male deer and a hunter will result in a dead deer. The value of P may be estimated from hunting data. In the northern part of the lower peninsula of Michigan in 1957, 306,000 hunters killed 41,000 of an estimated 75,000 adult male deer (Jenkins and Bartlett 1959). From these data, the value of P is found to be 0.681.

Equation 9.15 cannot be applied directly to an age-class model, because hunting mortality does not occur uniformly for all age classes of males.

Males of age-class-0 are protected by regulation. Hunters encountering a male of age-class-1 or -2 may not recognize it as a male because of its small antlers. Some hunters will pass up a younger male, expecting later to encounter more desirable older males. Table 9.3 indicates for each age class of males a hunter preference value, P_x, which is essentially the probability that an individual male deer will be taken during an encounter with a hunter. These values were developed as weighted averages of the P value above, based on reasonable expectations. For the different age classes of male deer, the annual hunting mortality is found with

$$(\Delta M_x)_H = \frac{M_x H \, P_x}{H + M_a} \qquad (9.16)$$

Hunting mortality must be combined with mortality occurring apart from hunting. However, birth and hunting in deer populations are discontinuous events, while other mortality is continuous through the year. The sequence of calculations for the model is therefore important. You may assume the year starts in the spring with births, that mortality is very low during the summer, that hunting occurs in the fall just after the breeding season, and that non-hunting mortality for both sexes is concentrated in the winter months. The values given in Table 9.3 describe a hypothetical population in the summer. The proper sequence for a deer-hunting simulation would require sequentially calculating (1) the proportion of females fertilized during the breeding season, (2) the hunting mortality of the males, (3) the number of births, (4) the mortality of females and remaining males, moving survivors to the next age class, (5) the proportion of births that are males and females, which are put into age-class-0 for each sex.

Exercise 9-5: A variety of simulations are possible using the deer-hunter model above. As a minimum exploration of the model, write a program to simulate the effects of different hunting pressures on the deer population. As measures of the performance of the deer population, have your program show the total number of deer, the number of births per year, and the proportion of males among adult deer. These measures should reach a steady-state condition after about 25 years at any given hunting pressure. The output of your program should show these three measures at steady-state plotted against the number of hunters. Vary the size of the hunter population from 0 to 10,000 in steps of 1000.

9.6 A Fisheries Age-Class Model

Age-class models are important to fisheries management because fecundity and fishing pressure both vary with the age of fish. Fisheries management has been based for several decades in the density-dependent stock-and-recruitment model attributed to Ricker. Hall (1988) reviews the model instructively. The key idea of this model is fundamentally that of the logistic model, based on the premise that there is some upper density of the population at which birth rate is balanced by death rate; below this density births exceed deaths, and above it deaths exceed births. Ricker's (1954) model is shown graphically in Figure 9.4, based on this equation:

$$F = P \cdot \exp\left[(P_r - P)/P_m\right] \tag{9.17}$$

where F is the density of recruited (i.e. next generation) adults and P is the density of parental stock adults. P_r is the density of parental stock which will produce just sufficient recruits for replacement, and P_m is the parental stock which produces maximal recruitment. (P_r is conceptually equivalent to K in the logistic.) Like the logistic, this model was formulated for homogeneous populations, but can be adapted for age-class models.

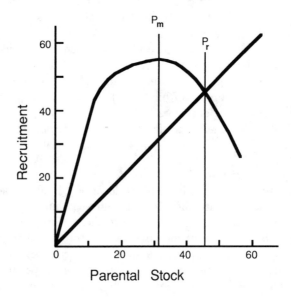

Figure 9.4. Density of recruitment of new adults in a fishery as a function of parental stock density. P_r is the upper limit of density for successful replacement. P_m is the parental density resulting in a maximum recruitment density.

The dynamics of a population of haddock in the North Sea have been modeled with a variation of the stock-recruitment model employing age classes (Jones and Hall 1973, Cushing 1977). A Ricker-type curve relating egg production of parental stock to recruitment of year-old fish was employed, which may be approximated with the following equation:

$$N_1 = F \cdot E \cdot \exp\left[(E_r - E)/E_m\right] \tag{9.18}$$

In this modification of Equation 9.17, N_1 is the density of one-year-old female recruits, and F is a survival factor relating eggs and larvae to number of female fish. E is the density measure of total egg production during a spawning season, E_r is the density of eggs required for replacement, and E_m is the density of eggs required for maximal recruitment. For the North Sea population of haddock, density values were based on 3 square kilometers of pelagic habitat. The constants were as follows:

$$F = 19.2 \times 10^{-6} \text{ females egg}^{-1},$$
$$E_r = 130 \times 10^6 \text{ eggs } (3 \text{ km}^2)^{-1},$$
$$E_m = 80 \times 10^6 \text{ eggs } (3 \text{ km}^2)^{-1}$$

The age-specific survival and fecundity information for the fish are summarized in Table 9.4, and describe the population just before spawning begins.

Age in years x	Density of females at age x N_x	Rate of survival x to $x+1$ s_x	Average number of eggs produced per female of age x m_x
1	2500	0.65	0
2	1630	0.42	0.58×10^5
3	700	0.35	1.4×10^5
4	250	0.33	2.2×10^5
5	90	0.33	3.0×10^5
6	30	0.33	3.8×10^5
7	10	0.30	4.4×10^5
8	5	0.20	5.0×10^5
9	2	0.00	5.5×10^5
10	0		

Table 9.4. Population characteristics of North Sea haddock, adapted from data in Jones and Hall (1973).

Exercise 9-6: With the information given above, develop an age-class simulation of the North Sea population of haddock. Use the fecundity data to determine the total production of eggs, then use the survival equations and data to determine the age classes for the next year, and finally use Equation 9.18 to find the number of first-year recruits. This simulation will show some interesting oscillations in the total population size. It should also show dominant age classes moving through the population. To demonstrate these phenomena, write your program to draw a graph of total population as a function of time for a period of at least 30 years. Then modify your program to produce histograms that show the size of each age class for a given year. Produce 10 such histograms for years 0 through 9 of your simulation.

9.7 Insect Life Stages: Flour Beetles

Flour beetles (*Tribolium* sp.) have been used frequently in laboratory investigations of population dynamics. At usual culturing temperature the stages of the beetle life cycle have the following durations in days: eggs, 4.6; larvae, 22.6; pupae, 6.6; adults, 120. Experiments are usually started with a few adults put into a vial containing 8 grams of whole wheat flour and brewer's yeast. The beetles are allowed to reproduce and grow in this confined volume at constant temperature and humidity. The flour is sifted at 30-day intervals to separate the larvae, pupae and adults; these are counted and put into a fresh supply of flour and yeast. Under these conditions, the data of Table 9.5 are typical of those collected in the classic work of Thomas Park (1948).

The data of the table show that adult beetles will intensely cannibalize two other life stages. Just after the population is founded, there are few adults, and there is little mortality in going from larva or pupa to adult (24 larvae and pupae become 24 adults). However, later in the experiment, there is a very low rate of successful transition from larva or pupa to adult (9-10 larvae and pupae become 5 adults, most of which were probably old surviving adults). Cannibalism appears to be occurring during the pupal stage. The other life stage that is cannibalized is the egg stage. For example, 20 adults should be laying about 250 eggs per day. Over a 30-day period this should result in about 7500 eggs and then 7500 pupae and larvae. However, after day 60, less than one larva and pupa per gram are found. Evidently the rate of cannibalism is intensely directed against the eggs. Park et al. (1965) indicate that cannibalism is the primary mechanism for limiting the population of flour beetles under these conditions. The model described below for cannibalistic control of

the confined flour beetle population illustrates some important features
of simulations involving insect life stages.

Culture age in days	Density of larvae + pupae number gram^{-1}	Density of adult beetles number gram^{-1}
0	—	3
30	24.0	3
60	3.0	24
90	1.0	24
120	0.7	23
150	0.7	22
180	2.0	20
210	2.5	14
240	4.0	10
480	9.0	5
960	10.0	5

Table 9.5. Typical data from laboratory cultures of flour beetles
(*Tribolium* spp.). See text for further details. Data are derived
from Park (1948).

The life cycle of the beetle is divided into four stages. The model
is based on an assumption that the egg and pupal stages are equally
long. The duration of the egg or pupal stage will represent one unit of
time for modeling purposes, so that a simulation based on the model will
use physiological time units. The larval stage lasts about three units of
physiological time, so for the model the stage will be broken into three
consecutive parts. The adult stage will consist of many parts in the model,
each a single interval in duration. Adults all will survive successfully
through three intervals, and then survive at a constant fractional rate S
for each succeeding time interval.

Egg-laying is assumed to be continuous throughout the adult stage, described by a constant R, measured as eggs adult^{-1} time-unit^{-1}. These
units presume that the male-female ratio is a constant. Adults are assumed to eat both eggs and pupae. Given the "natural history" of a flour
vial, with the adults tunneling through the flour and eating eggs when
they encounter them, the simple mass-balance Lotka-Volterra predation
term is an adequate descriptor of loss of eggs and pupae:

$$\frac{-dE}{dt} = k \cdot E \cdot A \tag{9.19}$$

$$\frac{-dP}{dt} = K \cdot P \cdot A \tag{9.20}$$

where A is adult density, E is egg density, P is pupal density, and K and k are cannibalism constants for eggs and pupae, and t is actual time, not physiological time. If we assume that the number of adults is constant over a unit of physiological time, then Equations 9.19 and 9.20 can be integrated (following procedures of Chapter 1) over one unit of physiological time:

$$E_u = E_o e^{kA} \tag{9.21}$$

where E_o and E_u are egg densities at the beginning and end of the time unit. Similarly,

$$P_u = P_o e^{KA} \tag{9.22}$$

describes pupal density after one time unit.

The sequential model system is described with the following:

$$A = A_1 + A_2 + A_3 \tag{9.23}$$

$$A_3 \quad \leftarrow \quad A_3 \cdot S + A_2 \tag{9.24}$$

$$A_2 \quad \leftarrow \quad A_1 \tag{9.25}$$

$$A_1 \quad \leftarrow \quad P e^{-KA} \tag{9.26}$$

$$P \quad \leftarrow \quad L_3 \tag{9.27}$$

$$L_3 \quad \leftarrow \quad L_2 \tag{9.28}$$

$$L_2 \quad \leftarrow \quad L_1 \tag{9.29}$$

$$L_1 \quad \leftarrow \quad E e^{-kA} \tag{9.30}$$

$$E \quad \leftarrow \quad R \cdot A \tag{9.31}$$

L_1, L_2 and L_3 are densities of the three parts of the larval stage. A_1, A_2 and A_3 are densities of the three parts of the adult stage. As in Exercise 9-1, the sequence of solution of the model is important, particularly the summation of number of adults before the calculations with Equations 9.26, 9.30 and 9.31.

Exercise 9-7: Implement the simulation described above for cannibalistic regulation of flour beetle populations. Assume the population begins with two adults placed in a closed container with one kilogram of flour (adult density = 0.002). Set the constants as follows:

$$R = 100 \qquad k = 0.30 \text{ gram adult}^{-1} \text{ time-unit}^{-1}$$
$$S = 0.85 \qquad K = 0.03 \text{ gram adult}^{-1} \text{ time-unit}^{-1}$$

Your output should plot number of (larvae + pupae) and number of adults over 100 units of physiological time.

The simulation should produce oscillating densities, with high adult densities depressing the abundance of immature stages. The simulation results are quite sensitive to the value of adult survival S. The oscillation is damped with values of S greater than about 0.9. With smaller values, the oscillations are undamped and some rather large excursions of larval densities are possible.

Conclusion

This chapter has covered several different models based on dividing a population into subclasses. The technique is used frequently in developing realistic models of economically important organisms, such as forest trees and insect pests. The increase in realism is obtained at the price of complexity. The modeler must be careful not to let the complexity cloud the central point of the particular model.

CHAPTER 10

SIMULATIONS OF POPULATION GENETICS

In the previous chapters, we have ignored the possibility that organisms might have different rates of survival or reproduction based on their genotype. This was explicit in our assumption of population homogeneity. In this chapter we will focus on the mechanisms of genetic variation, and ignore the size of the population.

Although we will not presume any great knowledge of genetics, we will use without definition a few terms that are commonly used in introductory biology courses and texts, including gene, genotype, allele, diploid, locus, etc.

10.1 The Hardy-Weinberg Principle

The classical models of population genetics are based on principles stated independently by the English mathematician G. H. Hardy and a German physician W. Weinberg. They published their principle within a few years after the rediscovery of Mendel's concepts of heritable factors, or genes. The basic requirement for their principle is random mating in a population. It will also require a large population and diploid individuals. In a population mating at random, matings between genotypes will occur as expected from the proportions of the different genotypes. (You may recognize this as another application of the law of mass action used in previous chapters.)

A population that follows the Hardy-Weinberg principle displays three interesting properties: (a) gene frequencies will not change from generation to generation; (b) after one generation, genotype frequencies will not change; (c) frequencies of genotypes and genes are related by a simple expression. These results can be illustrated using the simplified system we will be following for most of this chapter: a single locus with two alleles, A and a.

For our simple system, allele A will be dominant and allele a recessive. In a population, individuals can be of three types, AA, Aa, and aa, representing respectively the homozygous dominant, heterozygous, and homozygous recessive genotypes. The frequencies of the three genotypes in the population will be D, H, and R. Because no other combinations are possible, $D + H + R = 1$.

The frequency of the A gene is conventionally given as p, and the frequency of the a gene as q, so that $p + q = 1$. We will assume the gene frequencies are the same in both sexes. The notion of gene frequency is not quite as obvious as genotype frequency, because genes are not countable in the same way that individuals with genotypes are. It may be easiest to think of the genes as part of the gametes that go to make up the next generation. These gametes will carry either the A or the a gene. The genes will combine following the law of mass action, depending on their relative frequencies.

Given a population with D, H, and R known, the frequency of the A gene can be found with

$$p = 1 - q = \frac{D + 0.5H}{D + H + R} \tag{10.1}$$

Likewise, the frequency of the a gene is found with

$$q = 1 - p = \frac{.05H + R}{D + H + R} \tag{10.2}$$

It is possible to start out a population meeting Hardy-Weinberg assumptions with any combination of D, H and R. Then, the frequencies of A and a can be found with Equations 10.1 and 10.2.

If a Hardy-Weinberg population with known values of p and q reproduces, then the values of D, H and R in the next generation are found with

$$D = p^2 = (1 - q)^2 \tag{10.3}$$

$$H = 2pq = 2p(1 - p) = 2q(1 - q) \tag{10.4}$$

$$R = q^2 = (1 - p)2 \tag{10.5}$$

If the gene frequencies, p and q, of this generation are calculated with Equations 10.1 and 10.2, they will be found to equal the p and q of the preceding generation. This is the basis of the Hardy-Weinberg observation that gene frequencies will not change from generation to generation. Because these frequencies are constant, Equations 10.3-10.5 define the constant frequencies of genotypes after the founding generation. The simple relationship between gene frequencies and genotype frequencies is

$$(p + q)^2 = p^2 + 2pq + q^2 \tag{10.6}$$

The Hardy-Weinberg principle performs much the same function in population genetics that the exponential growth equation does for population growth: No real population follows the rule for very long if at all, but it provides a starting point for the study of real events. In the case of the Hardy-Weinberg equilibrium, a variety of processes serve to alter gene and genotypic frequencies. These processes form the basis of the simulations of this chapter. A more detailed discussion of the Hardy-Weinberg principle can be found in most genetics texts, and in various texts of population genetics. In particular, Ayala's (1982) textbook is clear and accessible at the elementary level of most of the simulations of this chapter.

10.2 Effect of Selection

Selection is a disturbance of the Hardy-Weinberg equilibrium caused by different genotypes reproducing at different rates. In our simple system, selection will take place whenever one of the genotypes contributes fewer gametes to the "mass-action" mating and as a result will be less frequent in the next generation. The measure of ability to contribute to the next generation is "fitness", and a value of 1 is usually given to the genotype or genotypes that reproduce most effectively. In a population following Hardy-Weinberg principles, each genotype has a fitness of 1. Relative fitness is usually symbolized by w. Any factor that alters relative survival and fertility of a genotype will alter its fitness, so that w is less than 1.

Initially, we will work with a simple, single-locus model, and assume that selection is operating against the recessive homozygous genotype, aa. Before selection operates in a generation, the aa genotype will have a frequency of R. After selection, it has a frequency of wR. The frequency of the a gene in the population will also be less after selection. In this case, the value of q in the next generation can be found with

$$q = \frac{0.5H + wR}{D + H + wR} \qquad (10.7)$$

where w is the measure of relative fitness. Because $p = 1 - q$, the frequencies of the three genotypes in the next generation can be found with Equations 10.3 through 10.5. Alternating calculations with these equations can show the decline in the frequency of the a gene through successive generations as a result of selection.

Exercise 10-1: Write and implement a program for simulating selection based on the elementary Hardy-Weinberg population genetics model. Flowcharts for the simulation are given in Figure 10.1. Determine if your program is operating properly by starting with

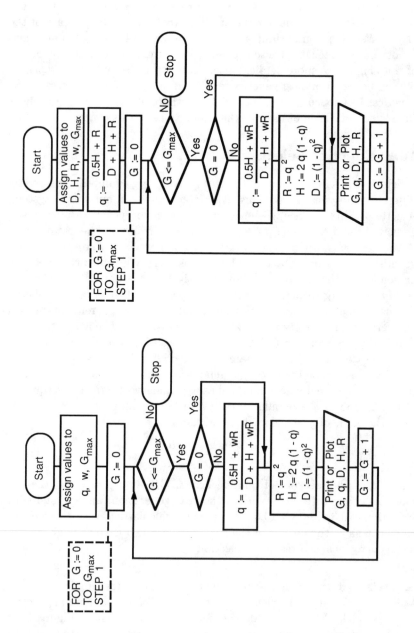

Figure 10.1. Two flowcharts for programs to simulate selection against recessive genotypes. Flowchart choice depends upon whether simulation begins with assigning gene frequency q

these values for the initial genotype frequencies: $D = 0.65$, $H = 0.27$, $R = 0.08$. First set up your simulation to plot the value of R and q as a function of generation, for generations 0 to 20.

This initial simulation should demonstrate that the value of q remains constant, and that the genotype frequencies stabilize immediately when there is no selection ($w = 1$). Next alter the fitness of R to show selection against the homozygous recessive genotype. Use $w = 0.16$. Observe the decline in both gene and genotype frequency, plotting R and q as before. Allow the simulation to proceed for about 80 generations.

This simulation is a valuable demonstration of the limited value of a eugenics program directed at a recessive gene, when selection can be exerted only on homozygous recessive individuals expressing the phenotype. When its frequency is small, the rate of removal of a recessive gene is very low. (In fact, when selection is maximal and fitness w is 0, the number of generations required to reduce q by half is $1/q$.)

10.3 Selection for Heterozygotes

The heterozygote is selectively favored in some circumstances, a condition known variously as heterosis, overdominance, or hybrid vigor. The model for selection favoring heterozygotes follows that for selection against the recessive homozygote, except that selection occurs against both homozygotes. Thus, given a population with frequencies D, H and R as before, the frequency of the a gene in the next generation is

$$q = \frac{0.5H + w_2R}{w_1D + H + w_2R} \tag{10.8}$$

where w_1 and w_2 are the relative fitnesses of the AA and aa genotypes, respectively. In this model, the gene frequencies p and q will come to an equilibrium that is independent of the starting frequencies.

The classical example of heterosis in human populations is sickle-cell anemia, a disease exhibited by native African populations. The anemia produced by the homozygous condition is severe, so that most aa individuals die before reaching maturity ($w_1 = 0$). However, in areas where malaria is prevalent, heterozygous individuals are more resistant to the effects of malaria than normal homozygotes. Heterozygous individuals have a greater fitness than either of the homozygotes, who are likely to die either of anemia or of malaria.

Exercise 10-2: Write and implement a program to simulate selection favoring the heterozygous condition, using our simple A/a model.

Test your program with $w_1 = 0.5$ and $w_2 = 0.2$. Begin your simulation with these values: $D = 0.68$, $H = 0.25$, $R = 0.07$. Your output should consist of a graph showing q for 20 generations. To demonstrate the independence of the equilibrium gene frequency and starting frequency, begin a second simulation with $D = 0.02$, $H = 0$, and $R = 0.98$.

Exercise 10-3: The basic Hardy-Weinberg model may be modified to simulate the effect of selection in a system which shows partial dominance. Write a program to simulate the result of homozygous recessive individuals having a fitness of 0.12 and heterozygous individuals having a fitness of 0.5, relative to the homozygous dominant individuals. Equation 10.7 will have to be modified appropriately. Implement your program with an initial value of $q = 0.10$. Your output should be in the form of a graph of q over generations from 0 to 20.

10.4 Simulation of Mutation

Although mutations occur very infrequently, they are the ultimate source of genetic variability in a population. Point mutation is fundamentally a chemical process involving the conversion of one base in the DNA code to another, thus causing a change in the genetic message at a particular locus. The chemical process is reversible.

We will consider here mutations that change one allele to another, which will obviously alter gene frequencies. We can suppose that allele A has a certain rate of mutation to allele a, and vice versa, as indicated in the following diagram:

$$A \underset{v}{\overset{u}{\rightleftarrows}} a$$

where u refers to the rate constant for forward mutation and v refers to the rate constant for reverse mutation. At equilibrium the forward and reverse mutation rates are equal. If q is the frequency of allele a in the population, then the equilibrium may be described by the equation

$$vq = up = u(1 - q) \tag{10.9}$$

This can be rearranged to give the frequency of a at equilibrium:

$$q = \frac{u}{u + v} \tag{10.10}$$

For any generation, the change in frequency of the a allele due to mutation can be found with

$$\Delta q = -vq + up = -vq + u(1 - q) \qquad (10.11)$$

It is useful to combine this model for effects of mutation with the model for selection, because selection frequently operates against mutations. Given a population described with genotype frequencies D, H and R, and with a fitness w less than 1 for the aa genotype, and mutation rates v and u, the net effect on the gene frequency of the next generation is given with

$$q = \frac{0.5H + wR}{D + H + wR}(1 - v - u) + u \qquad (10.12)$$

The result of selection and mutation acting simultaneously will be an equilibrium frequency. If the homozygous recessive genotype has a fitness of 0, then the equilibrium frequency of the a allele will be u, because it will be maintained in the population only by the mutation rate of A to a alleles.

Exercise 10-4: Using Equation 10.12, implement a simulation for the combined effect of mutation and selection against a recessive allele, and find whether an equilibrium frequency is obtained. Set $u = 0.00005$ and $v = 0.0000001$ and $w = 0.5$. Begin your simulation with $D = 0.9947$, $H = 0.005$ and $R = 0.0003$. Allow your simulation to proceed until a balanced condition is demonstrated. This may require several hundred generations. Plot q as a function of generation.

10.5 Selection Involving Sex-Linked Recessive Genes

Male organisms of many species have only a single X chromosome and are essentially haploid for genes carried on the X chromosome, while females have two X chromosomes. Traits which are carried on the X chromosome are said to be sex-linked, and males show a different pattern of inheritance for the genes carried on the X chromosome. A gene on the X chromosome that shows a dominant-recessive mode of inheritance in females will be expressed differently in males, with the recessive gene displaying a higher frequency of expression.

In this system, we will model a characteristic which is determined by a single pair of sex-linked alleles, A and a. We will assume that mating is random among the reproducing individuals. The females will have genotypes Aa, Aa and aa, with respective frequencies D, H and R. The frequency of allele a among the females in the population, q_f, may be determined as usual with

$$q_f = \frac{0.5H + R}{D + H + R} \qquad (10.13)$$

The frequency of allele a among males, q_m, is simply equal to the frequency of the a-genotype. Because a male receives the X chromosome only from his mother, q_m is equal to q_f of the previous generation.

To model selection in this sex-linked system, we will assume that aa, the homozygous recessive genotype in females, and $a-$, the (recessive) genotype in males, both produce phenotypes which are equally reduced in fitness compared to the AA, Aa or $A-$ phenotypes. If we are given a population with female genotype frequencies of D, H and R, and a measure of relative fitness w, then the contribution of females to the frequency of the a allele in the next generation is given with

$$(q_f)_{G+1} = \frac{0.5H + wR}{D + H + wR} \qquad (10.14)$$

The male contribution to the frequency of the a allele in the next generation is given by

$$(q_m)_{G+1} = \frac{wq_f}{wq_f + (1 - q_f)} \qquad (10.15)$$

We can now consider how to find the genotype frequencies of the next generation. First we move to the next generation with an update of gene frequency and let $q_m = (q_m)_{G+1}$ and $q_f = (q_f)_{G+1}$. Then we find for the females

$$R = q_m q_f \qquad (10.16)$$

$$H = q_f (1 - q_m) + q_m (1 - q_f) \qquad (10.17)$$

$$D = (1 - q_m)(1 - q_f) \qquad (10.18)$$

For the males, the genotype frequencies for the $A-$ and $a-$ genotypes are q_m and $(1 - q_m)$.

Exercise 10-5: Write a program using Equations 10.14-10.18 to simulate the effect of selection on sex-linked recessive genes. Set $w = 0.3$ and $q_m = q_f = 0.2$. Plot frequencies of recessive male and female phenotypes (q_m and R) for generations 0 through 8.

10.6 The Inbreeding Model of Sewell Wright

Inbreeding is the production of offspring by genetically closely-related individuals. Random mating among reproducing individuals is a basic assumption of the Hardy-Weinberg principle and inbreeding violates this assumption. In populations where inbreeding is significant, the Hardy-Weinberg equations may be modified by the inclusion of an inbreeding coefficient that was first described by Wright (1922). For the A/a model that we have been using, these are the appropriate equations:

$$D = (1 - q)^2(1 - F) + (1 - q)F \qquad (10.19)$$

$$H = 2(1 - q)q(1 - F) \qquad (10.20)$$

$$R = q^2(1 - F) + qF \qquad (10.21)$$

where F is the inbreeding coefficient. Inbreeding has no effect on the frequencies of genes in the population. However, inbreeding does alter genotype frequencies from Hardy-Weinberg expectations as described by these equations. If there is no inbreeding, $F = 0$ and these equations revert to the Hardy-Weinberg model. The inbreeding coefficient is the probability that an individual will receive, at a given locus, two alleles that are descended from a single allele carried by an ancestor. Thus, an individual with a heterozygous (Aa) genotype cannot be the result of inbreeding. However, an individual with the AA genotype may be the result of inbreeding if the AA genes were obtained from one allele of an ancestor. The same is true for a aa genotype.

For any population with inbreeding, the proportion of genes from a common ancestor will increase, and the inbreeding coefficient F will be altered with each generation. F will follow a pattern of change that depends upon the initial gene frequencies and the type of inbreeding. The change in F can be found using various recurrence relations for F under various standard systems of inbreeding.

Self-fertilization is the most radical sort of inbreeding. According to Li (1976), under conditions of self-fertilization the alteration in F from one generation to the next is given by

$$F_{G+1} = 0.5\,(1 + F_G) \qquad (10.22)$$

where G indicates generation number. If a population should shift from a system of random mating to self-fertilization, then $F_0 = 0$, $F_1 = 0.5$, $F_2 = 0.75$, $F_3 = 0.875$, etc.

For a system of sib-mating (brother-sister matings), the change in F is described with

$$F_{G+1} = 0.25 \left(1 + 2F_G + F_{G-1}\right) \tag{10.23}$$

and for a system of fixed-sire mating, the change in F is described with

$$F_{G+1} = 0.25 \left(1 + 2F_G\right) \tag{10.24}$$

Further discussion of the inbreeding coefficient under different mating systems is given in Crow and Kimura (1970) and Li (1976).

Exercise 10-6: Write a general program that will allow you to calculate D, H and R in populations with the patterns of inbreeding described for Equations 10.22, 10.23 and 10.24. Your program will have to retain F_G and F_{G-1} for use in the appropriate equations. Start each of your population simulations with $q = 0.5$. The output of your program should be a graph that shows the proportion of heterozygotes for generations 0 through 20 for each of the three patterns. This should demonstrate the loss of heterozygosity that is produced by systems of selfing, sib-mating, and fixed-sired mating.

Exercise 10-7: Modify the basic selection model of Section 10.2 to permit you to observe the effect of immigration on the frequency of the recessive allele in a population with selection against the allele. Using the A/a model, assume i is the fraction of the population that immigrates each generation, with a frequency q_i of allele a. After you have found the equations involved, write a program using your equations to simulate immigration. Set $w = 0.1$ and $q_i = 0.5$. Perform two simulations, first with no immigration ($i = 0$) and then with a moderate rate of immigration ($i = 0.2$). For each of these, plot q for generations 0 to 40. For initial values, use $D = 0.25$, $H = 0.50$, $R = 0.25$.

10.7 A Model of Competition with Selection

Levin's (1969) pioneering attempt to unite a simple model of selection with a model of competition resulted in a simulation that shows some unexpected properties. Levin combined the Volterra model of two-species competition (Section 7.7) with the simple selection model of this chapter. Here we will examine the simplest form of Levin's model, omitting several interesting features. We will work with two species, and consider only the effect of different genotypes in limiting resource use of the competing species. We will ignore selection for resistance to inter- and intraspecific competition.

The basic competition equation is slightly modified from the Volterra equations of Chapter 7. Here we will use the following equations:

$$\frac{dN_1}{dt} = r_1 N_1 \left(\frac{K_1 - \alpha_{11} N_1 - \alpha_{12} N_2}{K_1} \right) \tag{10.25}$$

$$\frac{dN_2}{dt} = r_2 N_2 \left(\frac{K_2 - \alpha_{22} N_2 - \alpha_{21} N_1}{K_2} \right) \tag{10.26}$$

These equations differ from those of Chapter 7 in the handling of the competition coefficients. The interspecific coefficients α and β are replaced with α_{12} and α_{21} respectively. In addition, these equations include intraspecific coefficients, α_{11} and α_{22}, which are equal to 1.0 in the equations of Chapter 7.

We will continue to work with the simple two-allele model, with A and a the alleles that are important in determining competitive ability. In this example, individuals possessing the AA or Aa genotype might be superior to the aa genotype in producing a chemical that inhibits growth of the other species.

Since there are three possible genotypes in each species, we will define them as before with D_1, H_1 and R_1 representing the AA, Aa and aa genotypes of species 1, and D_2, H_2 and R_2 the similar genotypes BB, Bb and bb for species 2. The inter- and intraspecific competitive abilities of these genotypes can be defined by a series of coefficients:

γ_{111} :	effect	of	the	AA	genotype	of	species	1	on	individuals	of	species	1
γ_{112} :	"	"	"	Aa	"	"	"	1	"	"	"	"	1
γ_{113} :	"	"	"	aa	"	"	"	1	"	"	"	"	1
γ_{121} :	"	"	"	BB	"	"	"	2	"	"	"	"	1
γ_{122} :	"	"	"	Bb	"	"	"	2	"	"	"	"	1
γ_{123} :	"	"	"	bb	"	"	"	2	"	"	"	"	1
γ_{211} :	"	"	"	AA	"	"	"	1	"	"	"	"	2
γ_{212} :	"	"	"	Aa	"	"	"	1	"	"	"	"	2
γ_{213} :	"	"	"	aa	"	"	"	1	"	"	"	"	2
γ_{221} :	"	"	"	BB	"	"	"	2	"	"	"	"	2
γ_{222} :	"	"	"	Bb	"	"	"	2	"	"	"	"	2
γ_{223} :	"	"	"	bb	"	"	"	2	"	"	"	"	2

The competition coefficients will be determined by these relationships, weighted according to their relative frequency within the populations:

$$\alpha_{11} = \gamma_{111}D_1 + \gamma_{112}H_1 + \gamma_{113}R_1 \tag{10.27}$$

$$\alpha_{12} = \gamma_{121}D_2 + \gamma_{122}H_2 + \gamma_{123}R_2 \tag{10.28}$$

$$\alpha_{21} = \gamma_{211}D_1 + \gamma_{212}H_1 + \gamma_{213}R_1 \tag{10.29}$$

$$\alpha_{22} = \gamma_{221}D_2 + \gamma_{222}H_2 + \gamma_{223}R_2 \tag{10.30}$$

Within a species, different genotypes will have different fitnesses depending upon their relative success in inhibiting the competing species, and in not inhibiting their own. Relative fitness is then a function of the number of individuals present of each species. Levin defined fitness for the different genotypes as follows:

Fitness of AA genotype of species 1 $= w_{11} = \gamma_{111}N_1 + \gamma_{121}N_2$

Fitness of Aa genotype of species 1 $= w_{12} = \gamma_{112}N_1 + \gamma_{122}N_2$

Fitness of aa genotype of species 1 $= w_{13} = \gamma_{113}N_1 + \gamma_{123}N_2$

Fitness of BB genotype of species 2 $= w_{21} = \gamma_{211}N_1 + \gamma_{221}N_2$

Fitness of Bb genotype of species 2 $= w_{22} = \gamma_{212}N_1 + \gamma_{222}N_2$

Fitness of bb genotype of species 2 $= w_{23} = \gamma_{213}N_1 + \gamma_{223}N_2$

The Volterra model has no provision for generation time, which is necessary for the usual procedures of finding changes in gene and genotypic frequency. Accordingly, Levin set the time unit Δt for the simple Euler solution of the Volterra equations to unity, and then allowed two units of time for each "generation" for both species. This technique was implemented by allowing selection to work to modify the genotypic frequencies of half the population for each time interval. To make calculations of post-selection genotypic frequencies easier, we first find the sum of the genotypic frequencies in that part of the population that is affected by selection:

$$S_1 = D_1w_{11} + H_1w_{12} + R_1w_{13} \tag{10.31}$$

$$S_2 = D_2 w_{21} + H_2 w_{22} + R_2 w_{23} \tag{10.32}$$

A new genotypic frequency at the end of each time unit is found by adding the selected half of each population and the unselected half:

$$D_1 = \frac{0.5 \, (D_1 w_{11})}{S_1} + 0.5 D_1 \tag{10.33}$$

The values for H_1, R_1, D_2, H_2 and R_2 are found similarly.

The gene frequencies for each time unit are calculated as before, following the Hardy-Weinberg rules:

$$q_1 = 1 - p_1 = \frac{0.5 H_1 + R_1}{D_1 + H_1 + R_1} \tag{10.34}$$

$$q_2 = 1 - p_2 = \frac{0.5 H_2 + R_2}{D_2 + H_2 + R_2} \tag{10.35}$$

The model assumes that the population follows a random pattern of breeding, so that genotypic frequencies for the next time interval are found with Equations 10.3-10.5.

This model employs many variables, each of which may be manipulated independently. It is reasonable to keep the values of the interspecific and intraspecific competition coefficients equal when $q_1 = q_2 = 0.5$. Similarly, K_1 is held equal to K_2.

Exercise 10-8: Write a program using the model above to simulate selection in a competitive situation for two species. Set $r_1 = r_2 = 0.25$, and $K_1 = K_2 = 100$. Let the resource-limiting constants be set as follows:

$$\gamma_{111} = 0.5 \quad \gamma_{112} = 1.0 \quad \gamma_{113} = 1.5 \quad \gamma_{121} = 1.5 \quad \gamma_{122} = 1.0 \quad \gamma_{123} = 0.5$$
$$\gamma_{211} = 1.5 \quad \gamma_{212} = 1.0 \quad \gamma_{213} = 0.5 \quad \gamma_{221} = 0.5 \quad \gamma_{222} = 1.0 \quad \gamma_{223} = 1.5$$

For initial values, use $q_1 = q_2 = 0.5$, $N_1 = 20$, and $N_2 = 40$. Plot the numbers of individuals in each population for 200 units of time. The flow of your program will be approximately as follows:

1. Assign constant and initial values to variables.
2. Go to (8) to find initial genotypic frequencies.
3. Calculate competition coefficients (Equations 10.27 - 10.30).
4. Find population sizes (Equations 10.25-10.26) with Euler methods ($\Delta t = 1$).

5. Calculate values of fitness for each genotype in the two populations.
6. Calculate post-selection genotypic frequencies (Equations 10.31 - 10.33).
7. Calculate gene frequencies with Equations 10.34 - 10.35.
8. Calculate genotype frequencies (Equations 10.3 - 10.5).
9. Print or plot the values of variable of interest.
10. Go to (3)

Your simulation should show oscillating and alternating population sizes of the two species. After your simulation is working well, alter the resource-limiting constants for species 2 by setting the γ_{2ii}-values all to unity. Rerun the simulation.

Conclusion

The field of populations genetics has been especially fertile for the development of simulation models. This chapter has presented only an introduction to some of the most fundamental models. Genetics is a particularly active area for stochastic models, some of which will be considered in the later sections on stochastic simulations.

CHAPTER 11

MODELS OF
LIGHT AND PHOTOSYNTHESIS

Many simulations of biological phenomena require an input of data about one or more environmental factors. These factors are considered to be significant in determining the results of the simulations, but are not directly controlled by the processes the models describe. In this chapter and the next, we will work with temperature and light intensity, two factors that appear in many models that use environmental inputs.

Light intensity and temperature vary deterministicly, because both follow well-defined cycles of daily and annual periodicity. However, both also display random variability caused by unpredictable factors such as cloud cover and storm front movements. This random variation will be considered in later chapters on stochastic models.

Some of the models that describe variation in light intensity and temperature in these two chapters will be used as components of larger models in later chapters. In the jargon of some modelers, light and temperature models used in this way are "forcing functions" for the larger models. Other environmental factors may be used similarly, for example annual patterns of precipitation, or the pattern of osmotic pressure changes caused by tidally fluctuating salt concentrations in estuaries.

In this chapter we will consider models of spatial and temporal variation in light intensity. We will also consider some simple models of the effect of light on photosynthetic rates of plants. The rate of photosynthesis, rather than light, is considered to be a forcing function in larger models of some systems.

11.1 Annual Cycles of Solar Radiation

The intensity of light that strikes a unit of the earth's surface is a function of several factors, including light intensity at the upper atmosphere, the angle of incidence, absorbing qualities of the atmosphere, altitude, and amount of light reflected by cloud cover. The theoretical maximal daily cumulative radiation at a given location is determined by the time of year, which fixes the angular height of the sun and the length of daylight. In the temperate zones, this daily total is a sinusoidal function of season (Gates 1962).

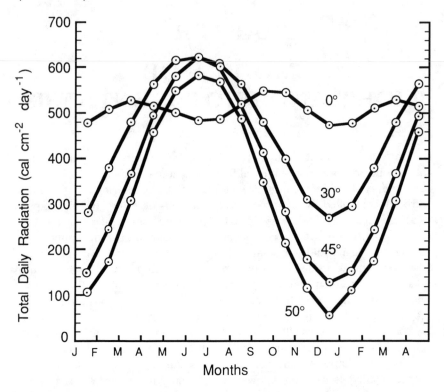

Figure 11.1. Annual variation in light intensity at different latitudes in the northern hemisphere. Points are from data in Hutchinson (1975).

Atmospheric absorption and scattering of radiation will depend upon the transparency of the atmosphere, and the distance sunlight must travel through it. Transparency will vary regionally and seasonally. If theoretical maximal values are corrected using estimates of atmospheric absorption, daily radiation for different latitudes can be estimated as in Figure 11.1. The curves are generally sinusoidal, and can be approximated with a sine function of time. Thus, the general equation for a weekly model of annual variation in daily total radiation in the northern hemisphere is

$$I_d = R \sin\left(2\pi \frac{W - 11}{52}\right) + M \qquad (11.1)$$

where M is the median daily total, and R is the range on either side of the median. I_d is total radiation in calories cm^{-2} day^{-1}. The time in weeks, W, is expressed relative to the calendar year, with week 1 beginning on the first day of January. The angle in radians is given as $2\pi(W - 11)/52$, with the denominator scaling the curve and the numerator providing the necessary phase shift. In Equation 11.1, the vernal equinox is assumed to occur in week 11. At each equinox, vernal and autumnal, sin(time) is 0, so that daily radiation equals M. Immediately following the vernal equinox, daily radiation will increase; it will decrease following the autumnal equinox. To model the daily total radiation at 45° North latitude, a suitable equation would be

$$I_d = 249 \sin\left(2\pi\frac{W - 11}{52}\right) + 378 \qquad (11.2)$$

In this equation, 378 is the median total radiation, and 249 is the range of variation on each side of the median, as estimated from Figure 11.1. The phasing of Equation 11.1 is proper for the northern hemisphere; for 45° South latitude, the correct phasing would be $2\pi(W - 37)/52$.

Exercise 11-1: Using Figure 11.1, find the constants needed to model total daily radiation by week at 30° North latitude using Equation 11.1. Write and implement a program to calculate total daily radiation based on the equation, and plot daily radiation as a function of time over a period of 52 weeks, beginning with the first week in January. Also find the equation for modeling the total daily radiation at 50° South latitude. Set up this equation to simulate the total daily radiation on a daily basis, rather than a weekly basis. For the southern hemisphere, assume the vernal equinox falls on 21 September (the 264th day of the year). Use this equation in a program to plot total daily radiation as a function of time over a 365-day year. (Assume the median and range values of Figure 11.1 apply equally to the southern and northern hemispheres.)

11.2 Modeling Daily Cycles of Solar Radiation

As the earth rotates daily on its axis, the intensity of direct sunlight on a horizontal unit of surface will be zero before dawn and after sunset, and will increase smoothly to a peak at solar noon. A variety of models have been developed to model light intensity over a 24-hour period. The following equation, based on a sine curve, has proven useful for temperate regions where the seasonal increase in daylight hours is accompanied by an increase in intensity due to increased angular height of the sun:

$$I_T = I_m \sin\left(2\pi\frac{T-6}{24}\right) + B \tag{11.3}$$

Here, I_T is the light intensity at hourly time T measured on a 24-hour basis beginning at midnight; the calculation of I_T is truncated to omit negative values. I_m is the daily maximum light intensity averaged over a year. B may be varied to model the combined seasonal effects of light intensity and daylength, being set to zero for the vernal or autumnal equinox. B will take on positive values for longer days and more intense noontime radiation, and negative values for shorter days with lower intensities; see Figure 11.2.

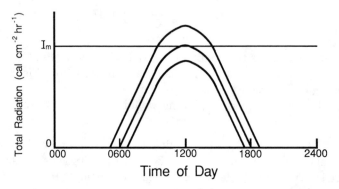

Figure 11.2. Truncated sine curve for simulating hourly light intensity on a 24-hour basis.

Exercise 11-2: Write and implement a BASIC program using Equation 11.3 to find light intensity for each hour over a 24-hour period. The output from your program should show light intensity vs. time over a 24-hour period. In addition, write your program so that it will numerically integrate the total radiation over the 24-hour period, and print out this sum as part of the output. Set $I_m = 30$ cal cm^{-2} hr^{-1}, and assume a 12-hour period of daylight.

11.3 Effects of Cloud Cover and Backscatter

The equations given above are based on direct radiation through clear skies. Cloud cover will have a variety of effects. If the sun is striking the ground directly while the sky is partly covered with white clouds, the total radiation can be increased significantly. If the sun is obscured by clouds, light intensity is decreased considerably. Gates (1962) estimates that the average cloud cover for the northern hemisphere is 52 percent. Half the light hitting a cloud is reflected back to space, about a fifth is absorbed, and the rest is transmitted to the earth's surface. On an annual basis, averaged over the northern hemisphere, direct radiation should therefore be reduced 35 percent by multiplying by a factor of 0.65. Of course, if an actual value for effective radiation is available for the particular locality being modeled, then this value should be used.

For some regions, cloud cover is highly seasonal. Anderson (1974) developed a complicated model of light intensity on the coast of southern California which included the heavy cloud cover that occurs during the early summer. For any given location, cloud cover may be modeled by fitting a polynomial equation to weekly weather data which may be available from several sources. Such an empirical equation may be used to supplement the basic sine curve of Equation 11.2. This method of fine adjustment is not required for the exercises of this text.

11.4 Reflection of Light from Water Surfaces

Light intensity is an important variable in many physical and biological models of marine and freshwater environments. For these models, the amount of light reflected at the surface of the water is important. Reflection is a function of the incident angle of light. Light penetrates almost without reflection when it is perpendicular to the water surface, while at very low angles reflection becomes almost total. The fraction of light reflected is obtained from Fresnel's law:

$$F_r = \frac{I_r}{I_i} = 0.5 \left(\frac{\sin^2(\phi_i - \phi_r)}{\sin^2(\phi_i + \phi_r)} + \frac{\tan^2(\phi_i - \phi_r)}{\tan^2(\phi_i + \phi_r)} \right) \qquad (11.4)$$

where F_r is the fraction of light reflected, I_i is the intensity of incoming radiation, and I_r is the intensity of reflected radiation. The angles involved are defined in Figure 11.3. For light entering water, the angles of refraction and incidence are related:

$$\sin \phi_r = \frac{\sin \phi_i}{1.33} \qquad (11.5)$$

This equation can be used to find angles of refraction needed to solve Equation 11.4. The fraction of incident light that enters the water is given by

$$F_w = 1 - F_r \qquad (11.6)$$

Precise estimates of variation in amount of reflected light may not be necessary for some models of aquatic systems. Ryther (1959) used a constant 5 percent loss to reflection of incident radiation. This constant loss was based on the observation that the fraction of total light entering the water is independent of angle of incidence; as angular height of the sun decreases, indirect skylight increases its proportion of total light, eventually exceeding the intensity of direct light from the sun.

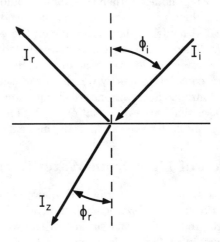

Figure 11.3. Reflection and refraction of incident light at the surface of water.

Exercise 11-3: Write a program to calculate the fraction of direct sunlight that enters a body of surface water for different angles of incident radiation between $0°$ and $90°$. ($\phi_i = 0°$ when the sun is directly overhead.) As output, produce a graph showing F_w for each $5°$ interval between $0°$ and $90°$. Note that the trigonometric functions built into most versions of BASIC require the expression of angles in radians. Many versions of BASIC do not have available the arcsin function to use with Equation 11.5 for finding the angles of refraction that are needed in Equation 11.4. The arctan function is usually present, so the following identity may be useful in this exercise:

$$\arcsin(x) = \arctan\left(\frac{x}{(1 - x^2)^{0.5}}\right)$$

11.5 Light Absorption in Lakes and Oceans

As light passes through water, its intensity declines at a rate that depends upon the length of the water column and the kinds and amounts of light-absorbing materials present. Beer's law provides an approximation for a simple model describing light attenuation as a negative exponential process:

$$I_z = I_o e^{-\eta bz} \tag{11.8}$$

where I_z is intensity of light at depth z, I_o is intensity of light at the surface of the column of water, and η is the coefficient of absorbance. Values of η depend on the nature and concentrations of absorbing materials in the water. If z is measured in meters, η will have units of m^{-1}. The factor b is a correction for the length of the actual light path through the water to depth z, related to the angle of refraction (see Equation 11.5) as follows:

$$b = \frac{1}{\cos \phi_r} \tag{11.9}$$

Interestingly, these two equations have been used also to model the decrease in the intensity of light passing through the leafy canopy of a forest. Light is absorbed by successive layers of tree leaves, so that the intensity of light reaching the ground depends upon the height of the canopy (Holm and Kellomaki 1984, Bachelet et al. 1983).

Exercise 11-4: Write and implement a computer program to model the intensity of light at a depth of 2 meters in a lake over a 24-hour period. Light intensity falling on the surface should follow the simple truncated sine-curve model of Exercise 11-2. Use values given for that exercise, and assume that 5 percent of the total light is reflected regardless of sun angle. Set b (Equation 11.8) equal to unity. For one-hour intervals over a 24-hour period, find and plot the intensity of the incident radiation, and the intensity of light at a depth of 2 m. Let $\eta = 0.4$.

11.6 Effect of Light on Rates of Photosynthesis

Physical models of light intensity become biologically interesting when they are used as parts of biological models. We will look at several models that describe the relationship between light intensity and the rate of plant photosynthesis. (Light intensity in photosynthetic work is measured in different units; here we will use calories per unit area per unit time.) One of the earliest of these models, due to Baly (1935), follows the Michaelis-

Menten saturation equation that we have discussed before:

$$P = \frac{P_{\max}I}{K_i + I} \tag{11.10}$$

Here, P is rate of photosynthesis, $P_{\max}$ is the maximum rate at saturating light intensities and I is a measure of intensity. K_i is the half-saturation constant ($K_i = I$ when $P = P_{\max}/2$) and has units of intensity. Thus, rate of photosynthesis is assumed to follow a hyperbolic curve, rising to the maximum rate when light is "saturating". Smith (1936) proposed the following formula as being more accurate:

$$P = \frac{P_{\max}KI}{\sqrt{1 + (KI)^2}} \tag{11.11}$$

This formula was adopted largely for empirical reasons. The curve resulting from this equation has a steeper slope at low light intensities and appears to fit experimental data better than Equation 11.10. K has units of intensity^{-1} in this equation.

Ryther (1956) observed that most aquatic algae (and perhaps most other plants, both terrestrial and aquatic) show a marked inhibition of photosynthesis at high light intensities. A model of the effect of light intensity on photosynthetic rate that incorporated the photoinhibition suggested by Ryther's (1959) data was proposed by Steele (1962):

$$P = P_{\max}kIe^{(1-kI)} \tag{11.12}$$

where the constant k is the inverse of the light intensity at which $P = P_{\max}$. This model appeared to fit the results of experiments fairly well. Vollenweider (1965) suggested that the fit could be improved by introducing another constant K, to give

$$P = P_{\max}KIe^{-kI} \tag{11.13}$$

The values for these constants may be estimated from experimental data using the maxima function with the techniques of Chapter 3. Figure 11.4 was obtained from such a curve-fitting procedure using the average of data collected by Ryther (1959) for 14 species of planktonic marine algae.

To account for photoinhibition, Vollenweider (1965) modified Equation 11.11 by adding another term:

$$P = \frac{P_{\max}KI}{\sqrt{1 + (KI)^2}} \cdot \frac{1}{\sqrt{\left(1 + (\alpha I)^2\right)^n}} \tag{11.14}$$

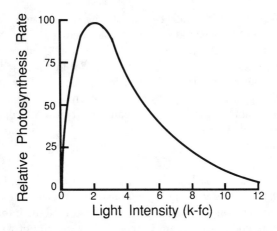

Figure 11.4. Photosynthetic response of plants to increased light, based on models of Ryther and Steele.

When $n = 0$, this equation becomes identical with Equation 11.11. Because it contains four constants that are determined by environmental conditions, this equation is extremely flexible. It can fit most data sets that are collected from experimental determinations of photosynthetic rate at different light intensities. As Fee (1969, 1973) points out, the difficult problem is to estimate accurately these constants ($P_{\max}$, K, α, and η); this usually requires some form of a nonlinear estimating procedure.

Exercise 11-5: According to Equations 11.10 and 11.11, light acts as a limiting factor for growth just as nutrients may. Chen (1970) suggested that Equation 11.10 could be used in combination with the multiple limiting nutrient models of Section 8.4; for example:

$$\mu = \mu_m \left(\frac{I}{K_i + I} \right) \left(\frac{N}{K_n + N} \right) \left(\frac{P}{K_p + P} \right) \qquad (11.15)$$

The terms in this equation are defined like those of Chapter 8: μ_m is maximum growth rate under saturating conditions of light intensity I, nitrogen concentration N, and phosphorus concentration P. The K values are half-saturation constants.

Write and implement a program using Equation 11.15 to simulate the growth of an algal population in an environment with constant total nutrients and seasonally varying light intensity. Assume total daily light varies on a sinusoidal annual cycle (Equation 11.2) having a maximum of 600 cal cm^{-2} day^{-1}, and a minimum of 100 cal cm^{-2} day^{-1}. Set K_i at 400 cal cm^{-2} day^{-1}. Use the

values given in Exercise 8-4 for the half-saturation constants, total nutrient concentrations, and fractional nutrient compositions
of N and P. Set the loss rate $R = 0.05$ day^{-1}, maximum growth
rate constant $\mu_m = 1.2$ day^{-1}, and initial biomass $= 9.25$ mg l^{-1}.
Use the simple Euler method of integration performed on a daily
basis to produce two years of simulation data. Plot your results
showing biomass, free nutrient concentrations, and daily radiation
values against time, to observe how varying light intensity affects
the system.

Exercise 11-6: Because Equation 11.14 uses 4 constants, it is not easy
to visualize how varying these constants changes the shape of the
curve describing photosynthetic response to light intensity. Write
and implement a program that will show how photosynthetic rate
varies with light intensity, using different values for K, α, and n.
Set $P_{\max} = 1$ so that your model describes relative photosynthesis,
and let I vary between 0 and 8 in steps of 0.1 units. Find the effect
of varying K by setting $\alpha = 1$ and $n = 1$, and plotting a curve of
P vs. I for $K = 0.5$; then on the same axes plot separate curves
for $K = 1.0$, 1.5, 2.0 and 2.5. Proceed further to investigate the
response by producing a set for curves for n varying between 0 and
2.5 with $K = 1$ and $\alpha = 1$. The produce another set for α varying
between 0 and 2.5 with $n = 1$ and $K = 1$.

11.7 Phytoplankton Production

The equations in the two previous sections may be combined to produce
a model of photosynthesis by planktonic algae at different depths in a
lake or ocean. Such a model would ignore species differences among the
algae making up the phytoplankton, and would assume that the algae are
uniformly distributed vertically. Other assumptions include an absorption
coefficient for light that is unvarying with depth, and a uniform response
of the algae to changes in light intensity.

One of the equations from the preceding section would be employed to
describe the photosynthetic rate based on light intensity, and the equations from Section 11.5 would be used to find light intensity at different
depths. If the surface light is sufficiently great to produce photoinhibition, then the curve describing photosynthesis at depth will show a peak
at the depth of optimal light intensity.

Exercise 11-7: Write and implement a program to describe photosynthetic rates at different depths below the surface of a lake. Use

Equation 11.8 to find light intensities, and Equation 11.14 to calculate photosynthesis. For values of constants in these equations, use those determined by Fee (1973) for Lake Michigan on 30 July 1970: $b = 1$, $\eta = 0.2$ m^{-1}, $\alpha = 1$ min cm^2 cal^{-1}, $K = 14.95$ min cm^2 cal^{-1}, $n = 0.531$, and $P_{max} = 4.30$ mgC m^{-3} hr^{-1}. Let the intensity of light penetrating the surface of the water be 0.5 cal cm^{-2} min^{-1}. Determine the rate of photosynthesis at each 0.1-meter depth increment from the surface to 20 meters. Plot light intensity and photosynthetic rate as a function of depth.

Exercise 11-8: Write a program to show 24-hour variation in photosynthetic rate at a given depth by combining Equation 11.14 with the sine equation for daily light intensity. Assume the maximum light intensity is 1.0 cal cm^{-2} min^{-1}, with other values taken from Exercise 11-7. Find the rate of photosynthesis for each 6-minute period (0.1 hour) of the 12 hours of daylight, for depths of 0.5, 5, 14, and 20 meters. Your output should consist of a graph showing rate of photosynthesis for each depth plotted against time of day. Because of the effect of photoinhibition, the shallower depths should display some interesting patterns, as in Figure 11.5.

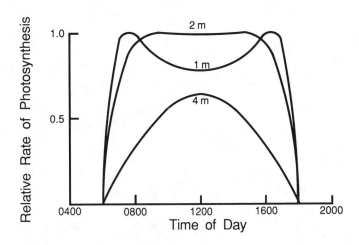

Figure 11.5. Simulated response of algal photosynthesis to daily changes in light intensity at different depths in a column of water.

Conclusion

This chapter has been concerned with temporal and spatial variation in light intensity, and the effect of light on photosynthesis. The photosynthetic models have been limited to the aquatic environment, and confined to very simple equations. Models that are considerably more complex than those given here have been developed for these phenomena, principally to escape the limits of the assumptions of the simple models. Models of light are used in many simulations from biological fields other than those in this chapter. For example, light models are used in simulations of photosynthetic biochemistry, of plant and animal behavior cued by photoperiodism, and in visual phenomena. The relationship between solar radiation and temperature will be discussed in the next chapter.

CHAPTER 12

TEMPERATURE AND BIOLOGICAL ACTIVITY

Models of temperature are important in most fields of biological inquiry because of the effect temperature has on rates of biological activity. This is produced primarily by effects of temperature on chemical reaction rates. At low temperatures reaction rates become vanishingly small, and biological activity all but ceases. At high temperatures competing reactions, such as the denaturation of vital proteins are favored, and organisms begin to lose their integrity. Biological activity is primarily confined to temperatures between 0 and 100° Celsius. This chapter will describe some methods for modeling seasonal and daily variation of temperature and heat balance, and for describing the effect of temperature on rates of biological processes.

12.1 Seasonal and Daily Variation in Temperature

Temperature of the environment is largely a function of solar radiation. For this reason environmental temperatures tend to follow sinusoidal patterns like those of light intensity in the previous chapter. Temperature also partly depends on the quantity of heat stored in the environment as a result of solar inputs. If solar radiation were stored and reradiated uniformly, the temperature curve would be the integral of the solar sine curve. Analytically, the integral of a sine function is a negative cosine function:

$$\int \sin x \, dx = -\cos x + c \qquad (12.1)$$

A negative cosine curve is identical to the sine curve, except that it lags the sine curve by 90° (i.e., it is out of phase by one-fourth wavelength). This relationship is diagrammed in Figure 12.1. The cosine approximation of temperature is inadequate, because the earth's surface is not homogeneous; air, land, and water can absorb, circulate and exchange heat energy at different rates. At most locations on the earth's surface, air temperature lags solar intensity by 3 to 4 weeks, rather than 13 weeks as

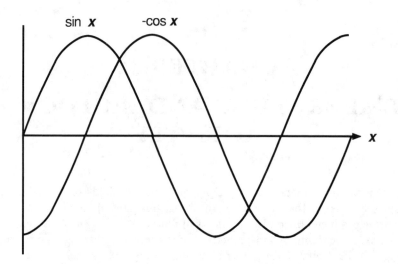

Figure 12.1. Comparison of sine curve with its integral, the negative cosine curve.

predicted by the negative cosine curve. If we use the 4-week figure and assume that average annual air temperature is 10°C with a range of 12°C for average daily temperatures on either side of the annual mean, we can find the average daily temperature for any week with

$$T = 10 + 12 \cdot \sin\left(2\pi \frac{W - 15}{52}\right) \tag{12.2}$$

This can be compared with a model for light intensity from Chapter 11:

$$I_d = 400 + 200 \cdot \sin\left(2\pi \frac{W - 11}{52}\right) \tag{12.3}$$

In many simulations, extremes of temperatures are more important than means. Webb et al. (1975) developed a model for air temperature of an Oregon forest which was composed of three elements: mean daily temperature, daily temperature range, and daily temperature excursion. Air temperature at any hour of any day is found with the relationship

$$\text{Air Temperature} = \text{Mean} + (\text{Range})(\text{Excursion}) \tag{12.4}$$

The following equations were used to find each of these components:

$$\text{Mean} = 10 + 8\sin\left(2\pi\frac{D - 107}{365}\right) \qquad (12.5)$$

$$\text{Range} = 7 + 2\sin\left(2\pi\frac{D - 107}{365}\right) \qquad (12.6)$$

$$\text{Excursion} = \sin\left(2\pi\frac{H - 17}{24}\right) \qquad (12.7)$$

Equation 12.5 indicates that mean daily temperature has an annual average of $10°C$, and a range of $8°C$. This is multiplied by the sine function which starts its 365-day cycle on day 107 (17 April); the peak mean temperature of $18°C$ will come on day 198 (17 July). Equation 12.6 indicates that the average daily fluctuation of $\pm7°C$ in temperature will take place on days 107 and 290; the maximum fluctuation of $\pm9°C$ will occur on day 198. Equation 12.7 provides the timing of the daily temperature variation, with the minimum at 0400 (4 am) and a maximum at 1600 (4 pm). In these three equations, time is expressed as D, Julian calendar days, and as H, hours, from 0 to 24 with solar noon at 12.

Exercise 12-1: Write a program to simulate air temperature in an Oregon forest, using Equations 12.4-12.7. Perform time increments in nested FOR-NEXT loops, using the outer loop to increment days, and the inner loop for hours. Set up your program to produce graphical output showing air temperature for each hour during the weeks of 1-7 January and 4-10 July. Then modify your program to produce a graph of air temperature over the 24-hour period of your birthday. (This will be useful for planning your birthday picnic in an Oregon forest.)

12.2 Heat Balance in Biological Systems

Heat balance determines the temperature of natural objects and organisms. If an object is absorbing or generating heat faster than it is giving off heat, obviously its temperature must rise. In this section we will consider the heat balance of an entire lake, and also of a desert-dwelling mammal.

The major source of heat energy input for a lake is solar radiation striking its surface. It is not surprising that the heat content and water temperature of a lake will follow an approximate sine function on an annual basis. The energy budget equation which describes the heat content of a lake is

$$Q_{t+\Delta t} = Q_t + (I_t + S_t - E_t - O_t) \tag{12.8}$$

where Q_t is the heat content of a lake at time t, and $Q_{t+\Delta t}$ is the heat content one week later. Heat content is measured on the basis of the exposed surface of the lake, as calories cm^{-2}. The other components of the equation are I_t, the effective radiation coming in through the lake's surface; S_t, the sensible heat transfer by conduction or convection to or from the atmosphere; E_t, the heat lost or gained by evaporation or condensation; and O_t, the heat lost as radiation to the atmosphere. These four components are rates, and are each expressed below as calories cm^{-2} $week^{-1}$.

Radiation input by weeks may be described by the following equation:

$$I_t = 2800 + 1400 \sin\left(2\pi\frac{w-11}{52}\right) \tag{12.9}$$

The form of this equation and the constants have been described above (e.g. Equation 12.2). Sensible heat transfer is based on Newton's law of cooling (see Chapter 1) and is given by

$$S_t = k\,(T_a - T_w) \tag{12.10}$$

where T_w is the surface temperature of the water and T_a is the temperature of the air overlying the water. Air temperature may fluctuate as in Equation 12.2. The transfer coefficient will vary with wind velocity; you may assume that it has a value of 100 calories cm^{-2} $week^{-1}$ $°C^{-1}$

Evaporation removes heat from the lake and may be described by the equation

$$E_t = RH\,(P_w - P_a) \tag{12.11}$$

The various terms in this equation are defined as follows. R is a rate constant that depends on wind velocity. Assuming an average wind velocity of 10 m sec^{-1}, R will have an approximate value of 1.0 gram $mmHg^{-1}$ cm^{-2} $week^{-1}$. H is the latent heat of evaporation and is a function of water temperature. Over the range of 0 to 50°C, H may be approximated by this equation:

$$H = 595.8 - 0.54T_w \tag{12.12}$$

P_w is the saturation water vapor pressure, which is also a function of surface water temperature. P_w is approximated by a cubic equation (Chapter 3):

$$P_w = 4.57 + 0.357T_w + 0.0065T_w^2 + 0.0004T_w^3 \tag{12.13}$$

P_a is the partial pressure of water vapor in the air and is quite variable. For purposes of this simulation we may assume a constant 80 percent relative humidity, so that P_a will be a function of air temperature as follows:

$$P_a = 3.66 + 0.286T_a + 0.0052T_a^2 + 0.0003T_a^3 \qquad (12.14)$$

According to the Stefan-Boltzmann law, thermal radiation of the lake to the atmosphere is a function of the absolute temperature of the lake surface, raised to the fourth power. The atmosphere will also radiate to the lake as a function of atmospheric temperature. The relationship between atmospheric temperature and radiation is not simple (Hutchinson 1957). We will employ a simple equation that relates the net thermal radiation between a lake and the atmosphere to the difference between air temperature and water temperature:

$$O_t = 77(T_w - T_a) \qquad (12.15)$$

where O_t is the net thermal radiation with units of calories cm^{-2} week^{-1}.

Exercise 12-2: Prepare a flowchart for a program to simulate water temperature of a lake based on the heat-balance model described with Equations 12.8-12.15. Then implement your program using the given transfer coefficients and rate constants. The output of your program should consist of a graph showing lake temperature for 104 weeks (2 years). Assume the lake has an initial temperature of 6°C on January 1, and that the lake circulates continuously throughout the year. Also assume the lake to have an average depth of 10 meters, so that

$$(T_w)_t = \frac{Q_t}{1000} \qquad (12.16)$$

based on the specific heat of water of 1 calorie gram^{-1} °C^{-1}. Although this simulation appears complex because of the number of equations involved, it is in fact greatly simplified and ignores many important features of the lake environment, including stratification. Section 12.7 below describes a model that includes stratification. For more information on models of heat balance for lakes, see Hutchinson (1957) and Wetzel (1983).

Exercise 12-3: Like most mammals placed in a hot environment, camels regulate their body temperatures partially by evaporative cooling (sweating) if they have enough drinking water. However, a partly dehydrated and thirsty camel exposed to heat will conserve water, and rather than sweating will allow its body temperature to increase. This takes advantage of the decreased rate of transfer of heat that occurs with increased body temperatures in warm environments (Newton's law). Camels rely on large body mass, about 460 Kg as an average, to moderate effects of increased heating. While the sun shines, a camel's body will absorb heat and increase in temperature, and during the relatively cool desert night it will lose heat.

Write a program to simulate the body temperature change of a partly dehydrated camel, based on the following considerations: (1) Internal metabolic heat production is constant at 250 Kcal hour^{-1}. (2) Air temperature varies sinusoidally on a 24-hour cycle, with a maximum of 46°C at 1400, and a daily mean of 33°C. (3) Between a camel's body and air, thermal conductance k is 62.5 kcal hour^{-1} °C^{-1}; heat loss and gain follow Newton's law (Equation 1.10; see also Equation 12.10). (4) T_t, body temperature at time t, is found with Q_t/M, where M is body mass in kilograms and Q_t is total heat content at time t.

Begin your simulation at midnight with a camel having a body heat content of 16500 kcal above 0°C. For each hour, find the total heat gained or lost by the camel (kcal), and plot the results of your simulation over a 96-hour period. Also produce a graph showing body temperature of the camel over the same time period.

12.3 Effect of Temperature on Chemical Reaction Rates

Biological activities such as feeding, digestion, respiration, photosynthesis, growth, and movement depend on chemical reactions which are catalyzed by enzyme proteins. Because of this dependency, temperature will affect biological activity in two ways. First, a temperature rise will increase the number of reactant molecules having an energy equal to or greater than the activation energy for the reaction. This effect tends to increase the reaction rate. Second, high temperatures may denature enzyme molecules, causing them to lose their catalytic activity. These two effects compete, resulting in the typical temperature optimum curve associated with essentially all biological activity. Some models of temperature dependency of biological systems are described in the next few sections.

One of the earliest models dealing with the effect of temperature on chemical reaction rates was developed by Arrhenius in 1889, based on the concept of activation energy. This equation defines the change in reaction rate with temperature as follows:

$$\frac{d(\ln k)}{dT} = \frac{(E_a)}{RT^2} \qquad (12.17)$$

where k is the reaction rate constant, T is absolute temperature in $°K$, R is the gas law constant, and E_a is the activation energy for the reaction. When integrated analytically, this equation can take the following form:

$$\ln k = \left(\frac{-E_a}{r} \cdot \frac{1}{T} \right) + \ln k_\infty \tag{12.18}$$

where $\ln k_\infty$ is an integration constant, representing the limiting value of k.

Notice that this is an equation for a straight line. This allows $\ln k$ to be graphed as a function of $1/T$, and therefore to obtain $-E_a/R$ as the slope. These Arrhenius plots are often used to find the activation energy for various biological reactions.

By integrating Equation 12.18 between limits and taking antilogs, the equation becomes

$$k_T = k_x \exp \left(\frac{E_a}{R} \cdot \frac{(T - T_x)}{T \cdot T_x} \right) \tag{12.19}$$

where k_T is the rate at temperature T, and k_x is the rate at some reference temperature T_x . Equation 12.19 can be made more useful by expressing temperature as $°C$ rather than $°K$. The reference temperature is set to $0°C$, and E_a/R is combined as a single constant:

$$k_C = k_0 \exp \left(\frac{A \cdot C}{C + 273} \right) \tag{12.20}$$

In this form, k_C is the rate at some Celsius temperature C, and k_0 is the rate at $0°C$. The value of the constant A may be found from the slope of the Arrhenius plot, where

$$A = \frac{\text{Arrhenius slope}}{273} \tag{12.21}$$

The value of A may also be estimated from the Q_{10} value (see below) and Equation 12.20. For further discussion of the development of the Arrhenius equation, see Hamil et al. (1966).

A simple and widely used method for describing the effect of temperature on chemical reactions is called the Q_{10} approximation. Q_{10} is the factor by which reaction velocity is increased for a temperature rise of $10°C$. The equation, in a form like Equation 12.20, is

$$k_C = k_0 Q_{10}^{C/10} \tag{12.22}$$

or

$$k_2 = k_1 Q_{10}^{(C_2 - C_1)/10} \tag{12.23}$$

where k_2 and k_1 are rates at temperatures C_2 and C_1. The equation is useful over relatively narrow ranges of temperature. Prosser and Brown (1961) discuss this problem further.

12.4 Effect of Temperature on Enzyme Activity

The amount of product produced by a reaction catalyzed by an enzyme depends on two reactions that have different rates. One reaction is the actual formation of product; the other is the denaturation of the enzyme. These reactions may be written together as follows:

$$S + E \xrightarrow{\ k_1\ } ES \xrightarrow{\ k_2\ } P + E$$

$$\downarrow k_d$$

$$D$$

where E is the active form of the enzyme, D is the denatured form, P is the product, S is the substrate, and ES is the enzyme- substrate compound. The rate of product formation is

$$\frac{dP}{dt} = k_2[ES] \tag{12.24}$$

If substrate is in excess and k_2 is the rate constant for the limiting reaction, we may assume that ES is in equilibrium with E and S, so that k_1 is given by

$$k_1 = \frac{[ES]}{[E][S]} \tag{12.25}$$

If this equation is rearranged and substituted into Equation 12.24, then

$$\frac{dP}{dt} = (k_2 k_1 [S])\,[E] = k_r[E] \tag{12.26}$$

where k_r is an overall constant involving k_1, k_2, and substrate concentration $[S]$, which will be essentially constant if in great excess. The reaction now simplifies to

$$D \xleftarrow{\ k_d\ } E \xrightarrow{\ k_r\ } P + E$$

The effect of temperature on k_r may be modeled using a modification of Equation 12.20:

$$(k_r)_T = (k_r)_0 \exp\left(\frac{A_r T}{T + 273}\right) \tag{12.27}$$

where T indicates temperature in °C.

The denaturation rate constant, k_d, also depends on temperature, and may be described by an equation similar to 12.27. Because of the very high activation energy involved, the reaction has a very high value of A_d, which is equivalent to a very high Q_{10} value. Here we will assume that denaturation is controlled totally by kinetic effects. The equation would be much the same if instead we were to assume a temperature effect on the equilibrium constant involved in the denaturation reaction.

Exercise 12-4: Using the equations above, write a program to simulate the effect of temperature on an enzyme-catalyzed reaction. Base your measurement of rate of product formation (i.e. activity of enzyme) on amount of product that is obtained after the reaction has proceeded for some arbitrarily specified period of time. Because denaturation of the enzyme depends on time, the optimum temperature shown by your simulation will partly be a function of length of this time period, as shown in Figure 12.3. A flowchart for the program is shown in Figure 12.2. Implement your simulation using the following values:

$$(k_r)_0 = 1 \quad (k_d)_0 = 0.0001 \quad A_r = 18$$
$$A_d = 55 \quad E_0 = 2 \quad t_{\max} = 20$$

Set $\Delta t = 0.01$. The output of your simulation should be a graph of the amount of product vs. temperature, for temperatures from 0 to 80°C. Your graph should resemble a typical optimum temperature curve like those of Figure 12.3.

12.5 Models of Temperature Effects on Biological Activity

Many models have been published to describe effect of temperature on rates of biological processes. The available literature is so large that Watt (1975) described it as "truly awesome". Many of these models attempt to use simple enzyme effects for complex biological processes involving many enzymes. In general, such models assume that a single enzyme will limit rates in complex systems like a developing embryo or feeding animal. The response of the whole organism is assumed to be similar to that of the limiting enzyme. This assumption is usually incorrect, because different enzymes can be limiting at different temperatures.

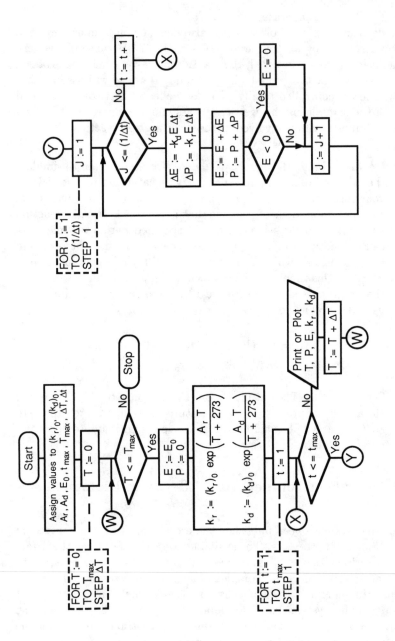

Figure 12.2. Flowchart for a program to simulate the effect of temperature on activity of enzymes.

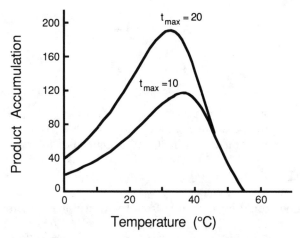

Figure 12.3. Curves showing simulation results of response of enzymes to different temperatures.

Overall response of a biological process to temperature is a composite of the response curves of all enzymes that might limit the rate within the temperature range for the organism. An overall response curve may be very different from that produced by a single enzyme. For example, Sharpe and DeMichele's (1977) model of development rate at different temperatures was based on Eyring's transition-state theory (Johnson et al. 1954, Eyring and Urry 1965). Assuming a single limiting enzyme, the model uses six constants that must be estimated from data. They note that the composite curve from multiple limiting enzymes could be described with their equation. The constants then would be simply regression coefficients for an empirical equation.

We present here two empirical equations which have been used to describe effects of temperature on biological rates. Below optimum temperature the rate rises approximately exponentially; above optimum it descends sharply.

O'Neill et al. (1972) developed an equation which results in a curve with the necessary form, and at the same time is based on constants which may be readily derived from most experimental data. The equation has the following form:

$$k_T = k_{\max} U^x \exp(xV) \tag{12.28}$$

where

$$U = \frac{T_{\max} - T}{T_{\max} - T_{\mathrm{opt}}} \tag{12.29}$$

and

$$V = \frac{T - T_{\mathrm{opt}}}{T_{\mathrm{max}} - T_{\mathrm{opt}}} \tag{12.30}$$

and

$$x = \frac{W^2 \left(1 + \sqrt{1 + 40/W}\right)^2}{400} \tag{12.31}$$

Here, k_T is the rate of the process at temperature T, k_{max} is the rate at the optimum temperature T_{opt}, and T_{max} is upper lethal temperature. Temperatures are all measured in °C. W is defined as follows:

$$W = (Q_{10} - 1)(T_{\mathrm{max}} - T_{\mathrm{opt}}) \tag{12.32}$$

The Q_{10} value should be estimated from the nearly linear increasing portion of the rate curve. The principal advantage of Equation 12.28 over other exponential equations is that its constants are readily found from plots of most real data. Its disadvantage is the obscure theory behind Equations 12.31 and 12.32. For temperatures above T_{max} it is necessary to set $k_T = 0$, because the equation is not designed to predict values beyond that point.

Logan (1988) reviewed a number of empirical equations that can produce temperature-rate curves resembling those of Figures 12.3 and 12.4. One of the more useful empirical models is partly based on an exponential function like the O'Neill equation above:

$$k_T = k_1 \left(\frac{\tau_1^2}{\tau_1^2 + k_2} - \exp\left(-\tau_2\right) \right) \tag{12.33}$$

In this equation, k_T is the rate at temperature T, k_1 and k_2 are empirical constants, and τ_1 and τ_2 are determined from experimental temperature data as

$$\tau_1 = T - T_{\mathrm{min}} \tag{12.34}$$

$$\tau_2 = \frac{T_{\mathrm{max}} - \tau_1}{T_{\mathrm{max}} - T_{\mathrm{opt}}} \tag{12.35}$$

The maximum and optimum temperatures for the reaction rate are defined as in the O'Neill model above. T_{min} is determined from the data, and is the low temperature at which the reaction rate drops effectively to zero. The bracketed part of Equation 12.33 contains two terms which compete to determine the rate. The first term (left-hand) resembles the Michaelis-Menten relationship, and will produce a sigmoid rise in the rate with increasing temperature. The second, exponential term describes the rate of denaturation of a limiting enzyme, which causes the rate to decline. As temperature increases, the two terms compete, and k_T reaches a maximum value at the optimum temperature. Above this temperature, the second term predominates because its rate of change per degree is much greater than the first term.

The constants k_1 and k_2 may be determined with nonlinear curve-fitting techniques. In actual practice, the values of the constants and the values of the temperature optimum, maximum and minimum may all be evaluated with techniques that permit the least-squares estimation of several coefficients.

Exercise 12-5: Write a program which determines k_T with Equation 12.28 for temperatures between 0 and 48°C . Use the following values for constants in the equation:

$$T_{\text{opt}} = 37 \quad T_{\text{max}} = 42 \quad Q_{10} = 2.5 \quad k_{\text{max}} = 100$$

Output for your program should be a curve showing k_T plotted against temperature in °C.

Exercise 12-6: Moner (1972) obtained the following values for growth rate (inverse of division time) for the ciliated protozoan *Tetrahymena pyriformis* at different temperatures:

°C	Rate	°C	Rate	°C	Rate
14	0.0023	21	0.0073	28	0.0129
15	0.0029	22	0.0087	29	0.0131
16	0.0036	23	0.0093	30	0.0132
17	0.0041	24	0.0100	31	0.0131
18	0.0048	25	0.0110	32	0.0111
19	0.0060	26	0.0117	33	0.0059
20	0.0066	27	0.0121	34	0.0000

Produce a graph of these data with a simple connect-the-dots routine, plotting rate vs. °C. From the plot, estimate the constants needed for the O'Neill model (Equation 12.28). Write a program as in Exercise 12-5 to draw a curve of k_T vs. temperature from 10°C to 40°C, using your estimates of the constants. On the same graph, plot the actual data points to examine the accuracy of the model equation. You will discover that the model is quite sensitive to small differences in the estimate of Q_{10}. You should try using various Q_{10} values to improve the fit of the plotted line to the data.

Exercise 12-7: As in Exercise 12-6, produce a rate vs. temperature graph of Moner's (1972) data. From the plot, estimate the values of T_{opt}, T_{max} and T_{min} needed for Logan's Equation 12.33. Then use the CURNLFIT program (Chapter 3) to estimate the constants k_1 and k_2. (To do this, you will need to read in the temperature as x-values and the rates as y-values. Put Equation 12.33 into CURNLFIT using k_T as Y, T as X, and your estimates of T_{min}, T_{max}, and T_{opt} as constants; use k_1 as coefficient A and k_2 as coefficient B.) For initial approximate values needed by CURNLFIT, use 3 times the measured maximum rate for k_1 (coefficient A), and 2 times $(T_{max} - T_{min})$ for k_2 (coefficient B).

12.6 Modeling Development in Variable Temperatures

Most plants, animals and microbes grow and develop in environments with fluctuating temperatures. Several methods are available for applying rate equations to development in varying temperatures. These models can have important practical application; for example, models for simulating the development of insect eggs or larval stages can permit agricultural controls to be applied most efficiently for maximum effectiveness. We will examine here a simple rate-summation model for predicting development in changing temperatures (Messenger and Flitters 1959, Grainger 1959, Tanigoshi and Browne 1978).

The rate-summation technique estimates duration of development based on the amount of time spent at a given temperature. This time is multiplied by the development rate, determined from constant temperature experiments. The time-rate product indicates how much development occurs during the period. Time-rate products are summed consecutively until the sum reaches unity, at which point the simulated development is complete. Consider a hypothetical example for eggs of a species of insect which experimentally have shown the following development rates at two temperatures:

Constant Temperature	Duration	Rate (1/duration)
10°C	100 hrs	0.01 hr^{-1}
20°C	33.3 hrs	0.03 hr^{-1}

Suppose that an egg is laid in an environment with temperatures alternating between 10°C and 20°C on a 12-hour cycle. If the egg begins to develop at the start of a 12-hour period of 10°C, development will proceed as follows:

Time period	Rate	Development	Sum	Total hrs
12 hrs at 10°C	0.01	0.12	0.12	12
12 hrs at 20°C	0.03	0.36	0.48	24
12 hrs at 10°C	0.01	0.12	0.60	36
12 hrs at 20°C	0.03	0.36	0.96	48
4 hrs at 10°C	0.01	0.04	1.00	52

Exercise 12-8: Write and implement a program based on the rate-summation method to find duration of development when temperature varies. Include in your program a sequence of DATA statements giving length of the period, temperature, and the rate at that temperature. Your program should read a DATA statement, and then calculate the proportion of development that occurs during the period, sum up the development to that time, and check to find if it exceeds 1.00. If not, data for the next period should be read. The portion of the final period required to complete development should be calculated. Output from your program should be in a tabular form, as given above. Test your program using a cycle of alternating 6-hour periods of rates of 0.022 and 0.034 at 12°C and 18°C respectively. Assume egg development begins at the start of a period with the higher rate.

12.7 Model of a Stratifying Lake

The heat-balance model of a lake in Section 12.2 was not very realistic because heat was assumed to be uniform throughout the water column. In fact, most heat is absorbed near the surface and carried down through the water column. One method for modeling lake temperature is to divide the lake vertically into separate compartments. For this method, heat transport is assumed to occur only in the vertical dimension between compartments. Each compartment represents one meter of depth below 1 cm^2 of surface, so the volume of each compartment is 100 cm^3. Heat content of each compartment is measured in calories; in this model, water at 0°C has a caloric content of 0. Thus, temperature of any compartment is its heat content divided by 100 cal °C^{-1}.

Solar radiation provides the principal heat input. Absorbed energy contributes to the heat content of a compartment at depth z as described by a modification of Equation 11.8:

$$(I_a)_z = I_z - I_{z+1} = I_o \left(e^{-\eta(z-1)} - e^{-\eta z} \right) \qquad (12.36)$$

where I_a is energy absorbed per unit time, I_z is amount of solar energy reaching the top of compartment z, I_{z+1} is the amount at the bottom, η is the absorption coefficient and I_o is the surface intensity.

Other types of heat transfer between the lake and the atmosphere are assumed to occur only at the surface. The remaining terms in the heat balance model (Equation 12.8) are

$$Q_0 = S - O - E \qquad (12.37)$$

where Q_o is the heat transport across the surface per unit of time, S is a sensible heat exchange (convective or conductive), O is heat radiated to the atmosphere, and E is heat transfer by evaporation or condensation.

In addition to the solar energy absorbed, each compartment will have its energy content altered by the transport of heat to the compartment from the adjacent warmer layer, and by the transport of heat from the compartment to the adjacent cooler layer. We will assume that the warmer layer is at a shallower depth, and the cooler layer is deeper. Amount of heat Q_z transported from compartment z to compartment $z + 1$ is assumed to be a function of the temperature difference and the eddy diffusivity coefficient, A_z, as follows:

$$Q_z = F A_z \left(T_z - T_{z+1} \right) \qquad (12.38)$$

where F is a proportionality constant. In finding heat transfer for the bottom compartment, assume the sediments have the same temperature as the bottom layer of water. The eddy diffusivity coefficient is a function of both the force that produces mixing, i.e. wind, and the force that resists the mixing, i.e. the density gradient; water of greater density lies beneath water of lesser density. The eddy diffusivity coefficient may be approximated with

$$A_z = A_{\min} + 0.9 A_{z-1} \exp \left[-G \left(\rho_{z+1} - \rho_z \right) \right] \qquad (12.39)$$

A_o will be the wind-induced surface value for eddy diffusivity. (To find A_z for the bottom compartment of water, assume the sediments to have a density value of 1.5). The constant 0.9 describes the exponential decrease in the surface value of eddy diffusivity with increasing depth. G is a constant of proportionality. $A_{\min}$ is the minimum eddy diffusivity, with a value of approximately 0.01 cal cm^{-2}. ρ_z and ρ_{z+1} are the densities of water in compartments z and $z + 1$. Water density is a function of temperature, and may be obtained from the following empirical quadratic equation which applies very well over the range of normal lake temperatures:

$$\rho_z = 0.999884 + \left(5.75 \times 10^{-5}\right) T_z - \left(7.27 \times 10^{-6}\right) T_z^2 \qquad (12.40)$$

Exercise 12-9: Write a program to simulate the warming of a lake in the spring, assuming an initial temperature of 4°C for all depths. Let the lake have a depth of 20 meters (i.e. 20 compartments), with compartment 1 representing the depth interval 0-1 m. The compartmental temperature values should be stored in a subscripted array. Losses and additions of heat in each compartment may be calculated with the equations above. The calculation of temperatures should be a two-part process. In the first part, the densities for all the compartments are calculated, and then the change in heat content and temperature is calculated for each of the compartments in turn, top to bottom:

$$\Delta T_z = \left[(I_a)_z + Q_{z-1} - Q_z\right]/100 \qquad (12.41)$$

with the values for I_a and Q coming from Equations 12.36 and 12.38. For compartment 1, the heat transfer across the surface can be considered to be Q_o (Equation 12.37).

In the second stage the changes are applied to each compartment to update its temperatures:

$$T_z \longleftarrow T_z + \Delta T_z$$

Remember to set the temperature of the sediments (compartment 21) to equal the temperature of compartment 20. Use the following values in setting up your simulation; some represent daily versions of weekly values in Exercise 12 2:

$$I_0 = 350 \quad A_0 = 0.5 \quad O = -20 \quad E = 70$$
$$S = 30 \quad \eta = 0.40 \quad F = 100 \quad G = 2000$$

Run your simulation for 30 days; the surface temperature should approach 26°C. Plot depth-temperature curves at least for times 0 and 30. Your output should consist of a graph showing temperature on the x-axis, and depth on the y-axis. Your graph may be clearer if you plot temperatures of each compartment as connected line segments, rather than as connected points. Graphs for comparison with actual lakes may be found in any textbook of limnology, for example Hutchinson (1957) or Wetzel (1983).

Conclusion

This chapter has introduced you to some representative models of temporal and spatial variation of temperature, and the effect of temperature on biological activity. Although they are interesting in themselves, you will find temperature models like these used in simulations of complex systems. For example, a model of lake temperature could be used in conjunction with a temperature-activity model and a model of photosynthesis as input for a model of algal productivity. Temperature-activity models are essential for simulations in which physiological rates of plants or poikilothermic animals are important components.

CHAPTER 13

COMPARTMENTAL MODELS
OF BIOGEOCHEMICAL CYCLING

In several previous chapters we considered various ecological models of growing and interacting populations. In this chapter we will study some ecological models for simulating flow of energy and material through organisms and their environment. These biogeochemical models focus on transfer through components of ecosystems, without considering individual organisms. Our approach will be to examine comparatively simple models that illustrate the principles of biogeochemical models. The techniques of compartmental modeling learned here will be useful in the next several chapters.

13.1 The Concepts of Material and Energy Flow

The concept of flow of materials between different components of the biosphere was worked out early in the twentieth century. For example, Lotka's (1925) box-and-arrow diagram for a global cycle of carbon would be at home (with slightly modified data) in current ecology textbooks. This analytical approach was evidently on Elton's mind as he worked out the first food web for an ecosystem in 1923 (Hutchinson 1978). The same approach influenced Lindeman (1942) in the first attempt to measure energy flow through an ecosystem.

One of the early attempts to obtain a complete, detailed description of energy content and flows in a single ecosystem was Odum's (1957) work on Silver Springs, Florida. The results of this research may be summarized in an energy-flow diagram (Figure 13.1). Similar diagrams are the basis of most ecosystem models.

The diagrams show the amount and direction of flow among components of an ecological community, and between the community and its environment. To develop a complete diagram, one must know for each component the standing crop, and the energy or material inputs and outputs. The organisms in the system may be grouped in a variety of ways; a common division is by trophic level. Depending on the interests and

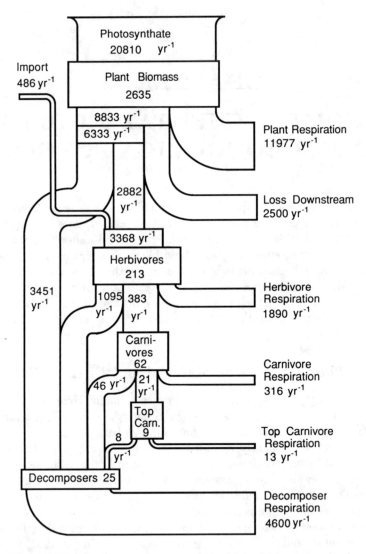

Figure 13.1. An energy diagram of the biological community in Silver Springs, Florida. Based on diagrams by Odum (1957) and by Patten (1971).

objectives of the modeler, a given trophic level may be subdivided in a variety of ways.

13.2 Block Diagrams and Compartment Models

Block diagrams are useful conceptual tools for developing simulations of flow, with each component of the system represented by a rectangular block. The blocks are connected by arrows showing flow of energy or material between components, or between components and the environment. The system may include all biological components of a community, or only a portion, depending upon the modeler's definition of the system. Once the system has been defined, everything else is "environment". Inputs to the system are termed driving or forcing functions. Figure 13.2 is an example of a block diagram.

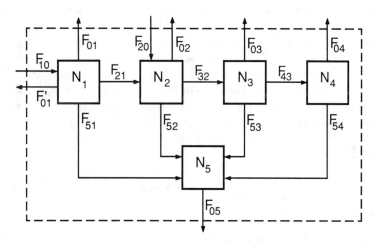

Figure 13.2. A simple block diagram of the Silver Springs energy flow compartmental model, modified from Patten (1971).

The blocks within a diagram may be defined with words. However, letters with subscripts to designate compartments, N_1 and N_2 for example, are more convenient because they may be used directly in writing computer programs. Between compartments, the flows F_{12}, F_{32}, etc., will have subscripts indicating receiving and donating blocks. Conventionally, a subscript of 0 indicates the environment. In this book, the first number in the subscript of a flow will represent the receiving compartment, and the second the source compartment. You should be aware that this subscript order may be reversed by different modelers working in different biological fields. No standardization exists.

In the Silver Springs model (Figure 13.2), N_1 = producers, N_2 = herbivores, N_3 = carnivores, N_4 = top carnivores, and N_5 = decomposers. F_{10} and F_{20} are the two driving flows from the environment. F_{01}, F_{02}, F_{03},

F_{04} and F_{05} represent respiratory losses to the environment from the system. F'_{01} represents the physical loss of plant biomass carried downstream and out of the system by water currents.

Under steady-state conditions, the sum of system inputs, F_{10} and F_{20}, must equal the sum of the system outputs, F_{01}, F_{02}, F_{03}, F_{04}, F_{05} and F'_{01}. Also at steady state, the sum of the inputs to any block must equal the sum of outputs from that block.

The units for the blocks will depend on the system. Usually they express mass or energy on the basis of volume or area. Typical units might be grams of dry weight per square meter, grams of organic carbon per cubic meter, or kilocalories per square meter. The flows will use the same units together with units of time, such as grams per square meter per day, or grams of organic carbon per square meter per year.

A compartmental model is a more formal statement of the block diagram, and includes the equations describing the flows between compartments. Sheppard (1948) apparently was the first to use "compartment" in this sense (Godfrey 1983). The flows or fluxes are used in the equations to provide an energy or material balance for each compartment. The rate of change for each compartment is found by adding the flows into the compartment and the flows out of the compartment. For the Silver Springs model these are

$$\frac{dN_1}{dt} = F_{10} - F_{21} - F_{01} - F_{51} - F'_{01} \tag{13.1}$$

$$\frac{dN_2}{dt} = F_{21} + F_{20} - F_{02} - F_{32} - F_{52} \tag{13.2}$$

$$\frac{dN_3}{dt} = F_{32} - F_{43} - F_{03} - F_{53} \tag{13.3}$$

$$\frac{dN_4}{dt} = F_{43} - F_{04} - F_{54} \tag{13.4}$$

$$\frac{dN_5}{dt} = F_{51} + F_{52} + F_{53} + F_{54} - F_{05} \tag{13.5}$$

Each flow is defined by an equation which usually involves a rate coefficient. We will designate the rate coefficients as f_{ij}. A large number of equations could be used to model flow between compartments, but only a few useful equations are encountered frequently. Some of these are described in Table 13.1. A variety of other examples may be found in Godfrey (1983).

$F_{ij} = k$	Flow from compartment j to i is constant, and is independent of time and system state.
$F_{ij} = f_{ij} N_j$	Flow to i is proportional to the content of j. This is a linear equation with control by the donor compartment only, with rate constant f_{ij}.
$F_{ij} = f_{ij} N_i$	Flow to i is proportional to the content of i. This is a linear equation describing control of flow by the receptor compartment.
$F_{ij} = f_{ij} N_i N_j$	Flow to i is controlled by both donor and receptor compartments in a cross-product manner. This is the mass-action approach. The equation is nonlinear.
$F_{ij} = f_{ij} \cdot f(\text{time})$	Flow from j to i is a function of time. An example is a sine function (Equation 11.2).
$F_{ij} = f_{ij} N_i (1 - g_{ij} N_i)$	Flow from j to i is controlled by a positive linear term and a negative non-linear term, much like the logistic equation.
$F_{ij} = \dfrac{f_{ij} N_i}{K + N_i}$	Flow from j to i is limited with a hyperbolic term as in Michaelis-Menten kinetics.

Table 13.1. Some useful equations for describing flow between compartments. F_{ij} is the instantaneous flow to compartment i from compartment j, and f_{ij} is the rate coefficient. N_i indicates the content of compartment i.

Compartment models of the type described here are often implemented on computers by solving the system of equations with matrix algebra. This method will be discussed in Chapter 16. For the present, we will solve the equations with a direct approach. Because most of the equations of flow are functions of the size of the compartments, it is necessary to use the two-stage numerical approach with Euler integration. For clarity of programming, the first stage should be separated into two parts as described in Section 6.2. That is, the various flows F will all be calculated, and then the net changes of content for each compartment will be determined to complete the first stage. Then, following the usual second stage procedures, values of the compartments will be updated.

13.3 The Silver Springs Model

We will continue working with the Silver Springs model as an example. It is not better than similar and more recent models. However, it is familiar and relatively uncomplicated, and it still appears in general ecology textbooks. The set of equations given in this section define the flows in the block diagram of Figure 13.2, and have been modified from Odum (1957) and Patten (1971):

(A. Forcing:)
$$F_{10} = M + R\sin\left(2\pi\frac{(T-11)}{52}\right)$$

$$F_{20} = k$$

F_{10} is the energy input from photosynthesis, assumed to be proportional to light intensity. M is the annual mean photosynthesis, and R is the range of the annual fluctuation around the mean. F_{20} is the energy in 70 loaves of bread fed daily to catfish by the tourist concession at the springs.

(B. Feeding:)
$$F_{21} = \tau_{21}N_1$$

$$F_{32} = \tau_{32}N_2$$

$$F_{43} = \tau_{43}N_3$$

Feeding is donor-dependent in this system, with τ_{ij} the linear coefficient having units of inverse weeks (wk^{-1}).

(C. Mortality:)
$$F_{51} = \mu_{51}N_1$$

$$F_{52} = \mu_{52}N_2$$

$$F_{53} = \mu_{53}N_3$$

$$F_{54} = \mu_{54}N_4$$

Here μ_{ij} is a coefficient of donor-dependent mortality, with units of wk^{-1}.

(D. Respiration:)
$$F_{01} = \rho_{01} N_1$$

$$F_{02} = \rho_{02} N_2$$

$$F_{03} = \rho_{03} N_3$$

$$F_{04} = \rho_{04} N_4$$

$$F_{05} = \rho_{05} N_5$$

Here ρ_{ij} is a coefficient of donor-dependent respiration, with units of wk^{-1}.

(E. Export:)
$$F'_{01} = \lambda_{01} N_1$$

λ_{01} is a coefficient of the loss downstream of small plants and pieces of plants that break loose from the plants that grow in the Silver Spring system. The coefficient has units of wk^{-1}.

After the F_{ij} values are calculated with the equations given above, the first stage in the two-stage Euler procedure is completed with the calculation of the ΔN_i values:

$$\Delta N_1 = (F_{10} - F_{21} - F_{51} - F_{01} - F'_{01}) \Delta t$$

$$\Delta N_2 = (F_{20} + F_{21} - F_{32} - F_{52} - F_{02}) \Delta t$$

$$\Delta N_3 = (F_{32} - F_{43} - F_{53} - F_{03}) \Delta t$$

$$\Delta N_4 = (F_{43} - F_{54} - F_{04}) \Delta t$$

$$\Delta N_5 = (F_{51} + F_{52} + F_{53} + F_{54} - F_{05}) \Delta t$$

The second stage of the Euler integration is completed as usual, with

$$N_i \longleftarrow N_i + \Delta N_i$$

Exercise 13-1: Write and implement a program to simulate energy flow through Silver Springs using the model equations above. For

the various compartment sizes and flows, use values given below
derived from Odum (1957). The values of mean annual standing
crop given in Figure 13.1 (units of kcal m^{-2}) are suitable for initial
compartment sizes:

$$N_1 = 2635 \qquad N_2 = 213 \qquad N_3 = 62 \qquad N_4 = 9 \qquad N_5 = 25$$

The values of rate coefficients (units of yr^{-1}) can be calculated by
inserting yearly values for the flows and for standing crops into the
flow equations for mortality, feeding, respiration and export. This
process yields coefficients as follows:

$$\mu_{51} = 1.310 \qquad \mu_{52} = 5.141 \qquad \mu_{53} = 0.742 \qquad \mu_{54} = 0.889$$
$$\tau_{21} = 1.094 \qquad \tau_{32} = 1.798 \qquad \tau_{43} = 0.339$$
$$\rho_{01} = 4.545 \qquad \rho_{02} = 8.873 \qquad \rho_{03} = 5.097 \qquad \rho_{04} = 1.444$$
$$\rho_{05} = 184.0 \qquad \lambda_{01} = 0.94$$

For the driving functions, use the following values:

$$k = 486 \text{ kcal m}^{-2} \text{ yr}^{-1} \qquad R = 175 \text{ kcal m}^{-2} \text{ wk}^{-1}$$
$$M = 400 \text{ kcal m}^{-2} \text{ wk}^{-1}$$

Most of the rate coefficients above are based on annual flows through
the system; weekly values may be found with division by 52. Use
weeks as the time unit for your simulation, with $\Delta t = 0.1$ week
as the unit for Euler integration. Larger Δt values will produce
unstable results.

Write your program to plot the contents of each compartment
for each week, including initial values. Your output will be a graph
with five separate lines, each showing the content of a model compartment.

After you have entered the program into your computer, make
an initial trial run, setting $R = 0$. This procedure will keep F_{10}
constant. The system should tend rapidly to a steady-state condition. If it does not, then check your program for errors.

After this initial check, carry out a 3-year simulation, beginning
with the first week in January. Set $R = 175$ kcal m^{-2} wk^{-1}.

Exercise 13-2: A crude simulation of ecological succession may be obtained with the model above by setting the solar input to a constant
value ($R = 0$) and reducing the starting size of all compartments to
some minimum value (such as 5 kcal m^{-2}). Allow the simulation

of the system to proceed for about 150 weeks to stabilize. Again use $\Delta t = 0.1$ week.

Exercise 13-3: Modify the Silver Springs model above by employing nonlinear feeding equations as follows:

$$F_{21} = \tau'_{21} N_1 N_2$$

$$F_{32} = \tau'_{32} N_2 N_3$$

$$F_{43} = \tau'_{43} N_3 N_4$$

Numerical values for the coefficients may be found by inserting the values for flows and standing crop sizes (Figure 13.1) and solving. This produces the following values for the coefficients:

$$\tau'_{21} = 0.00513 \qquad \tau'_{32} = 0.0290 \qquad \tau'_{43} = 0.0376$$

As in Exercise 13-1, run the simulation for a 3-year period, plotting the energy content of the five compartments. The larger excursions of the compartment contents produced by the nonlinear flows should be apparent.

13.4 Global Carbon-Cycle Model

The increased concentration of carbon dioxide that has been measured in the earth's atmosphere has caused concerns about global warming. There exist real controversies on the relative roles of burning fossil fuels and clearing tropical forests in producing the increase. The relative ability of terrestrial plants and the oceans to serve as reservoirs for the increase in carbon is also controversial. Complex compartmental models are important tools in working with analysis and prediction of the carbon cycle. Bolin (1981) has outlined a simple model of the carbon cycle that shows many of the expected characteristics of the cycle. We will use a modified version of this model as an introduction to the important subject of carbon cycle modeling, and as a further illustration of ecological compartment models.

In this model, carbon is distributed among seven compartments (Figure 13.3). The principal pool of the atmosphere can exchange carbon with both oceanic and terrestrial components. The oceans are represented by only two compartments, one for the mixed layers near the surface that come into contact with the atmosphere, and another for the deeper, more isolated waters. The terrestrial component is made up of four compartments: short-lived vegetation with a rapid carbon turnover, including annual plants and vegetative parts such as leaves; long-lived vegetation, particularly the woody trunks, stems and roots of trees; detritus, defined as dead and decomposing organic matter, sometimes called litter, duff, or humus; and organic material in the soil.

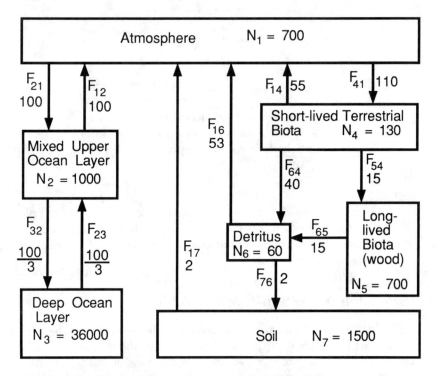

Figure 13.3. Block diagram of the equilibrium carbon cycle, modified from Bolin (1981). Pool sizes are in units of 10^{15} grams (=gigatons; GT). Transfers between compartments are in units of GT year^{-1}.

With two exceptions, the flows between compartments are linear, first-order, donor-controlled flows. That is, flow to compartment i from compartment j is described with

$$F_{ij} = k_{ij}N_j \qquad (13.6)$$

The two exceptions are the flows representing uptake of atmospheric carbon by terrestrial plants, and the transfer of carbon dioxide from the ocean to the atmosphere. In the first case, flow between plants and the atmosphere is controlled by both donor and receptor compartments. Within broad limits, a larger biomass of plants should be able to take up more atmospheric carbon than a smaller quantity of plants. Also, an increase in atmospheric CO_2 may stimulate plant growth, so that the transfer to plants may occur at a greater rate. However, the transfer is not directly proportional to the concentration of atmospheric CO_2; that is, a doubling of the CO_2 concentration would almost certainly not double the flow of carbon. Bolin (1981) suggested a linearized version of a nonlinear equation. Yearsley and Lettenmaier (1987) used the following nonlinear equation for this transfer:

$$F_{41} = k_{41} N_4 \left(1 + \beta \ln \frac{N_1}{(N_1)_0}\right) \tag{13.7}$$

Here, $(N_1)_0$ is the equilibrium value for the atmospheric carbon, N_1 is the current concentration, and β is a photosynthetic coefficient. The equation is semi-empirical, and indicates that uptake of atmospheric carbon is proportional to the amount of plant material when the atmosphere is at the equilibrium level of 700 GT. As atmospheric carbon rises above the equilibrium level, plants will be able to take up only a part of the increased atmospheric CO_2 .

The second instance of nonlinear flow is the transfer from the ocean to the atmosphere. This transfer is made complex by the chemical buffering system of the ocean. Bolin (1981) suggested that the flow could be described with

$$F_{12} = k_{21} \left((N_1)_0 + \xi \frac{(N_1)_0}{(N_2)_0} [N_2 - (N_2)_0]\right) \tag{13.8}$$

The equilibrium values for the atmosphere and mixed-layer are given with $(N_1)_0$ and $(N_2)_0$. The current value of the mixed layer is N_2, and ξ is a buffering constant. k_{21} (not a misprint) is the constant for the flow to compartment 2 from 1. This equation may be written in a much simpler form when appropriate constants are inserted. The result is:

$$F_{12} = N_2 - 900 \tag{13.9}$$

This form of the equation will hold for all reasonable values of an increase in the CO_2 content of the mixed layer. Elaborations of Bolin's basic model and more complex models may be found in papers by Yearsley and Lettenmaier (1987), Mulholland et al. (1987), and Detwiler and Hall (1988).

Exercise 13-4: On your computer, implement the global carbon model from the above information. As an early step in your program, solve for the transfer coefficients for the various flows (except F_{12} and F_{41}) using Equation 13.6. That is, find $k_{ij} = F_{ij}/N_j$. Use the whole-number equilibrium values in Figure 13.3. Because the model is sensitive to errors in the fourth or fifth significant digit for these constants, it is easiest to have the computer solve for these k_{ij} values and retain them.

Write your program using two-stage Euler integration, following the procedure described in Sections 13.2 and 13.3. That is, first calculate all the flows F_{ij} using Equation 13.6, but for F_{12} use Equation 13.9, and for F_{41} use Equation 13.7 with $\beta = 0.10$ and $k_{41} = 110/130$. Then find all the compartmental changes ΔN_i, and finally perform the updates for the new values of N_i. For this simulation set $\Delta t = 0.1$ year.

Begin your simulation with the equilibrium pool values from Figure 13.3. Plot the values of the various pools N_i through time for 20 years to be sure they remain at equilibrium.

After you are sure your program is working with the equilibrium values, alter your program so that a slug of 10 GT of carbon is added to the atmosphere at the beginning of year 10. This simulates the burning of a large amount of fossil fuel in that year. Follow the result of the perturbation for 90 years. For this simulation, the graphical output should show the departure of each pool from its equilibrium value, rather than pool size. That is, your x-axis should run from 0 to 100 years, and your y-axis from -10 to +10 GT.

Exercise 13-5: Modify the program of Exercise 13-4 to simulate a single incident of massive destruction of forest and release of carbon by burning of the wood. This is easily done by removing 10 GT from the compartment of long-lived biota at the same time that 10 GT is added to the atmosphere.

Exercise 13-6: For more than a century atmospheric carbon has been increasing principally because of combustion of fossil fuels including coal and oil. The amount of carbon added to the atmosphere each year has also increased. The amount of carbon from fossil sources added to the atmosphere each year by human activity may be described quite accurately with

$$\text{GT added each year} = 0.5 \ e^{0.03t}$$

where t is the number of years since 1900 (Rotty 1981).

Modify your program from Exercise 13-4 to simulate the addition of carbon from fossil fuel combustion for an 80 year period, 1900-1980. At least initially most of the added carbon accumulates in the atmospheric pool N_1. As output from your program, plot the cumulative carbon input to the atmosphere from fossil fuel combustion, and the change in the atmosphere from equilibrium. Then alter your program so that it will plot the distribution of the added carbon among the seven pools over the 80-year period. Initially, almost 100% of the added carbon will be in the atmosphere; this should decline to about 30% after 80 years, with 70% distributed among the other compartments.

13.5 Simulated Food Chain

Elliot et al. (1983) based a model of a planktonic grazing food chain on a general model of predation developed by Wiegert (1975, 1979). The model is instructive because it attempts to include a number of realistic factors in the transfer of material through the food chain. Some of these factors you have encountered previously in Chapters 7 and 8 for homogeneous populations, including self-limitation of population size, saturation of predators, and minimal prey density. The basic four-compartment model of the planktonic food chain is shown in Figure 13.4.

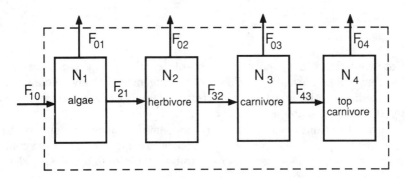

Figure 13.4. A block diagram for a simple 4-compartment model of a planktonic grazing food chain.

The model is simplified so that non-grazing losses from each compartment are combined in a single term. That is, rates of respiration, non-grazing mortality, etc., are all lumped together in a single factor to model loss to the environment. This flow to the environment from each of the four compartments is given by

$$F_{0i} = \lambda_{0i} N_i \tag{13.10}$$

where N_i is compartmental density (measured in units of mass) and λ_{0i} is the rate constant for loss, with units of time^{-1}.

The system is driven by a flow into compartment 1 from the environment, F_{10}. This flow will be a constant in the simulation described here. However, it may be made a function of time, for example a sine function as in the Silver Springs simulation above.

The predatory or grazing transfers of mass between all the compartments are modeled with equations having the same form. The equations include terms for both donor and receptor control:

$$F_{21} = \tau_{21} N_2 \left[D_{21} R_{21} \right]_+ \tag{13.11}$$

$$F_{32} = \tau_{32} N_3 \left[D_{32} R_{32} \right]_+ \tag{13.12}$$

$$F_{43} = \tau_{43} N_4 \left[D_{43} R_{43} \right]_+ \tag{13.13}$$

In these equations τ_{ij} is the feeding rate constant for grazing transfer to compartment i from j. It may also be thought of as the rate constant for exponential growth of N_i, whenever D_{ij} and R_{ij} are held constant.

The $+$ subscript for the bracketed term is used to indicate that this term must be positive or zero. That is,

$$[x]_+ = x \quad \text{if} \quad x \geq 0$$

$$[x]_+ = 0 \quad \text{if} \quad x < 0$$

(Note that this is not equivalent to absolute value.) In these equations, this convention prevents the flow from being negative, and having the prey feeding on the predator. (The terminology is easily programmed in BASIC with the statement IF X < 0 THEN X = 0.)

The explanations of the D and R components of these equations will be given for the specific example of F_{21}, the flow to compartment 2 from compartment 1. Only the subscripts need to be altered to find the equivalent terms for F_{32} and F_{43}.

D_{21} is a term for the donor control of flow to compartment 2 from 1. The term may take several forms, including $D_{21} = 1$ to indicate simple receptor control of flow. Wiegert (1979) suggests the following as a realistic expression:

$$D_{21} = \left[1 - \left(\frac{\alpha_{21} - N_1}{\alpha_{21} - \gamma_{21}} \right)_+ \right]_+ \tag{13.14}$$

The function of the $+$ subscripts is important and is described above. α_{21} is a constant that represents a saturating density of prey N_1. When prey density N_1 is greater than α_{21}, then D_{21} is equal to 1. Thus, F_{21} in Equation 13.11 will not be decreased because of a lack of prey. γ_{21} is a constant that represents some minimal density of prey. If prey density N_1 drops below γ_{21}, then D_{21} drops to zero. In this case, F_{21} in Equation 13.11 will become zero also, and there will be no feeding on the prey. (γ_{21} may represent for example the number of hiding places for prey, where they can be safe from predation.)

R_{21} is the term in Equation 13.11 for receptor control of the flow F_{21}. In this model it describes self-limitation of the predator population, with a modified logistic function. Weigert's (1979) expression was:

$$R_{21} = \left[1 - \left(1 - \frac{\lambda_{02}}{\tau_{21}}\right)\left(\frac{N_2 - \alpha_{22}}{\gamma_{22} - \alpha_{22}}\right)_+\right]_+ \qquad (13.15)$$

Again, the $+$ subscripts are important. λ_{02} is the rate constant for loss from compartment 2. α_{22} is a constant indicating the density at which predators begin to interact and to compete with each other. When N_2 exceeds α_{22}, R_{21} drops below 1 because of the predators' self-interference. γ_{22} represents the limiting maximum density of predators, so that R_{21} will decline as N_2 approaches γ_{22}.

This model system is extremely flexible, and can be used to simulate many features of the grazing food chain of aquatic ecosystems. The equations and constants may be modified to provide a variety of possible simulations, involving changes in refuge size, maximal densities, predation rates, and loss rates. These variations will produce widely varying results. The density of compartments may reach equilibrium, or oscillate in stable and unstable ways.

Exercise 13-71: Write a program for simulating planktonic grazing with the 4-compartment model of food chains described above. As described in Section 13.2, use a two-stage Euler integration, with the first stage subdivided into two parts. That is, for stage one first calculate the flow rates between components of the system with Equations 13.10 through 13.15, and then find the ΔN_i values by summing rates. For stage two, update the compartment values. It should be adequate to set $\Delta t = 0.1$ for the simulation of planktonic grazing.

The following coefficients were suggested by Elliot et al. (1983) as reasonable estimates for the grazing planktonic food chain in some freshwater systems:

$$\lambda_{01} = 0.10 \qquad \lambda_{02} = 0.46 \qquad \lambda_{03} = 0.37 \qquad \lambda_{04} = 0.20$$

$$\tau_{21} = 1.15 \qquad \tau_{32} = 0.74 \qquad \tau_{43} = 0.27$$

$$\alpha_{21} = 20.0 \qquad \alpha_{32} = 15.0 \qquad \alpha_{43} = 5.0$$

$$\gamma_{21} = 5.0 \qquad \gamma_{32} = 2.0 \qquad \gamma_{43} = 0.50$$

$$\alpha_{22} = 10.0 \qquad \alpha_{33} = 5.0 \qquad \alpha_{44} = 1.0$$

$$\gamma_{22} = 30.0 \qquad \gamma_{33} = 20.0 \qquad \gamma_{44} = 20.0$$

Use a constant input for this system of $F_{10} = 20.0$. For initial values of the compartments, use the following:

$$N_1 = 10.0 \qquad N_2 = 2.0 \qquad N_3 = 5.0 \qquad N_4 = 1.0$$

In this simulation, units of time and units of mass are arbitrary. As output for this simulation, plot values of the compartmental densities N_i against time for 0 to 240 units of time.

Exercise 13-8: Modify the simulation of Exercise 13-7 to show the effect of limiting the maximum density of the top carnivore. This modification might be needed for a simulation with a species that is more territorial, for example. Thus, in your program reduce the constant for maximum top carnivore density, γ_{44}, from 20 to 10 units and then rerun the simulation.

Conclusion

This chapter has introduced you to some of the concepts and techniques of working with models of material and energy flow through large-scale systems. In succeeding chapters you will work with compartmental models of smaller systems, and many of the methods you have learned here will be useful. The direct two-stage Euler approach has been described for implementing these models on the computer. The more elegant matrix approach will be taken up in Chapter 16. The three examples described in this chapter were relatively simple and designed to promote understanding of methodology. However, the same techniques may be applied to more complex, diverse systems. The simulation programs become longer, and the results less intuitively obvious, but the fundamental methods still apply.

CHAPTER 14

DIFFUSION MODELS

In the compartmental ecological systems of the previous chapter, the mechanisms of transport of material and energy between compartments were relatively straightforward. In physiological models, the mechanisms of transport are frequently the subject of interest. These mechanisms usually fall into three general categories: diffusion, active transport, and fluid flow. Each of these presents different problems requiring different modeling approaches. We will consider the first two mechanisms in this chapter. Fluid flow and other transfers among physiological compartments will be discussed in the next chapter.

A solution is made by dissolving some matter (solute) in a fluid (solvent). A solution may be described by its mass concentration, which is the mass of the solute per unit volume of solution. Molecules of solute are dispersed through the solvent by diffusion, a result of the thermal movements of the solvent molecules colliding with the molecules of solute.

14.1 Transport by Simple Diffusion

A simple model of transport of material by diffusion between a single compartment and an unchanging environment was discussed in Section 1.5. Frequently we are interested in diffusion between compartments, for example between two adjacent cells (Figure 14.1). Across a membrane separating the two compartments, net material transfer will proceed from the compartment of higher concentration to the compartment of lower concentration. The rate of transfer will be proportional to the difference in concentrations. For the simple model here, we will assume constant compartmental volumes and uniformity of concentration inside compartments.

The following equation describes forward diffusion from compartment i to j:

$$\left(\frac{dQ_i}{dt}\right)_{\text{f}} = \frac{-kQ_i}{V_i} \qquad (14.1)$$

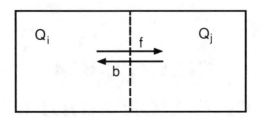

Figure 14.1. Diagram representing simple transport between two cells, with Q representing concentration. The direction of diffusion is indicated with f (forward) and b (back), relative to compartment i.

Q_i is the mass of some diffusing substance in compartment i, V_i is the volume of compartment i, and k is the rate constant for diffusion. Note that concentration is Q_i/V_i, and that k has units of volume (unit time)$^{-1}$. A negative is included to indicate a loss of material from compartment i.

When an amount of material exists in compartment j, there will occur a reverse or back diffusion from j to i. The effect of this on compartment i is described with

$$\left(\frac{dQ_i}{dt}\right)_b = \frac{kQ_j}{V_j} \tag{14.2}$$

The net rate of diffusion of the substance from compartment i is the sum of these two equations:

$$\left(\frac{dQ_i}{dt}\right)_{net} = \left(\frac{dQ_i}{dt}\right)_f + \left(\frac{dQ_i}{dt}\right)_b = k\left(\frac{Q_j}{V_j} - \frac{Q_i}{V_i}\right) \tag{14.3}$$

The change in amount of substance in each compartment may be solved with the usual procedure for two-stage simple Euler numerical integration. The first stage requires two equations:

$$\Delta Q_i = k\left(\frac{Q_j}{V_j} - \frac{Q_i}{V_i}\right)\Delta t \tag{14.4}$$

$$\Delta Q_j = k\left(\frac{Q_i}{V_i} - \frac{Q_j}{V_j}\right)\Delta t \tag{14.5}$$

The second stage involves the usual update procedure:

$$Q_i \leftarrow Q_i + \Delta Q_i \tag{14.6}$$

$$Q_j \leftarrow Q_j + \Delta Q_j \tag{14.7}$$

Exercise 14-1: Write and implement a program to simulate diffusion of a substance between two compartments of a hypothetical system. Initally only one of the compartments will contain a quantity of the diffusing substance. Use the following as constants in your simulation:

$$k = 0.75 \text{ ml min}^{-1} \qquad V_j = 60 \text{ ml} \qquad V_i = 20 \text{ ml}$$

For initial values, set $Q_j = 0$ mg and $Q_i = 100$ mg. Use a Δt value of 0.1 with simple Euler integration. As output, produce a graph which shows the concentration of substance in each compartment, from time 0 to a few minutes past the point at which steady-state occurs.

14.2 Linear Diffusion Gradient

The following simulation is an elaboration of the two-compartment model above. It provides a method for studying the pattern of concentration which results when solutes diffuse across a boundary that is initially quite sharp. Such a boundary might occur, for example, when a lump of sugar is dropped into a cup of tea and allowed to dissolve without stirring.

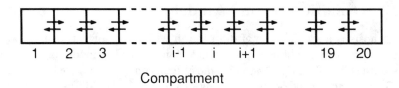

Compartment

Figure 14.2. Diffusive transport along a linear gradient of 20 compartments.

Assume that the diffusion system is represented by a series of compartments (Figure 14.2). For simplicity in simulation, assume also that the subcompartments are all of the same unit volume, i.e. $V_i = 1$. The net flux from compartment i to adjacent compartment $i + 1$ will be described by Equation 14.3, with slightly modified terminology:

$$F_i = k\,(C_i - C_{i+1}) \tag{14.8}$$

where F_i is net flux from compartment i to $i + 1$, C_i and C_{i+1} are the concentrations (Q/V) in compartments i and $i + 1$, and k is the diffusion rate constant as above.

The diffusion along the gradient of compartments can be found with two-stage Euler numerical integration. In stage one, all flows first are found between all compartments with Equation 14.8. Then as the second part of stage one, the net change is calculated for each compartment with

$$\Delta C_i = (F_{i-1} - F_i)\,\Delta t \qquad (14.9)$$

For the second stage of the Euler procedure, concentrations in each compartment are found with the usual Euler equation for up-dating:

$$C_i \leftarrow C_i + \Delta C_i \qquad (14.10)$$

Exercise 14-2: Write and implement the simulation of diffusion along a linear gradient of 20 compartments. Set $k = 0.1$. Use a Δt value of 0.1. Begin your simulation with $C_i = 100$ for values of i from 1 to 10, and $C_i = 0$ for $i = 11$ to 20. Output should consist of a graph showing compartment number (distance) on one axis and concentration on the other. Show these plots of concentration after 0, 20, 40, 60, 80 and 100 time units. The equations above are set up to permit easy use of subscripted variables. Set $F_0 = 0$ and $F_{20} = 0$.

14.3 Osmotic Pressure Model

Osmosis is a special case of diffusion in which the solvent, usually water, is the primary diffusing substance. The following model of osmotic pressure is based on the classical osmosis experiment, illustrated in Figure 14.3. The movement of water during osmosis is simply diffusion along a concentration gradient from high concentration of water towards a lower concentration of water. Only water can pass through the membrane; the large molecules of the solute are restricted to the osmometer chamber. Rate of water diffusion into the chamber is described by the equation

$$\text{Flow in} \ = k_i\,(W_e - W_i) \qquad (14.11)$$

The diffusion rate constant k_i is a function of the properties of the membrane, including its thickness and area. W_e is the concentration of water outside the compartment expressed as a mole fraction, usually 1.0, which indicates pure water. W_i is the mole fraction of water inside the membrane. If w is the number of moles of water and s is the number of moles of solute per unit volume inside the compartment, then

$$W_i = \frac{w}{(w + s)} \qquad (14.12)$$

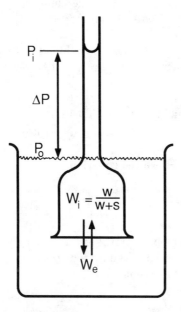

Figure 14.3. Diagram of the apparatus for the classic experiment for determining osmotic pressure. Terminology is defined in the text.

Substituting this into Equation 14.11 gives

$$\text{Flow in} = k_i \left(W_e - \frac{w}{w+s} \right) \tag{14.13}$$

The backflow of water across the membrane is the result of a difference in hydrostatic pressure, and is described by the equation

$$\text{Flow out} = k_h(\Delta P) \tag{14.14}$$

where k_h is a constant relating the flow rate to the difference in hydrostatic pressure across the membrane (ΔP; see Figure 14.3). k_h will vary with the area, thickness, and porosity of the membrane and with the fluid viscosity. Pressure difference ΔP will partly depend on the geometry of the manometer. Equation 14.14 may be simplified by basing the outflow on the amount of water in the chamber relative to the initial amount, which will be adjusted so that $\Delta P = 0$. This will give the following:

$$\text{Flow out} = k_o(w - w_0) \tag{14.15}$$

where w is the amount of water in the chamber at any time t, and w_0 is the initial amount of water when $\Delta P = 0$. k_o is k_h multiplied by a proportionality constant.

Equations 14.13 and 14.15 can be combined to produce an equation for net flow across the membrane as a result of two processes. The equation in words is:

Net flow of water = flow in (by diffusion)

- flow out (due to hydrostatic flow)

With the terms used above the equation is:

$$\frac{dw}{dt} = k_i \left(W_e - \frac{w}{w + s} \right) - K_o(w - w_0) \tag{14.16}$$

This equation can be solved by the usual two-stage Euler procedure for numerical integration. After the system reaches equilibrium, when net flow is zero (flow in = flow out),

$$k_i \left(W_e - \frac{w}{w + s} \right) = k_o(w - w_0) \tag{14.17}$$

Exercise 14-3: Write and implement a program to simulate a determination of osmotic pressure using the model above. Use the following constants and initial values:

$$k_i = 20 \qquad k_o = 0.30 \qquad W_e = 1 \qquad w_0 = 55 \qquad s = 1$$

For the Euler integration, let $\Delta t = 0.1$. The backflow from Equation 14.15 may be converted to pressure (atmospheres) with multiplication by a factor of 62.199. This in turn may be converted to inches of water with multiplication by a factor of 414, or to mmHg by a factor of 760.

Run your simulation until the net flow across the membrane is almost zero. At this point, the atmospheric pressure is the osmotic pressure. As output, plot osmotic pressure and net flow from time = 0 to equilibrium.

14.4 Countercurrent Diffusion

Countercurrent diffusion has evolved in a number of species as an efficient mechanism to transport materials or heat from one fluid-carrying vessel to another. One example is the special anatomical relationship between the loops of Henle in the kidney nephron and the medullary interstitial fluid (Guyton 1971). Another example is the system that whales, seals and birds use to transfer heat between arterial and venous circulation in their feet and flippers. This system, called the *rete mirabile*, is designed for the efficient conservation of heat. Both of these mechanisms have two counter-flowing vessels close together, so that materials or heat may be exchanged through simple diffusion. The countercurrent system maximizes the intensity of the gradient at all points where diffusion is occurring.

An effective model of countercurrent diffusion may be based upon two series of parallel compartments, separated by a membrane which permits transport. The A compartments represent one vessel through which fluid passes and the B compartments represent another vessel parallel to A.

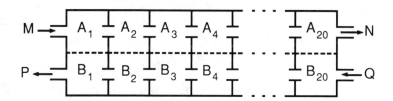

Figure 14.4. Conceptual diagram for a model to simulate countercurrent flow. The terminology is explained in the text.

Figure 14.4 illustrates the model system using terminology that simplifies programming. The diffusion between any pair of compartments A_j and B_j may be described by Fick's Law, which states that the rate of diffusion across a thin membrane is proportional to the area of the membrane, the difference in concentrations, and a constant of permeability:

$$F_{BA} = p\,a\,(A_j - B_j) \qquad (14.18)$$

where F is the flow to compartment B_j from compartment A_j, p is the permeability constant, and a is the membrane area. A_j and B_j designate not only the compartments, but represent the concentrations of the diffusing substance in the compartments. The simplifying assumptions associated with the model are (1) membrane area and permeability constants are the same for all compartments; (2) volume of each compartment is constant and equal to unity, so that concentrations and amounts in each compartment are equivalent.

Movement of fluid in the vessels is assumed to occur by plug flow. This flow is modeled by having the material in each of the A compartments move one compartment to the right for each time unit. Simultaneously material in each B compartment moves one compartment to the left for countercurrent flow. Simulation of countercurrent flow involves two sets of equations, one set for calculating diffusion flow, and a second set for the plug flow of fluids through the system. Diffusion may be simulated with the usual two-stage Euler procedure. As part of the first stage the diffusion between each pair of compartments is found with Equation 14.18, and then the changes in each compartment are found with

$$\Delta A_j = -F_{BA}\Delta t \tag{14.19}$$

$$\Delta B_j = +F_{BA}\Delta t \tag{14.20}$$

Each compartment is updated as usual with

$$A_j \leftarrow A_j + \Delta A_j \tag{14.21}$$

$$B_j \leftarrow B_j + \Delta B_j \tag{14.22}$$

The procedure for programming plug flow involves the use of the following sequence for the A compartments:

$$N \leftarrow A_n \tag{14.23}$$

$$A_j \leftarrow A_{j-1} \tag{14.24}$$

$$A_1 \leftarrow M \tag{14.25}$$

M represents the concentration of the input to the A compartments, and N the concentration of the output. The countercurrent flow through the B compartments will involve:

$$P \leftarrow B_1 \tag{14.26}$$

$$B_{j-1} \leftarrow B_j \tag{14.27}$$

$$B_n \leftarrow Q \tag{14.28}$$

Q is the input to the low concentration side, and P is the output. The terminology for these procedures is given in Figure 14.4.

In contrast with countercurrent flow, concurrent flow involves the flow of the solution through the B compartments in the same direction as through the A compartments. In Figure 14.4, concurrent flow through the B compartments would be from left to right. A simulation of concurrent flow would follow the same procedures as above, except that the sequence of Equations 14.26-14.28 would be reversed:

$$Q \leftarrow B_n \tag{14.29}$$

$$B_j \leftarrow B_{j-1} \tag{14.30}$$

$$B_1 \leftarrow P \tag{14.31}$$

Note that with concurrent flow, Q becomes the output and P the input for the B compartments.

Exercise 14-4: Program the model for countercurrent flow using a system of 20 pairs of A and B compartments ($n = 20$). Using subscripted variables will make your program much shorter than otherwise. This will allow you to write one FOR-NEXT loop to solve Equations 14.18-14.20 for all 20 pairs of compartments, and another loop to perform the updates with Equations 14.21-14.22. Then, a loop may be set up for the sequence involved with Equation 14.24, and finally another loop for Equation 14.27. Use constants of $a = 1.0$, and $p = 0.1$. Set $M = 100$ and $Q = 0$. For the Euler procedure let $\Delta t = 1$. Begin your simulation with A_1 through $A_n = 0$, and B_1 through $B_n = 0$. Allow your simulation to proceed for about 40 time intervals, and then plot concentration vs. compartment number for both A and B.

After you have produced the above output, modify your model to simulate concurrent flow, and run the simulation similarly. You can then compare efficiency of two flow types for lowering concentration of material in the A compartments.

14.5 A Model of Active Transport

A simple model for the active transport of some metabolite, B, into a cell may be developed from the following assumptions based on the diagram in Figure 14.5:

(1) Assume there are two compartments separated by a membrane with different permeabilities for substance B and a closely related compound C, with C having a higher permeability than B.

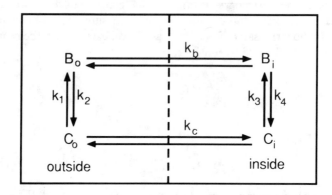

Figure 14.5. Diagram for a simple model of active transport. The terminology is explained in the text.

(2) Assume there is an enzyme on one side of the membrane which catalyzes the conversion of B to C. This might occur by a reaction such as:

$$B + W \rightleftarrows C + X$$

where W is an activating molecule, such as ATP. In this reaction, the equilibrium lies far to the right, so that the equilibrium constant $K >> 1$ and $\Delta G_o << 0$.

(3) Assume there is another enzyme which catalyzes the conversion of C to B. This might occur as:

$$C + Y \rightleftarrows B + Z$$

Again, the equilibrium is assumed to lie far to the right, and $\Delta G_o << 0$.

(4) Assume that concentrations of enzymes on each side of the membrane remain constant.

(5) Assume the two compartments have a unit volume and the separating membrane has a unit area. Following Equation 14.4, the rate of transport of B and C from inside the cell to the outside compartment due to diffusion would be described by these equations for the first stage of a two-stage Euler integration:

$$\Delta B_d = k_b \left(B_i - B_o \right) \Delta t \tag{14.32}$$

$$\Delta C_d = k_c \left(C_i - C_o \right) \Delta t \tag{14.33}$$

where k_b and k_c are diffusion rate constants, and B_i, B_o, C_i and C_o

represent concentrations of the two compounds inside and outside of the cell.

(6) Assume that the enzymatic reactions follow first-order kinetics, which will be the case if the reactants other than B and C have constant concentrations, and if the enzymes are well below saturation by substrate. The reactions will alter the concentrations of B and C inside and outside the cell according to the following first-stage Euler equations:

$$\Delta B_i = (k_3 C_i - k_4 B_i)\, \Delta t \tag{14.34}$$

$$\Delta Ci = (k_4 B_i - k_3 C_i)\, \Delta t \tag{14.35}$$

$$\Delta B_o = (k_1 C_o - k_2 B_o)\, \Delta t \tag{24.36}$$

$$\Delta C_o = (k_2 B_o - k_1 C_o)\, \Delta t \tag{14.37}$$

where k_1, k_2, k_3, and k_4 are the rate constants for the reactions. In the system based on the above assumptions, the concentration of B and C will be altered by diffusion and by the enzymatic reactions. The update expressions for the Euler integration of the equations will be

$$B_i \leftarrow B_i + \Delta B_i - \Delta B_d \tag{14.38}$$

$$B_o \leftarrow B_o + \Delta B_o + \Delta B_d \tag{14.39}$$

$$C_i \leftarrow C_i + \Delta C_i - \Delta C_d \tag{14.40}$$

$$C_o \leftarrow C_o + \Delta C_o + \Delta C_d \tag{14.41}$$

Exercise 14-5: Using the equations in Section 14.5, write and implement a program to simulate active transport of B against a diffusion gradient. Use the following rate constants, all with units of $\min^{-1}$:

$$k_1 = 0.005 \qquad k_2 = 0.5 \qquad k_3 = 0.5$$
$$k_4 = 0.005 \qquad k_b = 0.001 \qquad k_c = 0.1$$

Begin your simulation with these initial values for concentration (mM l^{-1}):

$$B_o = 50 \qquad B_i = 50 \qquad C_o = 1 \qquad C_i = 1$$

Use a Δt value of 1.0 minute. Your output of the simulation should be a graph of the concentration of B_i and B_o against time. Allow the simulation to proceed to near steady-state.

14.6 Simple Approach to Active Transport

In the previous section we discussed a possible mechanism for active transport of a substance against a diffusion gradient. This resulted in a steady-state concentration gradient across the membrane. This same result could be obtained by simply employing different constants for the diffusion rates in each direction across the membrane.

In this case, rate of diffusion is described by a modification of Equation 14.3:

$$\frac{dQ_j}{dt} = k_1 \frac{Q_j}{V_j} - k_2 \frac{Q_i}{V_i} \tag{14.42}$$

where Q and V are defined as before, and k_1 is greater than k_2 if active transport is toward compartment i. At equilibrium the rates in each direction are equal, so that

$$k_1 \frac{Q_j}{V_j} = k_2 \frac{Q_i}{V_i} \tag{14.43}$$

and

$$\frac{k_1}{k_2} = \frac{Q_i V_j}{Q_j V_i} = \frac{C_i}{C_j} \tag{14.44}$$

Thus, the ratio of the two constants is equal to the equilibrium or steady-state ratio of the two concentrations, C_i and C_j.

Neither of the active transport processes described above account for mediated or facilitated transport processes. These are membrane transport processes, either active or passive, that show saturation-type kinetics because only a limited number of transport sites exist. In addition, they may show specificity for a particular chemical species being transported.

The distinction between simple diffusion and mediated-transport processes is seen in Figure 14.6. This shows graphically that the carrier or transport sites of the mediated-transport system become saturated at high concentrations of the diffusing substance, and that the rate does not exceed T_{max}. This may be represented with a model equation of the Michaelis-Menten type:

$$T_c = T_{max} \frac{C_i}{K_c + C_i} \tag{14.45}$$

where T_c is the rate of carrier-mediated diffusion, T_{max} is the maximum rate, C_i is the concentration of diffusing substance, and K_c is the half-saturation constant.

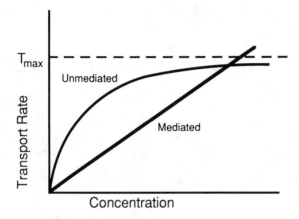

Figure 14.6. Graph showing general relationship between rates of transport and external concentration for mediated and unmediated (passive diffusion) transport processes. Based on a graph in Lehninger (1975).

Conclusion

As part of the coverage of compartmental models in physiology, this chapter has looked at simple models of diffusion. The modeling and simulation of diffusion can become quite complex; see Crank (1956) for example. The next chapter considers some different physiological models of fluid flow among compartments.

CHAPTER 15

COMPARTMENTAL MODELS IN PHYSIOLOGY

The passage of a substance between compartments has been the fundamental characteristic of the two previous chapters. In this chapter we again consider compartmental models, but for a different class of problems. These models are usually encountered in physiological descriptions of flow and circulation of fluids, for example the circulation of blood, or the passage of drugs or tracers among different internal sites of an organism.

For purposes of modeling, a useful definition of compartment was given by Milhorn (1966) as follows:

"If a substance is present in a biological system in several distinguishable forms or locations, and if it passes from one form or location to another form or location at a measurable rate, then each form or location constitutes a separate compartment for the substance."

An almost identical but more pithy definition of compartment was offered by Atkins (1969): "...a quantity of a substance which has a uniform and distinguishable kinetics of transformation or transport."

There are two general classes of models of compartmental flow in physiology, distinguished by an assumption of constancy of volumes of the compartments. Models with varying compartmental sizes involve submodels of pressure. Those with a constant volume resemble the compartmental models of the previous two chapters.

The distinctions between compartmental models of physiology and ecology are not in the concepts of the models themselves, but in their use. In ecology the quantities or concentrations in the compartments and transfer rates are measured directly. In many physiological systems, the number of compartments may be unknown. The number, volumes and concentrations of physiological compartments, and the flow rates between them must be estimated from experimental data.

In many instances, the physiological situation is a better mimic of the modeling approach to biology discussed in the Introduction to this book than is ecological research. On the other hand, ecological models are more likely to result in counter-intuitive simulation results because of their complexity. In general, the structure and flow of physiological models are comparatively simple.

15.1 A Single-Compartment Model

A model involving only a single compartment can provide a useful description for some important physiological mechanisms. The compartment may represent the body of an entire organism, or perhaps an organ, for example the liver. Suppose a drug is being administered at regular intervals to build up an effective concentration in a diseased compartment. The dose is constant, with D being the increase in concentration in the compartment immediately after drug administration, and T the dose interval. If loss from the compartment is linear with concentration, then the usual equation for exponential decay will apply for a description of concentration:

$$\frac{dX}{dt} = -aX \tag{15.1}$$

with X being the concentration in the compartment, and a the rate constant of loss for the compartment. (Because only a single compartment with a constant volume is involved, here X may represent either the concentration or the mass of drug in the compartment.) Loss may occur by transport of fluids from the compartment, either through excretion, or metabolic breakdown of the drug. This results in exponential decay of the drug concentration between doses, producing the response of drug concentration graphed in Figure 15.1. As the number of doses becomes large, the system will reach a fluctuating plateau. The amount excreted during the dose interval will just equal the amount given in the dose (Winter et al. 1980). At this plateau, the maximum concentration just after a dose will be

$$X_{\max} = \frac{D}{1 - e^{-aT}} \tag{15.2}$$

and the minimum just before a dose

$$X_{\max} = X_{\max} e^{-aT} \tag{15.3}$$

The gradual build-up of drug concentration in the compartment is the reason that many courses of drug therapy call for a large initial "loading" dose, to raise the effective concentration to $X_{\max}$.

In most cases, the assumption of linearity used above is probably less realistic than assuming that there exists some maximum rate of removal.

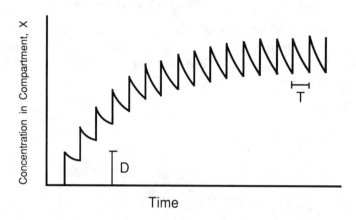

Figure 15.1. Generalized result of repeated doses of a drug or tracer into a single compartment. D indicates the size of the dose, and T the dose interval.

This alternative assumption presumes that the mechanism operating to remove the drug can become saturated if the concentration of the drug becomes sufficiently great. The familiar Michaelis-Menten formulation can be used as an approximation of this situation:

$$\frac{dX}{dt} = -k_{\max} \frac{X}{K + X} \tag{15.4}$$

where $k_{\max}$ is the maximum rate of removal from the compartment when concentration X is very great, and K is the half- saturation constant. With this modification the overall system behavior is much like that of Figure 15.1, but is not as easily analyzed.

Exercise 15-1: Write and implement a simulation for drug dosage of a single compartment, using Equation 15.4 to describe the elimination of the drug. Assume that drug dosage results in an immediate increase in concentration of 1 mg (kg body wt)$^{-1}$, and that the dosing interval is 8 hours. Begin with the concentration of drug $= 0$, and add the first dose at time 0. For Equation 15.4, let $k_{\max} = 0.2$ and $K = 10$. Find concentrations using simple Euler integration with $\Delta t = 0.1$. The output of your simulation should show concentration plotted against time for at least 288 hours (12 days).

15.2 Terminology and Bucket Models

The models used in study of flow rates in physiological systems with fixed-volume compartments are sometimes termed "bucket" or "bathtub" models. We can imagine compartments to be containers of fluid, connected by pipes and pumps (Figure 15.2).

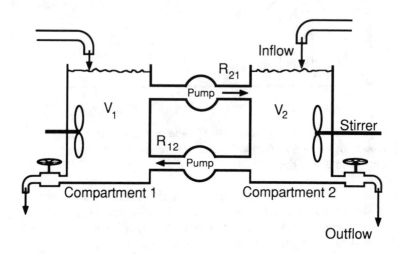

Figure 15.2. Diagram of a two- compartment bucket model.

The amounts of fluid in the buckets labeled 1 and 2 are constant at volumes V_1 and V_2 ml. R_{21} is the constant rate of flow (ml min^{-1}) to bucket 2 from bucket 1, and R_{12} is the constant rate to 1 from 2. The buckets are considered to be stirred completely so that any dissolved material is uniformly distributed within the bucket. If a material, for example a drug or tracer, is introduced into solution in compartment 1, it will be mixed instantaneously throughout the compartment. If the amount introduced is Q_1 mg, its concentration will be $C_1 = Q_1/V_1$. The rate of transfer of material to compartment 2 from compartment 1 will be

$$F_{21} = R_{21}C_1 = R_{21}\frac{Q_1}{V_1} \tag{15.5}$$

with units of mg min^{-1}. This equation is formally similar to Equation 14.1 but conceptually different, because the fluids do not flow between compartments in the systems described with Equation 14.1.

The rate constant for transfer from compartment 1 to 2 is

$$a_{21} = \frac{R_{21}}{V_1} \tag{15.6}$$

with units of min^{-1}. Like the example of compartmental models of Chapter 13, changes of amount of material in a compartment may be found with numerical integration. The flow between compartments is donor-controlled in most physiological models. Hence the rate of flow from compartment j to compartment i is

$$F_{ij} = R_{ij}C_j = a_{ij}Q_j \qquad (15.7)$$

The two-stage Euler integration will proceed as usual, finding for the first stage of changes in amount (ΔQ_i) or concentration (ΔC_i) of material in compartment i:

$$\Delta Q_i = \text{(sum of flows into } i \text{ - sum of flows out of } i) \, \Delta t$$

or

$$\Delta C_i = \text{(sum of flows into } i \text{ - sum of flows out of } i) \, \Delta t$$

As the second stage, the amount or concentration is updated as usual: $Q_i \leftarrow Q_i + \Delta Q_i$ or $C_i \leftarrow C_i + \Delta C_i$.

15.3 Liver Function - A Two-Compartment Model

The assumption of donor-controlled linear relationships between compartments in physiological systems has produced a number of useful models, particularly in mammalian physiology. As an example, in this section and the next we will consider two compartmental models with different behaviors that have been used with the same experimental technique.

The state of health and functioning of the liver is often assessed with dye-tracer techniques (Winkelman et al. 1974). The dye that is used most frequently in these tests is bromosulfophthalein (BSP). The conventional procedure calls for a single injection of the dye directly into the bloodstream, followed by taking samples of the blood at intervals. With a normally functioning liver, the dye will be removed from circulation after it is conjugated with cysteine or glutathione. Outside of the liver there appears to be little removal of the dye. Dye concentrations in the plasma can be measured after the blood cells are removed. In alkaline solution the dye becomes purple and can be measured with a spectrophotometer (Mitoma 1985).

The usual clinical procedure to test humans for liver function involves a single injection of 2 to 5 mg BSP (kg body wt)$^{-1}$, with a determination of concentration in the blood after 45 minutes. Liver function is assessed by comparing the experimental concentration with the range of normal values. The model being used in this case is a two-compartment model (Figure 15.3). If a series of blood samples is taken at close intervals over a period of 90 minutes after injection, then more information may be obtained about rates of transfer between liver and plasma (a_{12} and a_{21}) and of excretion (a_{20}). Some elementary techniques for estimating transfer coefficients from data are described clearly in Cullen (1985).

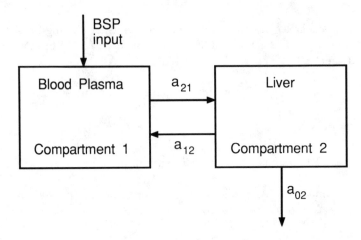

Figure 15.3. Diagram of two- compartment model for dynamics of elimination of bromosulfophthalein (BSP) after injection of trace amounts.

Exercise 15-2: For the two-compartment model of BSP dynamics given in Figure 15.3, Carson et al. (1983) give the following values of rate constants of mass transfer between compartments:

$$a_{12} = 0.0038 \qquad a_{21} = 0.14 \qquad a_{02} = 0.032$$

These coefficients were determined from experiments using adult humans with normal liver function. The units of the coefficients are $\min^{-1}$. Write and implement a program to simulate the loss of dye from compartment 1 (plasma) in a study of liver function that begins with an intravenous injection of 350 mg BSP. Use the given rate constants to find transfers of the dye between the two compartments and the environment. Set up the usual two-stage Euler integration procedures with a $\Delta t = 0.1$. Your program should find the concentration of dye in the plasma compartment from time 0 (time of injection) to 120 minutes. Assume a plasma volume of 3.5 liters. Plot the output as concentration against time. Also produce a graph showing the log of concentration vs. time; this should be a concave curve, indicating the interaction of two rates. (A single rate would produce a straight line with the log plot.)

15.4 Multi-Compartment Models of Liver Function

The two-compartment model above is not very detailed with respect to known dynamics of BSP in experiments. The ability of the test to detect certain types of inadequate liver function is therefore limited. Rates of transfer among compartments can be changed in ways that will not produce abnormal concentration values in the conventional test. It is possible to measure separately the conjugated and unconjugated BSP as a four-compartment model, which results in clearer separation of some liver problems (Molino et al. 1978).

A better model of BSP kinetics was obtained with measurement of some human subjects whose gall bladders had been removed previously (Molino and Milanese 1975, Milanese and Molino 1975). These circumstances permitted a simultaneous determination of unconjugated and conjugated BSP in the bile tracts inside the liver. For this case, a six-compartment model was used to describe the movement of BSP (Figure 15.4). All the rate constants were determined from experiments. There was a delay of several minutes before either conjugated or unconjugated BSP began to appear in the bile.

Exercise 15-3: For the six-compartment model of BSP dynamics given in Figure 15.4, one of the experiments by Molino and Milanese (1975) produced the following values for rate constants of transfer between compartments (min^{-1}):

$$a_{12} = 0.0300 \qquad a_{21} = 0.1130 \qquad a_{32} = 0.0043$$
$$a_{03} = 0.0480 \qquad a_{52} = 0.0090 \qquad a_{45} = 0.9900$$
$$a_{54} = 0.0078 \qquad a_{65} = 0.6280 \qquad a_{06} = 0.0580$$

The time delay before appearance of either form of BSP in the bile is 20 minutes. Write and implement a program to simulate the transport of dye among the different compartments after a single initial intravenous injection of 300 mg BSP. Use the rate constants above. Set up the usual two-stage Euler integration with a Δt value of 0.1. Simulate the time delay by setting a_{03} and a_{06} to zero until after 19 minutes have lapsed. Find the amounts of BSP in each of the six compartments and the amount excreted for a period of 160 minutes after injection. Plot amounts (mg) of BSP in all six compartments and in the environment. (An easy way to do this is to plot 7 lines on the same graph: the first showing Q_1, the second $(Q_1 + Q_2)$, the third $(Q_1 + Q_2 + Q_3)$, etc. The seventh line should show $(Q_1 + Q_2 + Q_3 + Q_4 + Q_5 + Q_6 + Q_0)$.)

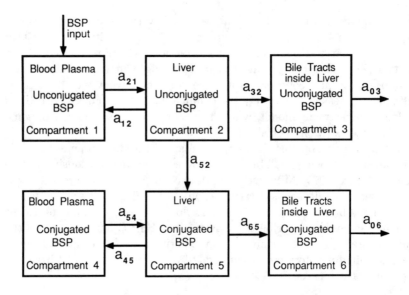

Figure 15.4. Diagram of six-compartment model for the dynamics of elimination of BSP tracer by the liver.

15.5 A Multicompartment Model of Calcium Metabolism

Some interesting oscillatory behavior is displayed by a model developed for the dynamics of calcium in mammals. Staub et al. (1981) studied the flow of a calcium tracer, ^{45}Ca, in calcium-deficient rats. Calcium excretion in such rats is much below normal, so that the rats' calcium metabolism functions as an essentially closed system. Calcium level in metabolic compartments fluctuates much more than in normal rats, permitting more accurate determinations of rate constants.

A five-compartment model was minimally required to simulate the system (Figure 15.5). Compartment 1 represents calcium in plasma, and the other compartments are associated with metabolism of calcium by bone. The flows between compartments are all linear and donor-controlled except the flow from bone compartment 2 to 3. That is, except for F_{32} all the flows of material between compartments are given by

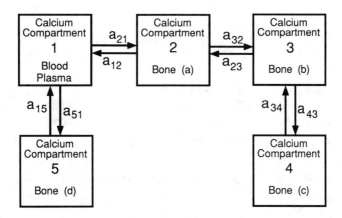

Figure 15.5. Diagram of multicompartment model for calcium metabolism in mammals. Compartment 1 represents calcium in blood plasma; the other compartments are associated with bone.

$$F_{ij} = a_{ij}Q_j \qquad (15.7A)$$

Transfer from compartment 2 to 3 is:

$$F_{32} = a_{32}Q_3^2Q_2 \qquad (15.8)$$

This nonlinear flow is typical of autocatalytic components. The system is assumed to be closed completely, with no environmental input of calcium. The values for the mass transfer coefficients of this particular system were determined from a series of experiments to be

$$a_{12} = 8.506 \times 10^{-2} \qquad a_{23} = 6.019 \times 10^{-1} \qquad a_{34} = 7.190 \times 10^{-3}$$
$$a_{21} = 6.534 \times 10^{-1} \qquad a_{32} = 7.533 \times 10^{-3} \qquad a_{43} = 2.816 \times 10^{-1}$$
$$a_{15} = 1.951 \times 10^{-2} \qquad a_{51} = .2922 \times 10^{-1}$$

With these coefficients and the single nonlinear function, the system is self-regulating and displays an oscillation with a period of approximately 24 hours. The simulation provides a good approximation of the cycling which has been observed experimentally.

Exercise 15-4: Implement a program for the flow of calcium in this system, using the following initial values for amount of calcium (mg) in each compartment:

$$Q_1 = 1.558 \qquad Q_2 = 11.792 \qquad Q_3 = 6.079$$
$$Q_4 = 224.537 \qquad Q_5 = 25.534$$

Because of the nonlinearities of the system, use a Δt of 0.01 with simple two-stage Euler integration. Your output should consist of a graph showing the amount of calcium in each compartment over a period of 96 hours (4 days).

15.6 A Simple Model of Compartmental Fluid-Flow Processes

The study of ways in which fluids respond to mechanical forces in biological systems involves topics in physics and complex applied mathematics considerably beyond the scope of this book. We will look at some simple compartmental models of fluid flow which can produce interesting biological simulations.

A basic compartmental fluid-flow model is diagrammed in Figure 15.6. This model involves a chamber with a variable volume, V, which is supplied with fluid from a single input, and which loses fluid through a single output. The input has a conductance, C_i, which under pressure P_i, results in an input flow of F_i. The output has a conductance, C_o, which results in an output flow of F_o. The pressure in the compartment, P, is the result of the volume of fluid, V, acting against the elastic wall of the compartment having a stiffness, S.

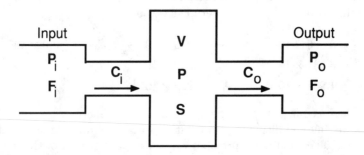

Figure 15.6. Diagram of compartmental model for a generalized fluid-flow process. Terminology is discussed in the text.

Assuming that the chamber walls will stretch in direct proportion to the force applied, compartmental pressure will be given by the equation

$$P = VS \qquad (15.9)$$

Alternatively, if the relationship between pressure and volume is not a linear proportion, then an empirical equation may be used:

$$P = a + bV + cV^2 \qquad (15.10)$$

Here, a, b, and c are stiffness coefficients derived from measurements.

Flow into the chamber is defined by

$$F_i = C_i(P_i - P) \qquad (15.11)$$

Conductance of the input, C_i, is the flow that results from a single unit of pressure difference, and P_i and P are the existing pressures. C_i could also include the effect of a simple passive valve having flaps that close under a given back pressure. The valve would close when the pressure difference $(P_i - P)$ reached some value less than zero, at which point C_i would become 0. Using the notation of Section 13.5, this mechanism would also be written as

$$F_i = C_i(P_i - P)_+ \qquad (15.11A)$$

Similarly, outflow is regulated by a conductance value, C_o, and a pressure differential:

$$F_o = C_o(P - P_o) \qquad (15.12)$$

The fundamental mass balance equation for the change in compartmental volume is

$$\frac{dV}{dt} = F_i - F_o \qquad (15.13)$$

Integrating this equation will show the volume of the compartment. The two-stage Euler solution of this equation would involve finding change in volume with

$$\Delta V = (F_i - F_o)\,\Delta t \qquad (15.14)$$

and then updating with

$$V \leftarrow V + \Delta V \qquad (15.15)$$

Exercise 15-5: Write and implement a program to simulate volume change of a compartment using the simple model above, with the internal pressure given by Equation 15.9. Use the following values in setting up your simulation:

$$P_i = 15 \text{ mmHg} \qquad\qquad P_o = 6 \text{ mmHg}$$

$$C_i = 0.8 \text{ ml min}^{-1} \text{ mmHg}^{-1} \qquad C_o = 1 \text{ ml min}^{-1} \text{ mmHg}^{-1}$$

$$S = 0.15 \text{ mmHg ml}^{-1}$$

A two-stage Euler integration will be suitable for this simulation with Δt set to 0.1. Begin the simulation with $V = 50$ ml. As output, show F_i, F_o and V plotted against time from 0 to 20 minutes, to observe how the system approaches steady state.

15.7 Mechanical Operation of the Human Aorta

An interesting simulation of ventricular and aortic blood pressure has been developed by Sias and Coleman (1971). The simulation is based on the model of flow presented in the previous section, but with the addition of a variable input pressure, a flap valve to prevent backflow, and a complex output pressure. The explanation that follows is somewhat modified from their basic model.

The aorta is assumed to be a single compartment of variable volume. Blood flows into this compartment through the aortic valve when ventricular pressure is greater than aortic pressure. Blood flows out of the compartment (aorta) at a rate which is a function of aortic pressure. The driving function of the model system is the blood pressure in the left ventricle of the heart. This pressure is periodic with a frequency of 72 beats min^{-1}, and may be simulated by a truncated sine curve having a maximum pressure of 120 mmHg at peak systole. Aortic pressure is a function of blood volume in the aortic compartment. The aortic volume in ml is found by integrating the equation for flow into and out of the aorta.

Ventricular pressure (input pressure for the aorta) is defined by the following equation

$$P_v = \left[P_{max} \sin\left(2\pi \frac{t}{0.8333} \right) \right]_+ \qquad (15.16)$$

where t is time in seconds, and P_{max} is 120 mmHg. The constant 0.8333 is the duration of one heartbeat in seconds ($1/1.2$). The $+$ subscript (see Section 13.5) indicates that the ventricular pressure always is assumed to be non-negative. (If a calculated P_v should fall below zero, then P_v is set to zero.)

Aortic pressure, P_a, is a function of aortic volume and stiffness of the walls of the aorta. Because of the nonlinear relationship between internal pressure, volume and stiffness, P_a is described by an empirical equation:

$$P_a = c_o + c_1 V_a + c_2 V_a^2 + c_3 V_a^3 \qquad (15.17)$$

where the coefficients c_i are obtained by fitting experimental data to a cubic polynomial (Chapter 3).

Blood flow into the aorta from the ventricle is described by

$$F_v = C_a \left(P_v - P_a \right)_+ \qquad (15.18)$$

where C_a is input conductance through the aortic valve. C_a has a value of 16.67 ml sec^{-1} mmHg^{-1} when the valve is open. The valve may be assumed to close when the pressure differential $(P_v - P_a)$ is less than zero. In this case, flow becomes zero as indicated by the $+$ subscript.

The outflow of blood from the aorta to the peripheral vascular system is a function of aortic pressure, and the pressure and arterial conductance of the peripheral system, as in Equation 15.12. Values for the peripheral system factors are complex, so that outflow may be described by an empirical function of aortic pressure:

$$F_p = b_o + b_1 P_a + b_2 P_a^2 + b_3 P_a^3 \qquad (15.19)$$

where the coefficients b_i are obtained by fitting a polynomial equation to experimental data.

Changes of aortic volume are found with the equation

$$\frac{dV_a}{dt} = F_v - F_p \qquad (15.20)$$

This equation is integrated to find aortic volume. The two-stage Euler procedure used with Equation 15.13 may be used here also.

Exercise 15-6: Develop and implement a simulation for flow in the aorta using the equations in Section 15.7. Use the values given for constants. The following data from Sias and Coleman (1971) may be used with a polynomial curve-fitting program (e.g. POLYFIT) to find the values for the coefficients in the empirical equations for aortic pressure (Equation 15.17) and peripheral blood flow (Equation 15.19):

Aortic Volume ml	Aortic Pressure mmHg		Aortic Pressure mmHg	Peripheral Blood Flow ml sec^{-1}
0	0		20	0.00
50	20		50	16.67
100	70		100	83.33
150	175		150	166.67
200	220		200	250.00

Plot P_a, P_v and F_v through a time period of 2.40 seconds. Output from your program should resemble that of Figure 15.7. Begin your simulation with $t = 0$ and $V_a = 86.22$. It is convenient to use 0.01 second as the basic time unit, with $\Delta t = 0.01$ second for simple Euler integration of Equation 15.20.

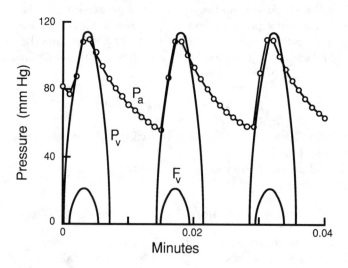

Figure 15.7. Some simulation results for a model of aortic segment, showing pressure and rates of blood flow.

Conclusion

This chapter has introduced some elementary compartmental models of physiology that are useful in the study of many metabolic systems. The concepts involved have important practical applications in most branches of physiology and in medicine. The problems of finding the number and characteristics of compartments in tracer studies have important theoretical implications in mathematical structural identifiability. A key assumption of these models is that the metabolic processes are not controlled actively by hormonal regulation. The system is specified by storage of materials in any compartment, the physical passage of materials between compartments, and by reaction rates of the materials in compartments. The rate of transfer is determined only by concentrations of mass of material in the donor and receiving compartments. In a subsequent chapter we will consider models in which control is exerted by a variety of regulating mechanisms.

CHAPTER 16

APPLICATION
OF MATRIX METHODS
TO SIMULATIONS

Up to this point, all of our computer programs have used what might be called a direct or brute-force approach. These methods are most easily understood by individuals just learning how to develop computer simulations. However, the direct approach can become rather cumbersome when dealing with large numbers of variables and the equations which define them. This problem was exemplified by the age-class models of Chapter 9, and the compartment models of the previous three chapters. As you will see, some of the exercises presented in those chapters are much more conveniently programmed using matrix methods.

Matrix algebra often provides an efficient way of manipulating a large amount of data. A few computer instructions can result in the solution of a great number of related linear equations. Knowledgeable use of matrix manipulations can make programming simpler. In this chapter you will learn some matrix techniques that can be applied directly to modeling problems like the exercises of this book. For additional information, the textbook on matrix methods for biologists by Searle (1966) is particularly useful.

16.1 Some Brief Definitions

A matrix is an array of numbers or symbols, and is usually named by a single letter printed in boldface. To show some notation of matrix use, we will arrange nine numbers in a 3×3 matrix. The numbers will be designated with a subscripted variable a_{ij}.

$$\boldsymbol{A} = \begin{bmatrix} a_{11} & a_{12} & a_{13} \\ a_{21} & a_{22} & a_{23} \\ a_{31} & a_{32} & a_{33} \end{bmatrix} = \begin{bmatrix} 52 & 28 & 76 \\ 95 & 63 & 34 \\ 87 & 42 & 23 \end{bmatrix} \tag{16.1}$$

The matrix contains three columns (running up and down) and three rows (running across). For a number a_{ij}, i indicates the row, and j indicates the column. Often the diagonal from top left to bottom right is important in using the matrix. (The numbers in the matrix may be somehow related; they may be coefficients involved in a biological system, for example, but that is not relevant just yet).

In addition to matrices like A above, you will also use in this chapter some one-dimensional matrices. In our simulations, the usual form of a one-dimensional matrix is a vertical column of numbers or symbols, called a column vector, having the following form:

$$B = \begin{bmatrix} b_1 \\ b_2 \\ b_3 \end{bmatrix} \qquad (16.2)$$

16.2 The Matrix Approach to Linear Compartmental Models

Let us assume that the diagram of Figure 16.1 represents a 3-compartment model of some physiological or ecological system, such as those encountered in the previous chapters.

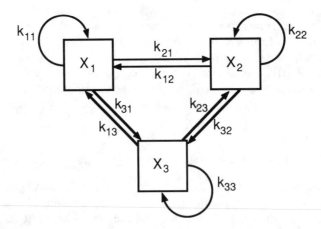

Figure 16.1. Block diagram of model for interactions among three compartments.

Suppose we are working with a model in which all possible transfers occur and are donor-controlled, with rate constants of k_{11}, k_{12}, k_{21}, $k_{13}, \ldots , k_{33}$.

The notation here is the same as in previous chapters: for rate constant k_{ij}, i designates the receiving compartment and j the donor compartment.

If we use the direct approach that has been used previously to model this system, we would write the usual set of equations describing the change in contents of the compartments as follows (ignoring the auto-effect coefficients of k_{11}, k_{22}, and k_{33}):

$$\frac{dX_1}{dt} = -k_{21}X_1 - k_{31}X_1 + k_{12}X_2 + k_{13}X_3 \qquad (16.3)$$

$$\frac{dX_2}{dt} = k_{21}X_1 - k_{12}X_2 - k_{32}X_2 + k_{23}X_3 \qquad (16.4)$$

$$\frac{dX_3}{dt} = k_{31}X_1 + k_{32}X_2 - k_{13}X_3 - k_{23}X_3 \qquad (16.5)$$

If we were working with a model having more compartments there would be more equations, and each would be longer. The methods of writing a program using the two-stage Euler technique for solving equations like these should be quite familiar to you by now.

To use matrix techniques in writing a computer program for this model, we will have to alter these three equations. This will require two separate, elementary steps of algebra. First, for each of the equations 16.3-16.5 above we can combine terms which involve the same compartment, and obtain the following three equations:

$$\frac{dX_1}{dt} = (-k_{21} - k_{31})X_1 + \qquad k_{12}X_2 + \qquad k_{31}X_3 \quad (16.6)$$

$$\frac{dX_2}{dt} = \qquad k_{21}X_1 + (-k_{12} - k_{32})X_2 + \qquad k_{23}X_3 \quad (16.7)$$

$$\frac{dX_3}{dt} = \qquad k_{31}X_1 + \qquad k_{32}X_2 + (-k_{13} - k_{23})X_3 \quad (16.8)$$

Second, we will redefine the coefficients and produce a new set of equations:

$$\frac{dX_1}{dt} = a_{11}X_1 + a_{12}X_2 + a_{13}X_3 \tag{16.9}$$

$$\frac{dX_2}{dt} = a_{21}X_1 + a_{22}X_2 + a_{23}X_3 \tag{16.10}$$

$$\frac{dX_3}{dt} = a_{31}X_1 + a_{32}X_2 + a_{33}X_3 \tag{16.11}$$

In these three equations, the values for each a_{ij} are taken from the values of the k coefficients in the preceding set of equations (Equations 16.6-16.8). For example, $(-k_{21} - k_{31})$ of Equation 16.6 is equal to a_{11} of Equation 16.9, etc.

These a_{ij} coefficients are put into a matrix:

$$A = \begin{bmatrix} a_{11} & a_{12} & a_{13} \\ a_{21} & a_{22} & a_{23} \\ a_{31} & a_{32} & a_{33} \end{bmatrix} \tag{16.12}$$

As a final step, the values of X_1, X_2, and X_3 are arranged in a column vector as in Equation 16.2:

$$X = \begin{bmatrix} X_1 \\ X_2 \\ X_3 \end{bmatrix} \tag{16.13}$$

With the coefficients in matrix form, the product of the matrix A and the column vector X of X_i values is written as

$$\begin{bmatrix} dX_1/dt \\ dX_2/dt \\ dX_3/dt \end{bmatrix} = \begin{bmatrix} a_{11} & a_{12} & a_{13} \\ a_{21} & a_{22} & a_{23} \\ a_{31} & a_{32} & a_{33} \end{bmatrix} \begin{bmatrix} X_1 \\ X_2 \\ X_3 \end{bmatrix} \tag{16.14}$$

In matrix notation this is expressed by

$$\dot{X} = A \cdot X \tag{16.15}$$

This matrix operation is described as post-multiplication of a square matrix by a column vector. The result is another column vector. The operation consists of the following two steps:

(1) Each column of the matrix is multiplied by the corresponding vector element. In this case, each element of column 1 of the A matrix is multiplied by X_1, each element of column 2 by X_2 , and so on.

(2) The multiplication products in each row are summed across to generate the corresponding elements of the product vector. In this case, the sum of the products of row 1 = dX_1/dt.

The key point of the matrix manipulation is that Equation 16.15 or 16.14 are equivalent to the previous sets of equations. That is, Equation 16.15 indicates the same operations as Equations 16.3-16.5 or Equations 16.6-16.8 or Equations 16.9-16.11.

A simple two-stage Euler integration is easily accomplished with matrix multiplication. The instantaneous rates of change in the X_i variables are found with Equation 16.15 for the first part of stage one. For the second part of stage one, these instantaneous rates need to be multiplied by Δt to find the ΔX_i values. That is,

$$\begin{bmatrix} \Delta X_1 \\ \Delta X_2 \\ \Delta X_3 \end{bmatrix} = \Delta t \cdot \begin{bmatrix} \dot{X}_1 \\ \dot{X}_2 \\ \dot{X}_3 \end{bmatrix} \tag{16.16}$$

or

$$\Delta \boldsymbol{X} = \Delta t \cdot \dot{\boldsymbol{X}} \tag{16.17}$$

In matrix terminology, a matrix with a single element is a scalar. Multiplying a matrix by a scalar involves multiplying each element in the matrix by the value of the scalar.

The second stage of the simple Euler integration is the updating of the X_i variables, with the ΔX_i values added to the X_i values:

$$\begin{bmatrix} X_1 \\ X_2 \\ X_3 \end{bmatrix} \leftarrow \begin{bmatrix} X_1 \\ X_2 \\ X_3 \end{bmatrix} + \begin{bmatrix} \Delta X_1 \\ \Delta X_2 \\ \Delta X_3 \end{bmatrix} \tag{16.18}$$

or equivalently

$$\boldsymbol{X} \leftarrow \boldsymbol{X} + \Delta \boldsymbol{X} \tag{16.19}$$

Most compartmental models will have some driving or forcing function involving inputs from the environment. These functions are easily accommodated with the matrix approach. Let F_i represent the instantaneous forced input to compartment i. This forcing function input may be incorporated in the two-stage Euler integration, Equations 16.16 and 16.17, with

$$\Delta \boldsymbol{X} = \Delta t \left(\boldsymbol{F} + \dot{\boldsymbol{X}} \right) \tag{16.20}$$

where $\boldsymbol{F}$ represents a column vector of the forcing functions:

$$\boldsymbol{F} = \begin{bmatrix} F_1 \\ F_2 \\ F_3 \end{bmatrix} \tag{16.21}$$

The addition of two column vectors of the same size requires adding each of their corresponding elements together.

Some dialects of BASIC have built-in functions which permit the calculations described above to be accomplished with simple MAT statements. If these are not available on your computer, it is relatively easy to write programs for matrix operations using the routines listed in Appendix 1.

The following BASIC statements would accomplish the matrix operations for the two-stage Euler integration described above. Assuming the existence of forcing functions, it will produce the results of Equations 16.15, 16.20 and 16.19 in sequence. The statements begin after preliminary operations such as dimensioning arrays, assigning values to coefficients, and reading data already have been completed:

```
10 REM N = # OF ROWS & # OF COLUMNS
20 REM I,J = LOOP AND SUBSCRIPT COUNTERS
30 REM DT = EULER TIME INCREMENT, DELTA-T
40 REM X(I) = STATE OF COMPARTMENT I
50 REM XD(I) = X-DOT(I)
60 REM DX(I) = DELTA-X(I)
70 REM F(I) = FORCING FLOW RATE
80 REM A(I,J)= COEFFICIENTS FOR THE TRANSFER MATRIX
90 :
100 REM STAGE ONE-PART 1 (EQN 16.15)
110    FOR I = 1 TO N
120       XD(I) = 0
130       FOR J = 1 TO N
140          XD(I) = XD(I) + A(I,J) * X(J)
150       NEXT J
160    NEXT I
170 REM STAGE ONE-PART 2 (EQN 16.20)
180    FOR I = 1 TO N
190       DX(I) = DT * (F(I) + XD(I))
200    NEXT I
210 REM STAGE TWO UPDATE (EQN 16.19)
220    FOR I = 1 TO N
230       X(I) = X(I) + DX(I)
240    NEXT I
250 END
```

As indicated in Statement 10, N is the dimension of the square matrix. In the case of our 3-compartment example above, $N = 3$. The power of this method is clear when you realize that the same statements would be just as effective for a 30-compartment system as for the system with three compartments. (The listing above is set up for clarity and for consistency with the discussion. Sightly faster computer execution will result if Line 230 is combined with Line 190, and the remaining Lines 210-240 are deleted.)

Exercise 16-1: Carry out Exercise 13-1, the linear Silver Springs model, using matrix operations rather than the direct approach of Chapter 13. The graphical output from the matrix approach should be identical with the original exercise.

Exercise 16-2: Carry out Exercise 15-3, the 6-compartment linear liver model, using matrix multiplication techniques in your program. The output from this approach should be identical to that of the direct approach of Chapter 15 if your program is written correctly.

16.3 Matrix Solutions of Nonlinear Functions

Matrix multiplication is best adapted for solving linear equations. To write programs for models that contain nonlinear equations, it is necessary to modify the approach used in the previous section. You have encountered examples of models with nonlinear transfers including the nonlinear feeding version of the Silver Springs model in Exercise 13-3 and the calcium model of Section 15.5.

Suppose that all the fluxes between compartments in Figure 16.1 were functions of both the donor and acceptor compartments. Using the direct approach, the rates of change using nonlinear transfers would be expressed with the following set of equations:

$$\frac{dX_1}{dt} = k_{12}X_1X_2 + k_{13}X_1X_3 - k_{21}X_1X_2 - k_{31}X_1X_3$$

$$(16.22)$$

$$\frac{dX_2}{dt} = k_{21}X_2X_1 + k_{21}X_2X_3 - k_{12}X_2X_1 - k_{32}X_2X_3$$

$$(16.23)$$

$$\frac{dX_3}{dt} = k_{31}X_3X_1 + k_{32}X_3X_2 - k_{13}X_3X_1 - k_{23}X_3X_2$$

$$(16.24)$$

Factoring out the common X_i variable from each equation and combining the remaining variables will give this set of equations:

$$\frac{dX_1}{dt} = X_1 [\quad 0 \quad + X_2 (k_{12} - k_{21}) + X_3 (k_{13} - k_{31})]$$

(16.25)

$$\frac{dX_2}{dt} = X_2 [X_1 (k_{21} - k_{12}) + \quad 0 \quad + X_3 (k_{23} - k_{32})]$$

(16.26)

$$\frac{dX_3}{dt} = X_3 [X_1 (k_{31} - k_{13}) + X_2 (k_{32} - k_{23}) + \quad 0 \quad]$$

(16.27)

The matrix used in calculating nonlinear transfer rates therefore becomes

$$\boldsymbol{A}_{nl} = \begin{bmatrix} 0 & (k_{12} - k_{21}) & (k_{13} - k_{31}) \\ (k_{21} - k_{12}) & 0 & (k_{23} - k_{32}) \\ (k_{31} - k_{13}) & (k_{32} - k_{23}) & 0 \end{bmatrix} = \begin{bmatrix} a_{11} & a_{12} & a_{13} \\ a_{21} & a_{22} & a_{23} \\ a_{31} & a_{32} & a_{33} \end{bmatrix}$$

(16.28)

Note that multiplying this $\boldsymbol{A}_{nl}$ matrix by the $\boldsymbol{X}$ vector does not produce the $\boldsymbol{X}$(dot) vector as in the linear system. Instead, multiplication results in a vector that is equivalent only to the bracketed parts of Equations 16.25-16.27. Multiplying this resultant vector by the $\boldsymbol{X}$ vector will produce the required $\boldsymbol{X}$(dot) vector. Using this matrix to solve nonlinear systems involves only minor modification of the technique used for linear systems. In order to produce the $\boldsymbol{X}$(dot) vector, the following lines of code must be inserted into the BASIC listing above:

```
155 REM PRODUCE X-DOT VECTOR (EQS 16.25-7)
156 XD(I) = XD(I) * X(I)
```

Exercise 16-3: Using the above matrix multiplication method, carry out Exercise 13-3, the nonlinear-feeding form of the Silver Springs energy flow model. Notice that only the feeding coefficients are nonlinear. The other transfers are simple donor-dependent linear relationships. You will have to set up two matrices of coefficients. One, $\boldsymbol{A}_{nl}$, should consist of the nonlinear feeding coefficients with zeros making up the other elements; the other, $\boldsymbol{A}_l$, should consist of the linear coefficients.

16.4 Multicomponent Interspecific Action

The equations of Volterra in Section 7.7 are used to describe the interaction of two populations. The basic approach may be extended as a simple model of the interactions of an unlimited number of populations within an ecosystem. The extension requires merely adding terms to the two-species equations. The two-species equations are

$$\frac{dN_1}{dt} = r_1 N_1 - r_1 \frac{N_1^2}{K_1} - r_1 N_1 \frac{\alpha N_2}{K_1} \tag{16.29}$$

$$\frac{dN_2}{dt} = r_2 N_2 - r_2 \frac{N_2^2}{K_2} - r_2 N_2 \frac{\beta N_1}{K_2} \tag{16.30}$$

Factoring out common terms gives

$$\frac{dN_1}{dt} = r_1 N_1 \left(1 + a_{11} N_1 + a_{12} N_2\right) \tag{16.31}$$

$$\frac{dN_2}{dt} = r_2 N_2 \left(1 + a_{22} N_2 + a_{21} N_2\right) \tag{16.32}$$

where

$$a_{11} = \frac{-1}{K_1} \qquad a_{22} = \frac{-1}{K_2} \qquad a_{12} = \frac{-\alpha}{K_1} \qquad a_{21} = \frac{-\beta}{K_2}$$

In these equations, the coefficients indicate the influence of one population on another or the ability of a population to utilize its environment, rather than material or energy transfers between compartments. Similar equations were used in Section 10.7.

A block diagram which shows all the interactions between four populations is presented in Figure 16.2. The differential equation expressing the change in size of population 1, for example, is

$$\frac{dN_1}{dt} = r_1 N_1 + r_1 N_1^2 a_{11} + r_1 N_1 N_2 a_{12} + r_1 N_1 N_3 a_{13} + r_1 N_1 N_4 a_{14} \tag{16.33}$$

The equations for the other populations will be similar. Factoring out the common terms in the equation will give

$$\frac{dN_1}{dt} = r_1 N_1 (1 + N_1 a_{11} + N_2 a_{12} + N_3 a_{13} + N_4 a_{14}) \tag{16.34}$$

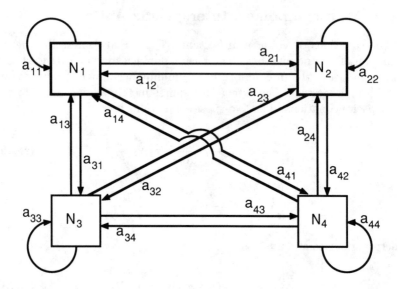

Figure 16.2. Block diagram for model of interactions among four compartments.

By selecting appropriate values for the coefficients, a_{ij}, almost any kind of positive or negative interaction between the four populations may be expressed.

Exercise 16-4: The form of Equation 16.34 suggests that this four-population model would lend itself to the programming techniques of nonlinear matrix multiplication discussed in the previous section. Begin by writing out the four equations required to describe changes in population density of the four species. Then write and implement a program using matrix techniques to simulate the interaction of the four populations. Use the following coefficients:

$$a_{11} = -0.0005 \qquad a_{12} = -0.001$$
$$a_{13} = -0.001 \qquad a_{14} = -0.001$$
$$a_{21} = +0.00003 \qquad a_{22} = -0.1$$
$$a_{23} = 0 \qquad a_{24} = 0$$
$$a_{31} = +0.0007 \qquad a_{32} = 0$$
$$a_{33} = -0.11 \qquad a_{34} = 0$$
$$a_{41} = +0.001 \qquad a_{42} = 0$$
$$a_{43} = 0 \qquad a_{44} = -0.115$$

$$r_1 = 0.5 \qquad r_2 = 0.2$$
$$r_3 = 0.3 \qquad r_4 = 0.3$$

Use $\Delta t = 0.1$ with simple Euler integration. Your output should consist of a graph of population sizes over at least 50 time units. Begin your simulation with all the populations having initial sizes of 10.

16.5 The Leslie Age-Class Matrix

The technique of using matrices for age-class models of populations (see Chapter 9) was developed principally by Leslie (1945). He devised a way to include in a single matrix all the coefficients necessary to describe reproduction and survival for each age class. This procedure allows the updating of all the age-classes of a population using a single matrix-vector multiplication. A column vector, N_t, is set up showing the numbers of individuals in each age class, N_x, at time t; for example:

$$\begin{bmatrix} N_0 \\ N_1 \\ N_2 \\ N_3 \\ N_4 \\ N_5 \end{bmatrix}_t$$

A square matrix, L, is set up with age-specific reproduction values m_x entered in the top row of the matrix. Age-specific survival values s_x are entered as a diagonal in the matrix. For example:

$$\begin{bmatrix} m_0 & m_1 & m_2 & m_3 & m_4 & m_5 \\ s_0 & 0 & 0 & 0 & 0 & 0 \\ 0 & s_1 & 0 & 0 & 0 & 0 \\ 0 & 0 & s_2 & 0 & 0 & 0 \\ 0 & 0 & 0 & s_3 & 0 & 0 \\ 0 & 0 & 0 & 0 & s_4 & 0 \end{bmatrix}$$

If the matrix L is multiplied by the vector N_t, the result is a vector showing the age-classes at time $t + 1$:

$$N_{t+1} = L \cdot N_t \tag{16.35}$$

As an example, the Leslie matrix procedure applied to the sowbug population from Table 9.1 would result in the following matrix multiplication, and would produce a vector with the population age-class sizes for the first year:

$$
\begin{bmatrix}
0.000 & 3.13 & 42.53 & 100.98 & 118.75 & 0 \\
0.114 & 0 & 0 & 0 & 0 & 0 \\
0 & 0.1042 & 0 & 0 & 0 & 0 \\
0 & 0 & 0.1391 & 0 & 0 & 0 \\
0 & 0 & 0 & 0.1250 & 0 & 0 \\
0 & 0 & 0 & 0 & 0.0000 & 0
\end{bmatrix}
\cdot
\begin{bmatrix}
10000 \\
1104 \\
115 \\
16 \\
2 \\
0
\end{bmatrix}_{t=0}
=
\begin{bmatrix}
N_0 \\
N_1 \\
N_2 \\
N_3 \\
N_4 \\
N_5
\end{bmatrix}_{t=1}
$$

$$(16.36)$$

The Leslie matrix approach is a very flexible method for modeling the dynamics of populations with fixed schedules of birth and death. As such, it finds more use in the investigation of human populations than in natural populations of plants or animals, which rarely have constant values for survival and reproduction. The method can be modified for simultaneous solutions of two-sex populations, for harvesting of exploited populations, etc. Usher (1972) and Cullen (1985) provide good reviews of some of the extensions of the method. As defined and used in this section and in Chapter 9, m_x is equivalent to F_x used in other formulations of the Leslie matrix; see Pollard (1973) and Jenkins (1988).

Exercise 16-5: Perform Exercise 9-4 for the U.S. population, using the procedures of Leslie matrix multiplication. The output from both exercises should be identical.

Conclusion

This chapter has introduced a few ways in which matrix algebra can be used to simplify the programming of some biological simulations. These methods find wide application and will be included in examples of some subsequent chapters.

CHAPTER 17

PHYSIOLOGICAL
CONTROL SYSTEMS

Much of the theory about control of physiological systems has been developed by analogy to technological control systems of engineers. As a result, the literature on physiological control mechanisms has kept much of the engineering terminology and analytical approach. While this specialized material will have to be mastered by anyone doing serious study of the subject, the principles of physiological control can be learned with only a few new terms. This chapter presents a brief introduction to the subject with some elementary examples.

17.1 The Generalized Feedback Control System

A block diagram with the necessary features of a feedback control system is shown in Figure 17.1.

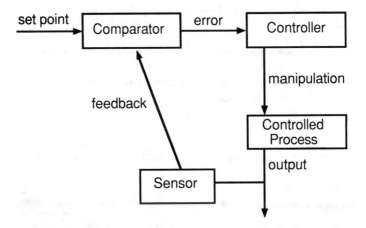

Figure 17.1. Diagram of a generalized feedback control system.

The basis of the system is a controlled process that has an output such as heat production or fluid flow. The intensity of output is measured by a sensor, which sends a signal to the system comparator. This feedback signal is a function of output intensity. The comparator adds or compares the feedback signal with a reference or set-point value, and then sends an error signal to the controller. The controller changes the error signal to a manipulating function which alters the output of the process. The overall result of feedback is that the output is controlled so that it tends toward an intensity set by the reference value.

In some physiological systems, these components can be identified explicitly. For example, body temperature in most warm-blooded animals is apparently set by a reference value. In many other systems, it is impossible to identify either a reference point or an error detector. However, these systems may still tend toward a steady-state or an equilibrium condition. The interactions of the system components may result in a more or less passive control; an example is the calcium flow system of Section 15.5. The general relationship between components of passive systems may be shown in "function blocks". These graphically describe the input-output behavior of one or more system components (Figure 17.2).

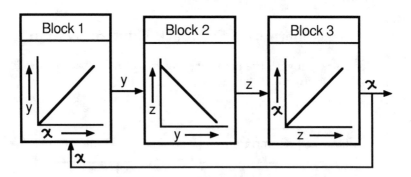

Figure 17.2. Diagram of a control system for a hypothetical physiological process with output x.

Jones (1973) defines such a negative feedback passive-control system in the following way:

(1) It is made up of a set of processes each with its own input and output variables.

(2) The processes are coupled in sequence, so that the output of one provides the input for the next to form a complete loop.

(3) Included in the loop is an odd number of processes (usually one) which exhibit sign reversal.

The example shown in Figure 17.2 meets all of these requirements. Block 2 has a negative slope, while the others are positive. Notice specifically that there is no reference value specified in this system. Instead, the equilibrium or steady-state value results from the interaction of the functions. The numerical value of this point may be found by the simultaneous solution of the function equations making up the control loop. This may be demonstrated graphically with the following steps.

First, we lump together the system components that are external to the function block having the reversed sign. In this example, Block 2 has the reversed sign. Lumping the equations of Blocks 1 and 3 gives a curve relating y to z. This curve has the form shown in Figure 17.3a. Next, this curve is inverted by plotting it on axes of z against y (Figure 17.3b), which are also the axes of Block 2. Finally, the curve from Block 2 is plotted on the same graph, as in Figure 17.3c. The steady-state or equilibrium point for the system is defined by the intersection of the two curves.

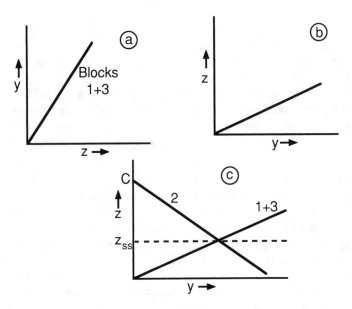

Figure 17.3. Graphical evaluation of the equilibrium point for the hypothetical system of Figure 17.2. See text for explanation of the evaluation procedure.

Mathematically, the steady-state point is defined by the simultaneous solution of all the function equations. As an example, assume that the following three equations define the functions for the system diagrammed in Figure 17.2:

$$\text{Equation 1:} \quad y = k_1 x \qquad (17.1)$$

$$\text{Equation 2:} \quad z = k_2 y + C \qquad (17.2)$$

$$\text{Equation 3:} \quad x = k_3 z \qquad (17.3)$$

The steady-state value for z would be given by

$$z_{ss} = \frac{C}{1 - k_1 k_2 k_3} \qquad (17.4)$$

The steady-state value for x would be given by

$$x_{ss} = \frac{k_3 C}{1 - k_1 k_2 k_3} \qquad (17.5)$$

An examination of Equation 17.4 shows that the product of the slopes $k_1 k_2 k_3$ must be less than 1 if z_{ss} is to have a value that falls between $z = C$ and $z = 0$; see Figure 17.3c. This condition will be met if one of the values of k should have a sign opposite the other two, as in Figure 17.2. If the product of the slopes in Equation 17.4 is not less than one, then steady-state solution of the equations falls outside the positive values for x and z, and thus outside the domain of control. The result is positive feedback, which characterizes runaway or vicious-circle systems.

Exercise 17-1: An elementary example of a control mechanism is a greatly simplified two-component model of the system which regulates the concentration of carbon dioxide (CO_2) in the blood of mammals (Milhorn 1966). In this model, the controller consists of the medullary respiratory center together with the lungs and other parts of the body that are used to remove CO_2 from the blood. The controlled process is the removal of CO_2 from the blood, which is a function of breathing rate. The system demonstrating the feedback loop is diagrammed in Figure 17.4.

The two input-output functions may be described with the following approximate relationships:

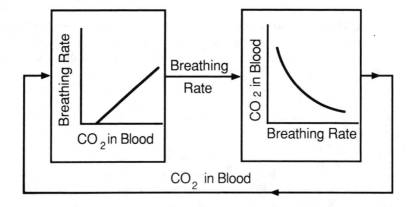

Figure 17.4. Diagram for feedback system for regulation of carbon dioxide levels in the blood.

$$B = c_1 + c_2 D \tag{17.6}$$

$$D = c_3 + c_4 \frac{1}{B} \tag{17.7}$$

Here, B is a measure of breathing rate or blood ventilation, and D is a measure of CO_2 concentration. For this system the units will be arbitrary.

The system may be simulated simply by repeatedly taking the output of one of these equations and using it as the input of the other. The system can be extremely unstable unless one builds into it some inertia. In this case, an increase in breathing rate does not normally produce an instantaneous decrease in CO_2 concentration throughout the body. The lag in this situation may be simulated by averaging the previous three values of D with the new value from Equation 17.7, and using this average value for D in Equation 17.6. Write and implement a computer program to simulate this feedback system, using the following values for constants:

$$c_1 = -5 \qquad c_2 = 0.25 \qquad c_3 = 0 \qquad c_4 = 200$$

Begin your simulation with an organism that has suffered some respiratory distress (e.g. near-drowning) so that $D = 80$, which is far from equilibrium. Plot D and B as a function of the number of cycles between the two equations, which is equivalent to some arbitrary time.

Your simulation should converge on the steady-state values for B and D. (These values can be obtained from the algebraic solution of Equations 17.6 and 17.7. For the given constants, the steady-state values are $D = 40$ and $B = 5$.)

17.2 Regulation of Thyroxine Secretion by the Pituitary

The following model provides a good example of a simple feedback control system in which the primary function blocks describe changes in variables rather than the static control of the previous example. The example is drawn from human endocrinology.

The rate of secretion of thyroxine (TH) from the thyroid gland is regulated by the concentration of thyroid stimulating hormone (TSH) in the plasma of blood and tissues. At the same time, the release of TSH from the pituitary is regulated by the concentration of TH in the plasma. In the plasma, both TH and TSH are slowly degraded to inactive products so that the concentration of each in the plasma at any time is a function of net rates of change (rate of secretion minus rate of degradation).

In this elementary model, we will neglect the effects of the related hormone, triiodothyronine, and will assume only a single form of thyroxine occurs in the plasma. We also assume that the sole effect of TSH is on thyroxine secretion, neglecting the role of the hypothalamus. This greatly simplified model is diagrammed in Figure 17.5. As indicated by the diagram, the input to the pituitary function-block is the plasma concentration of TH. The output is the rate at which TSH is released by the pituitary. Likewise, the input to the thyroid block is the plasma concentration of TSH, and the output is the rate of TH release by the thyroid to the plasma.

At any instant in time, rate of secretion of TSH by the pituitary is described by the relationship:

$$TSH \text{ secretion } (\mu\text{gr day}^{-1}) = \frac{b_1}{1 + b_2[TH]^k} \qquad (17.8)$$

where b_1, b_2, and k are coefficients for the reverse sigmoidal function (Figure 17.5) that describes empirically the pituitary output of TSH as a function of $[TH]$. Removal of TSH from the plasma may be described by an ordinary first-order decay process. Thus, net rate of change of the amount of TSH in plasma is given by:

$$\frac{dQ_{TSH}}{dt} = \frac{b_1}{1 + b_2[TH]^k} - a_{TSH}Q_{TSH} \qquad (17.9)$$

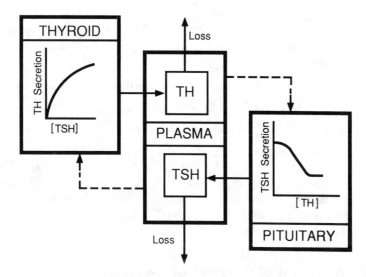

Figure 17.5. Diagram of feedback regulationof thyroid hormone concentration, $[TH]$, by the concentration of thyroid stimulating hormone, $[TSH]$.

where Q_{TSH} is the mass of TSH in the plasma, and a_{TSH} is the rate constant for loss due to degradation.

Secretion of TH by the thyroid is given by the familiar hyperbolic (Michaelis-Menten) equation

$$TH \text{ secretion } (\mu\text{gr day}^{-1}) = \frac{b_3[TSH]}{b_4 + [TSH]} \qquad (17.10)$$

where b_3 represents the maximal rate at which the thyroid can secrete TH and b_4 is the concentration of plasma TSH that produces half of the maximal rate. A first-order decay equation describes the loss of TH from plasma, so that the net rate of change in $[TH]$ is given by:

$$\frac{dQ_{TH}}{dt} = \frac{b_3[TSH]}{b_4 + [TSH]} - a_{TH}Q_{TH} \qquad (17.11)$$

Here, Q_{TH} is the mass of TH in the plasma, and the rate constant for decay is a_{TH} .

Exercise 17-2: Write and implement a computer program to simulate the thyroid-pituitary feedback control system described above. A simple two-stage Euler integration will be adequate for this simulation with a Δt value of 0.1. Set up the first Euler stage to solve for ΔQ_{TSH} and ΔQ_{TH}. The second Euler stage will involve the usual updating of Q_{TSH} and Q_{TH}. Use the following values for the coefficients in the model:

$$k = 3 \qquad b_1 = 8\mu\text{gr day}^{-1} \qquad b_2 = 0.000008 \text{ liters}^k \mu\text{gr}^{-k}$$

$$b_3 = 100\mu\text{gr day}^{-1} \qquad b_4 = 1.5\mu\text{gr liter}^{-1}$$

$$a_{TH} = 0.1 \text{ day}^{-1} \qquad a_{TSH} = 0.25 \text{ day}^{-1}$$

For volume of plasma V use a value of 10 liters. Concentrations are found as usual with Q/V. For initial concentrations, set $[TSH] = 2\mu\text{gr liter}^{-1}$ and $[TH] = 30\mu\text{gr liter}^{-1}$. Generate about 50 days of simulated data to watch the system go to a steady-state condition. Plot concentrations of TSH and TH over the time period. There should be little oscillation in this system.

17.3 A Model of Control of Eating

Many technological feedback systems are set up with a controller made of a simple switch, so that the controlled process is either on or off. A familiar example is the thermostat-furnace system used for heating most buildings. An biological example of this type of control is a model of feeding in laboratory rats (Booth and Toates 1974, Gold 1977). In this rudimentary model the controller acts only as an on-off switch to start and stop feeding. The simplified control model is diagrammed in Figure 17.6.

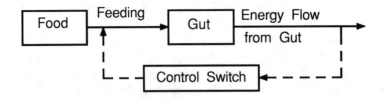

Figure 17.6. Diagram of model for control of feeding in rats.

The gut serves as a storage compartment for food energy. The rate of transfer of energy from the gut to the body is proportional to surface area of the gut contents; the surface area varies as the square root of the gut content. An equation describing this relationship is

$$M = k_M \sqrt{G} \qquad (17.12)$$

where M is rate of energy transfer from the gut to the body at any instant, k_M is the uptake coefficient, and G represents the energy content of food in the gut. k_M may vary during a 24-hour period. Because rats are nocturnal, the coefficient will be higher during the nighttime.

Feeding will fill the gut rapidly in comparison with the rate of energy removal. The rate of filling, F, is given by

$$F = \varepsilon I \qquad (17.13)$$

where I is the rate of eating, and ε is a "switching" coefficient. If the organism is eating, then the controlling switch is "on" with $\varepsilon = 1$, and the gut is filling at rate F. If the animal is not eating, then $\varepsilon = 0$.

In this model, the switch is turned on and off by the level of available energy, M. If M drops below some minimum level, M_L, then the rat will start to eat. When M reaches some upper level M_H, the animal will stop eating, turning the switch off with $\varepsilon = 0$. In more formal notation,

$$\varepsilon = 1 \qquad \text{if} \quad M \leq M_L$$

$$\varepsilon = 1 \qquad \text{if} \quad M < M_H \text{ and } \varepsilon = 1$$

$$\varepsilon = 0 \qquad \text{if} \quad M \geq M_H$$

This model assumes that rats are neither gaining nor losing weight on average, so that normal rat activities will use energy as it becomes available. The energy is neither metabolically stored, nor supplemented with energy from storage. The rats are assumed to have food available at all times. For this model the feedback system is defined by Equation 17.12 and by

$$\frac{dG}{dt} = F - M \qquad (17.14)$$

The control of feeding is of course far more complex than this system suggests, and involves social and sensory factors as well as physiological energy balance. The system of feeding control is also sensitive and efficient, because animals rarely gain or lose large proportions of their body mass.

Exercise 17-3: Write and implement a program for simulating the control of feeding using the system described above. Use the following values in the simulation:

$$I = 1000 \text{ cal min}^{-1} \qquad M_H = 60 \qquad M_L = 18$$

Let $\Delta t = 1$ min. Allow your simulation to proceed for 48 hours (2880 minutes). Begin your simulation at midnight with $G = 4000$ and $k_M = 0.9$. At 8 am, change k_M to a daytime value of 0.6, and at 8 pm let k_M revert to 0.9. For each time unit, your simulation should do the following in sequence: (1) check to see whether it is day or night and set k_M, (2) calculate M, (3) check to see whether feeding needs to be turned on or off, (4) find ΔG and update G with the simple Euler procedure.

Output for your program should be a plot of values of G and of M over the time period of your program.

17.4 Sweating and Temperature Control

Under an external heat stress, the human body will absorb heat until it reaches a point where it is unable to maintain a constant body temperature with cooling by convection and radiation. Above this point, body temperature will be controlled primarily by sweating. The general feedback control pathway for this process is diagrammed in Figure 17.7.

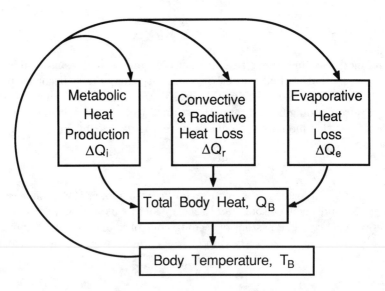

Figure 17.7. Block diagram for model of control of temperature above thermal neutrality.

The heat production of metabolism, H_m, is a function of body temperature. Rate of heat production at any instant may be found with this equation:

$$H_m = P_H \left(Q_{10}^{T_d/10} \right) \tag{17.15}$$

where P_H is metabolic heat production and Q_{10} is the factor by which heat production increases for a 10°C rise in temperature (Section 12.3). P_H has an average value of 1.25 kcal min^{-1} at 37°C. A typical value for Q_{10} is 2. T_d is the temperature above normal body temperature (37°C) defined by

$$T_d = (T_B - 37°C)_+ \tag{17.16}$$

where T_B is body temperature. In this model T_d must be a positive number or zero, as indicated by the + subscript (Section 13.5).

Convective and radiative loss of heat, H_r, is a function of body temperature and environmental temperature. The instantaneous rate of loss may be found with:

$$H_r = k_r (T_B - T_A) \tag{17.17}$$

where k_r is the specific rate of heat loss, T_B is body temperature, and T_A is environmental temperature. Typically, k_r will take on values of about 0.167 kcal min^{-1} °C^{-1}.

The rate of heat loss due to evaporation may be described with the following hyperbolic saturation equation:

$$H_e = \frac{E_{\max} T_d (1 - R_H)}{k_e + T_d} \tag{17.18}$$

where $E_{\max}$ is the maximum evaporative heat loss, typically 23.3 kcal min^{-1}. T_d is defined above, R_H is relative humidity expressed as a decimal fraction, and k_e is the half- maximum value for the hyperbolic equation. k_e is approximately 0.2 °C. Because of insensible evaporation, H_e has a minimum value of 0.083 kcal min^{-1}. Thus, if H_e calculated with Equation 17.18 is less than 0.083, then H_e is assumed to be 0.083.

The total change in heat content of the body, ΔQ_B, over a small interval of time Δt may be found with the standard Euler equation

$$\Delta Q_B = (H_m - H_r - H_e) \Delta t \tag{17.19}$$

The Euler expression for updating body heat content is therefore

$$Q_B \leftarrow Q_B + \Delta Q_B$$

In this model, the heat capacity of the body is assumed to be about equal to that of water, 1 kcal kg^{-1} °C^{-1}. The average adult human body mass can be assumed to be 70 kg. Accordingly, the relationship between temperature of the body and its heat content is

$$T_B = Q_B/70 \qquad (17.21)$$

The zero point for heat content is fixed arbitrarily at 0°C. The value of Q_B for an average human body at normal temperature is therefore about 2590 kcal.

The model described by these equations illustrates some interesting elementary feedback properties. The feedback is passive, with the exception of the 37°C minimum established for body temperature. The model will display negative feedback properties until a combination of temperature and relative humidity exceeds the capacity of evaporative cooling. Then the system will switch to positive feedback, generating heat by raising the body temperature and metabolic rates in a upward cycle.

Exercise 17-4: Develop a computer simulation of the control of human body temperature during heat stress using the above information. Use the given constants, and start your simulations with $T_B = 37$. Use $\Delta t = 1$ min. Set $R_H = 0.80$ and $T_a = 55°C$. Produce a graph showing T_B over 1200 minutes (20 hours). T_B should approach steady-state within this time limit.

Exercise 17-5: Based on the simulation of Exercise 17-4, find the body temperatures obtained after 300 minutes at ambient temperatures of 30°, 35°C, 40°C, 45°C, 50°C, 55°C, and 60°C, with $R_H = 0.5$. Start each case at $T_B = 37°C$. Plot the body temperatures at 300 minutes vs. ambient temperature. Also produce results and plot them on the same graph for the same series of temperatures with $R_H = 0.80$ and with $R_H = 0.90$. With some combinations of temperature and humidity, the evaporative cooling mechanism will be overwhelmed; assume that body temperatures above 41.4°C result in death.

17.5 Temperature Control below Thermal Neutrality

When a small mammal is cooled below its thermal neutral point, body temperature change is resisted by the insulating effects of fur, and by increasing the production of metabolic heat, also called chemical thermogenesis. A simulation of the cooling of a bat is interesting because of their unusual thermal responses, large naked wings, and ready ability to become torpid (Bakken and Kunz 1988, Kurta and Fujita 1988).

Heat loss, ΔQ_L, from the body of a bat may be described as usual by the equation for Newton's law of cooling (Chapter 1):

$$\Delta Q_L = k_L \left(T_B - T_A \right) \Delta t \qquad (17.22)$$

where T_B is body temperature, T_A is ambient temperature, and k_L is thermal conductance. The value of k_L depends on body surface area and insulation effects. In a series of experiments, values for k_L at various temperatures for a common species, the little brown bat (*Myotis lucifugus*), were determined as follows (Holyoak and Stones 1971):

$T_A(^\circ C)$:	35	30	25	20	15	10	5
k_L :	1.00	0.70	0.60	0.50	0.46	0.44	0.42

Note that k_L is expressed in energy equivalent units, cc O_2 (gram body wt)$^{-1}$ hr^{-1}. This dependency is best modeled with an empirical equation of the polynomial type (Chapter 3).

Chemical thermogenesis, ΔQ_R, varies as a straight-line function of body temperature:

$$\Delta Q_R = [2 + 2 \left(T_N - T_B \right)] \, \Delta t \qquad (17.23)$$

where T_N is the body temperature at thermal neutrality. For this species of bat, thermal neutrality is $35^\circ C$. If the difference $T_N - T_B$ is very large, the rate of metabolism predicted by the equation is also going to be large. However, this species has been found to have a maximum rate of energy output equal to about 10 cc O_2 gr^{-1} hr^{-1}. Thus, ΔQ_R will follow Equation 17.23 between an upper limit of 10 and a lower limit of 0.

Temperature is assumed to affect the rate of chemical thermogenesis, following the Q_{10} rule, with a factor of 2 for each $10^\circ C$. This effect can be described by an equation of this type:

$$\Delta Q'_R = \Delta Q_R \exp\left[-A \left(T_N - T_B \right)\right] \qquad (17.24)$$

For any given environmental temperature, there is some body temperature at which heat loss and generation are balanced. The computer may be programmed to find this body temperature by a stepwise lowering of body temperature until heat loss equals heat generation. Using this procedure, it is possible to simulate the bat's response to lowered temperatures. A flowchart for this process is given in Figure 17.8.

With this simulation it is possible to observe the following characteristic features of temperature control below thermal neutrality (see Figure 17.9):

(1) A plateau region of thermoregulation where body temperature is relatively constant with declining ambient temperature above the critical temperature.

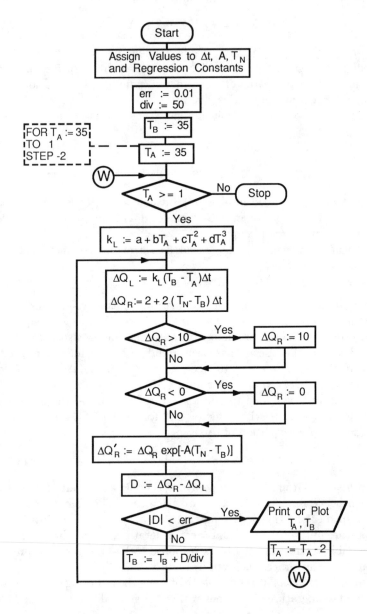

Figure 17.8. Flowchart for program to simulate temperature regulation of a bat being cooled.

(2) Ambient temperatures below the critical temperature result in positive feedback because of the Q_{10} effect on thermogenesis. The organism therefore comes to equilibrium at a body temperature considerably below that of the thermoregulatory zone.

(3) In the zone of hypothermic cooling, the body temperature declines rapidly with ambient temperature maintaining a slight differential all the way down to $T_A = 0$

(4) Heat production approaches the limit of 10 cc O_2 gr^{-1} as the bat approaches the critical temperature.

Exercise 17-6: Using the information provided above, produce a program to simulate the temperature control characteristics of a bat at temperatures below thermal neutrality. The program should proceed from an ambient temperature of 35°C to 1°C in steps of -2°C. As output, plot your results as shown in Figure 17.9. Set $A = 0.0693$, $T_N = 35$, $\Delta t = 1$. You will have to find the coefficients of the polynomial equation describing the relationship between k_L and temperature, using the techniques of Chapter 3.

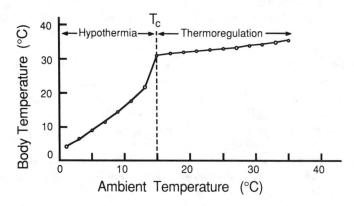

Figure 17.9. Results of simulation of cooling a bat.

17.6 Hormonal Control with Two Feedback Mechanisms

In average adult human males, testosterone levels in the blood are known to cycle with a period of about three hours. The following model of this periodicity is interesting because it uses double feedback circuits. It also combines several modeling techniques you have studied previously.

Production of testosterone is stimulated by luteinizing hormone (LH) from the pituitary. Production of LH is stimulated by luteinizing hormone releasing hormone (LHRH) which is produced by the hypothalamus. The cycling of testosterone is a result of pulsing of the release of LHRH.

The cause of pulsing of LHRH is not definitely known; some models have used an internal, independent clock. Release of LHRH is known to be regulated at least partially by testosterone levels in a negative feedback loop. Cartwright and Husain (1986) proposed a feedback model for the system, with both testosterone and LH acting to control LHRH production from the hypothalamus. Based on reasonable assumptions and realistic parameters, this model closely mimics important features of the testosterone cycle. However, the summation-comparison mechanism, a unique feature of this model, has not been identified either structurally or biochemically and remains hypothetical.

The model is based on the following assumptions which appear to agree with the known biology of the system:

(1) The rate of release of LH depends linearly on the concentration of LHRH at the pituitary.

(2) Similarly, the release of testosterone depends on the concentration of LH at the testes.

(3) The hypothalamus releases LHRH either at a constant rate, or not at all. The release is triggered by a combination of low levels of LH and testosterone at the hypothalamus.

(4) LH, LHRH, and testosterone are all eliminated from the system at a rate that depends on their concentration. This process therefore follows the familiar first-order kinetics of exponential decay.

(5) In response to LH concentration, production of testosterone by the testes is subject to a delay of about 25 minutes, caused by the time required for biochemical synthesis of testosterone.

(6) In the whole system, there are delays of a few minutes required for each hormone to be transported between the site of production and the site of action.

Based on these assumptions, Cartwright and Husain (1986) proposed the following model to account for testosterone cycling. The simplest equation models change in LH:

$$\frac{dL_t}{dt} = -q_L L_t + r_L R_{t-3} \tag{17.25}$$

where L_t is the concentration of LH at the pituitary at time t. q_L is the rate constant for clearance of LH from the system. r_L is the rate constant for LH production as a function of LHRH. R_t is the concentration of LHRH at the hypothalamus at time t. About three minutes are required for LHRH level at the hypothalamus to begin to affect production of LH at the pituitary. Thus, LHRH concentration at the pituitary is equal to its concentration at the hypothalamus three minutes previously. Hence, a subscript of $t-3$ will be used to represent a 3-minute time delay for effective transport of LHRH from the hypothalamus to the pituitary.

Change in testosterone is found with

$$\frac{dT_t}{dt} = -q_T T_t + r_T L_{t-30} \tag{17.26}$$

where T_t is testosterone concentration in the blood at the testes at time t. q_T is the rate constant for clearance of testosterone from the blood. r_T is the rate constant for testosterone production. There are 2 delays combined in the time delay of $(t-30)$. The first delay is for LH transport from the pituitary to the testes (about 5 minutes). The second delay is caused by the time required for adjusting testosterone synthesis (about 25 minutes).

The equation for change in concentration of LHRH requires an off-on switching function to model the production of LHRH:

$$\frac{dR_t}{dt} = -q_R R_t + r_R[S(Q)] \tag{17.27}$$

Here, R_t is the blood concentration of LHRH at the hypothalamus at time t, and q_R is the rate constant for bloodstream clearance of LHRH. r_R represents a constant rate of production of LHRH. Q is evaluated with

$$Q = 2 - \frac{L_{t-5}}{L_S} - \frac{T_{t-5}}{T_S} \tag{17.28}$$

L_{t-5} represents LH concentration with a 5-minute delay for transport from the pituitary to the hypothalamus. T_{t-5} is testosterone concentration, with a similar 5-minute transport delay, testes to hypothalamus. L_S represents half the level of LH required to switch off the production of LHRH. Likewise, T_S is half the level of testosterone needed to switch off the production of LHRH. The switching function is given by $S(Q)$ with

$$S(Q) = 0 \qquad \text{if} \qquad Q < 0$$

$$S(Q) = 0.5 \qquad \text{if} \qquad Q = 0$$

$$S(Q) = 1 \qquad \text{if} \qquad Q > 0$$

Exercise 17-7: Using the model above, write and implement a program to simulate testosterone cycling in adult human males. Cartwright and Husain (1986) proposed the following values as being realistic constants for use with the model:

$q_L = 0.015$ min^{-1} $r_L = 5$ ng LH (ng LHRH)$^{-1}$ min^{-1}

$q_T = 0.023$ min^{-1} $r_T = 0.01$ ng testosterone (ng LH)$^{-1}$ min^{-1}

$q_R = 0.10$ min^{-1} $r_R = 0.1$ ng LHRH ml^{-1} min^{-1}

$L_S = 30$ ng LH ml^{-1} $T_S = 8$ ng testosterone ml^{-1}

Begin your simulation with these initial concentrations:

$L = 30$ ng ml^{-1} $T = 16$ ng ml^{-1} $R = 1.5 \times 10^{-4}$ ng ml^{-1}

A simple two-stage Euler integration will be adequate for this simulation, with $\Delta t = 1$ minute. To permit simulation of the time delays of the model, you will have to include a method of retaining values of concentrations. Time delay will be easier to program for this exercise using subscripted variables rather than using the direct approach of Exercise 7-3. Set the starting levels for the time-delay concentrations at the initial levels given above.

The simulation will show some erratic system behavior initially for about 100 minutes, until the time-lag values stabilize. Your output should include plots of values of R_t, L_t, and T_t for at least 800 minutes. The cyclic behavior of testosterone levels should be evident.

Exercise 17-8: It is possible to simulate the effects of castration on the hormonal cycle with the model above. This can be accomplished by modifying your program from Exercise 17-7 so that at times greater than 300 minutes, $r_T = 0$, indicating nonfunction of the testes. You should be able to see the effects of this manipulation on levels of LH and cycle frequency as T_t drops to zero.

Conclusion

This chapter has presented some fundamental concepts that are used to describe the feedback controls exhibited by physiological systems. Some further examples of such feedback will be used in later chapters with biochemical reaction pathways.

PART THREE
PROBABILISTIC MODELS

Up to this point, all the models studied have been deterministic. However, random processes also play a significant role in most biological systems. There are several reasons for their importance, including:

1) Many important biological processes are the result of discrete events which appear to occur more or less randomly through time. Some examples from a population viewpoint are births of individuals in continuously breeding populations, deaths of individuals, and determination of individual genotypes.

2) Many systems are strongly influenced by environmental factors which are themselves subject to much random variation in space and through time. Although light intensity and temperature are deterministic in their effects, these may fluctuate in a random fashion, so that randomness may play an important role in determining the net effect.

3) Individual organisms respond differently to the same stimulus because of differences in genotype and previous environmental conditioning.

Although some of the factors mentioned above might be averaged by extremely large populations, the finite size of real populations, living in a real environment, results in apparently random fluctuations from the normal or deterministic behavior of the populations. The type of model which includes random fluctuations, in addition to deterministic behavior, is called a stochastic or probabilistic model. Stochastic models are generally more complex than deterministic models because of the mathematics involved in describing random processes. However, in the next several chapters, we will emphasize the direct approach, that of using the Monte Carlo method for simulating stochastic processes. The material included in these chapters, as in the previous chapters, must be considered as only a brief introduction to an important area of modeling. Further material may be found in the classic book by Bailey (1964), in Goel and Richter-Dyn (1974), and in appropriate chapters in Poole (1974) and Swartzman and Kaluzny (1987).

From a purely philosophical view, one may question whether truly random processes really exist. The boundary between what we consider random behavior and what we know to be deterministic has been steadily pushed back throughout the history of science. On closer examination, each of the examples of random processes given above may be shown to be the result of deterministic causes. They appear to occur randomly because we are unable to predict their occurrence from the information available. Thus, there is a close relationship between unpredictability and what we call randomness. Randomness is a useful concept for characterizing the behavior of unpredictable systems or system components. The stochastic model and the Monte Carlo technique provide an alternate modeling approach which allows us to gain new insight into certain biological systems.

You should study the following chapters with the objectives of learning how the Monte Carlo technique may be applied in simulating random processes, and understanding how random processes may affect the understanding of biological systems.

CHAPTER 18

MONTE CARLO MODELING OF SIMPLE STOCHASTIC PROCESSES

Several simple nonbiological and biological processes will be examined in this chapter. The purpose of the chapter is to introduce Monte Carlo methods and to provide some simple familiar systems which permit easy analysis of the simulation results. The following chapters will provide more extensive biological applications of the methods you learn in this chapter.

18.1 Random Number Generators

Monte Carlo simulations depend on the generation of random numbers by the computer. The random numbers generated by the computer are actually pseudorandom, because they are created by arithmetic operations performed on numbers by the computer. The same "random" numbers will be produced in the same order if the generating process is started in the same way. To avoid this problem, most dialects of BASIC initiate their random number generators with a starting number (seed) that is itself likely to be random, for example the time of day.

Truly random numbers are difficult to obtain. Their production requires some inherently random physical process such as radioactive decay, or noise in an electronic circuit. Such sources may be difficult or expensive to obtain. For the purposes of the Monte Carlo simulations in this textbook, the random number generators built into BASIC are adequate, and a more elaborate source is not needed. However, before you use a BASIC random number generator for serious scientific work, you should have a very good idea of its characteristics. Some may have serious limitations (Gleason, 1988). The theory that underlies the generation of random numbers by computers is an active field of mathematical research, as are the methods for detecting nonrandomness in a series of numbers.

18.2 The Fundamental Monte Carlo Simulation

The Monte Carlo technique of computer simulation uses a random number to decide whether or not a particular event occurs. The event may relate to an individual atom, molecule, organism, gene, etc. This is called the Monte Carlo approach because a random number takes the place of a coin flip, dice toss, or wheel spin used in non-computer simulations of this sort. The number generated by the computer is usually in the form of a decimal fraction between zero and one so that it may be compared easily with the probability of a particular event, also typically expressed as a decimal fraction.

The simplest example is a coin flipping simulation in which "heads" occur with a probability of 0.5. To use a Monte Carlo simulation to determine whether any flip produces a head or a tail, one simply compares the random number with the 0.5 probability value. If the random number is greater than or equal to 0.5, the result can be called a "head"; if it is less than 0.5, it may be called a "tail". The flowchart for a simulation of a series of coin flips is given in Figure 18.1.

The Monte Carlo technique may be used with equal ease for a number of yes-no or on-off circumstances in biological situations, such as

1) whether an organism dies or not during a given period of time;
2) whether a heterozygous individual produces an A gamete or an a gamete
3) whether a rat in a 'T' maze turns right or left at the next intersection.

The probability value that is compared with the generated random number may be fixed at some value such as 0.5 or 0.95, or it may change during the simulation in some defined way.

In some simulations, the random number may be used to decide among a number of alternative actions. For example, suppose you wished to simulate a game that uses a pair of dice, and the BASIC on your computer can generate a random number R that falls between 0.00000 and 0.99999. The random number R may be converted into random digits between 1 and 6 by the BASIC statement

```
D = INT(R*6 + 1)
```

For the game of dice, the operation of obtaining a random number and converting it to a digit would have to be performed twice, once for each die. The outcome of such a simulation is like that of a real dice game (or perhaps better, since the computer dice cannot be manipulated.) Similar

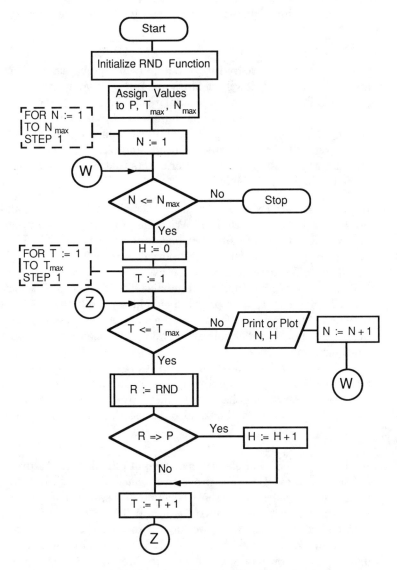

Figure 18.1. Flowchart for a coin-tossing simulation.

R is a uniform random number generated by the computer's RND function.

P is the probability of heads in a given toss of the coin.

T_{max} is the number of tosses per set of simulated tosses.

N_{max} is the number of simulated sets of tosses.

operations on the random numbers generated by the computer may be used to provide random numbers along a continuum between any two limits.

Exercise 18-1: Write a program to simulate coin-flipping as described above, based in the flowchart of Figure 18.1. Set your program to provide 8 flips in a set, and to simulate 100 such sets. Write your program so that it prints out the number of heads in each set (as shown on the flowchart). In addition, have your program produce a print out of the total number of heads and tails for the 800 flips, and also a tabulation of the number of sets that produced 0 heads, 1 head, 2 heads, ..., 8 heads.

Exercise 18-2: Modify Exercise 18-1 so that it makes 5 flips in a set, and produce at least 200 such sets. Write the program to provide chi-square values for the following two goodness-of-fit tests: (1) Total number of heads and tails (expected values of 500 for each). (2) Frequency of occurrence of heads in five flips, with expected probability values coming from the binomial distribution for 5 tosses. As in Exercise 18-1, your program should print out the observed numbers of heads and tails; it should also print the expected numbers, and the calculated chi-square values. Check your computed chi-square value against a value from a table of chi-square values to find how well the simulation data fit the expected binomial pattern. Methods of chi-square analysis and tables may be found in most elementary statistics and biometry textbooks, for example Zar (1984). Most of the books also discuss the binomial distribution.

Exercise 18-3: Write and implement a program to simulate the throwing of a pair of dice. Use your program to generate enough data (at least 150 throws) so that the frequency of 2's, 3's, 4's, etc may be compared with the expected frequency. Those who are familiar with dicing know that with two dice there is only 1 way to throw either 2 or 12, 2 ways to throw 3 or 11, 4 ways to throw either 5 or 9, 5 ways to throw 6 or 8, and 6 ways to throw a 7. Since there are 36 possible ways to throw the two dice, the expected proportion of 2's is 1/36, for 3's is 2/36, etc. Include in your program a calculation of a chi-square value for a goodness-of-fit test for the simulated results. Your program should print out the observed and expected numbers, and the calculated chi-square value. Check the chi-square value from your program against the appropriate values (10 degrees of freedom) from a table of chi-square.

18.3 Stochastic Simulation of Radioactive Decay

Radioactive disintegrations proceed at a rate which cannot be modified by any known chemical or physical means. For a given element this rate at any instant is proportional only to the number of nuclei present, N, and is described by the differential equation

$$\frac{dN}{dt} = -kN \tag{18.1}$$

The integration of this equation, yielding the classical decay equation, was discussed in Chapter 1.

Because the atoms act independently of one another during radioactive decay, the process may be simulated by using a procedure which looks one at a time at each radioactive atom. The given probability of decay applies equally to all undecayed atoms during one interval of time. The Monte Carlo technique may be used to decide whether or not an atom is going to decay during a given time interval. For each atom, in turn, a random number is generated. If the number is greater than the probability of decay, the atom is considered undecayed. If the random number is less than the probability of decay, the atom is considered to have decayed. The basic technique may be combined in a variety of ways with a time counter to simulate the decay of a number of radioactive atoms.

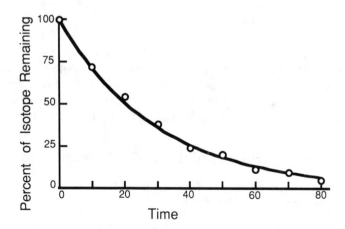

Figure 18.2. Results of stochastic simulation of decay of ^{11}C isotope. The solid line represents deterministic expectation. The circles show results of simulated decay.

Exercise 18-4: Use the Monte Carlo technique described above to write
a program for simulating radioactive decay. Implement the model
using data from one of the isotopes included in the following list.
The tabulation is based on half-life, $t_{1/2}$, where

$$k = \frac{\ln 2}{t_{1/2}}$$

$$\text{decay probability } = 1 - \exp(-k)$$

The units of time are selected to provide reasonable probability
values for the simulation.

Isotope	t (unit of time)	half-life (t)	k (t^{-1})	probability (t^{-1})
^{11}C	10 min	2.05	0.33812	0.28689
^{14}C	1000 yr	6.00	0.11552	0.10910
^{13}N	10 min	1.01	0.68628	0.49656
^{24}Na	10 hr	1.48	0.46834	0.37396
^{124}I	1 day	4.00	0.17329	0.15910

Write your program to produce simulation data on the decay of at
least 200 atoms. Compare the results of the Monte Carlo simulation with those of the deterministic model by plotting the number
remaining undecayed vs. time. It is instructive to draw the line
predicted by the deterministic solution; see Figure 18.2.

18.4 Monte Carlo Simulation of a Monohybrid Cross

With the usual terminology from Chapter 10, the monohybrid cross
may be diagrammed as follows:

$$Aa \times Aa \rightarrow 1aa : 2Aa : 1AA$$

This 1:2:1 ratio for the genotypes of the offspring is subject to the random
processes involved in gametogenesis and zygote formation. When a single
gene locus is involved, there will be only two kinds of gametes formed,
each occurring with a probability of 0.5.

This situation may be simulated by generating a uniform random number R between 0 and 1, and testing it against 0.5. If R is equal to or greater than 0.5 assume that the gamete from the first parent contains the allele A; if less than 0.5 assume it contains a. Then examine a second random number representing the second parent. If R is equal to or greater than 0.5, again we will say the gamete contains the A allele, and if less than 0.5, the a allele. If both are less than 0.5 then the offspring is aa and if neither are less than 0.5 then the offspring is AA. If only one R of the pair is less than 0.5, then the offspring is either Aa or aA. A series of random number pairs may be generated and examined in this way to simulate the production of a number of offspring.

Exercise 18-5: Write and implement a program for the Monte Carlo simulation of the production of offspring by monohybrid parents. Simulate the production of at least 100 offspring, with the computer keeping a tally of the number of the three possible genotypes. Write your program so that at the end of the simulation it will print out the tallied numbers, the expected numbers and a calculated chi-square value for a goodness-of-fit test of the observed and expected numbers. Use the calculated value to see whether the simulated reproduction deviates significantly from the expected 1:2:1 ratio.

18.5 Monte Carlo Model of a Dihybrid Cross

A dihybrid cross involves two individuals heterozygous for two non-linked genes. The cross may be diagrammed with

$$AaBb \times AaBb \rightarrow 9AxBx : 3aaBx : 3Axbb : 1aabb$$

where A and B are dominant alleles, a and b are recessives, and x may be either. The phenotypic ratio results from random processes involved in both gametogenesis and in zygote formation. The probability of any given gamete containing the A gene is 0.5 and the probability of any given gamete containing the B gene is 0.5. The genotype of any zygote is the result of four random events, two involving gametogenesis, and two involving zygote formation.

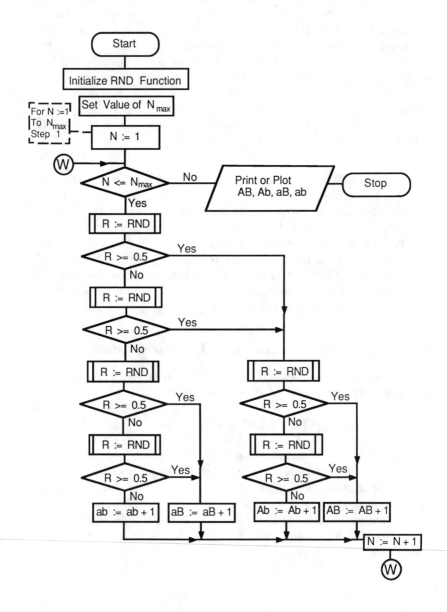

Figure 18.3. Flowchart for program of stochastic simulation of a dihybrid cross.

The dihybrid cross may be simulated by generating uniform random numbers and testing these against 0.5. From the flow diagram in Figure 18.3 it will be apparent that the *aabb* genotype (*ab* phenotype) occurs when four random numbers in succession are less than 0.5. Because it is only necessary to have one *A* gene to express the *A* phenotype, it is not necessary for the random number to exceed 0.5 more than once to fall into the *Ax* category. The same is true for the *B* allele and the *B* phenotype.

Exercise 18-6: Write and implement a program to simulate the production of at least 60 offspring by dihybrid crossing. Your program should print the tally of the resulting phenotypes, and find a chi-square value from a goodness-of-fit test based on the expected phenotypic ratio from the dihybrid cross.

18.6 Tests of Randomness

In order to appreciate the true nature of random numbers, and also to see just how random they really are, it is quite interesting to perform some standard tests on a set of pseudorandom numbers which have been generated by the computer. Table 18.1 shows the results of sorting 1200 pseudorandom two-digit numbers.

The 1200 random numbers were taken from a series generated by a computer with a RND function in BASIC intended to produce positive decimal fractions uniformly distributed between 0 and 1. The numbers making up the table are the first two digits following the decimal point, because these are the more important digits from the standpoint of probability simulation. The occurrences of different pairs within the test series of 1200 numbers are shown in the table. The table indicates that in this series of numbers, 9 began with 0.00..., 11 began with 0.01..., 13 began with 0.99..., etc.

These numbers were examined with tests similar to some of those employed by the Rand Corporation (1955) in its classic publication on random numbers. These goodness-of-fit tests are informative and can detect various kinds of nonrandomness. You should be aware that they cannot detect many other kinds. The tests and the results for this particular set of numbers follow:

First				Second Digit							
Digit	0	1	2	3	4	5	6	7	8	9	Totals
0	9	11	15	9	15	6	16	13	10	16	120
1	9	13	11	20	7	15	13	11	14	10	123
2	11	17	20	10	10	13	10	7	15	12	125
3	13	19	6	9	10	4	16	14	11	17	119
4	10	11	15	11	7	11	9	11	10	11	106
5	9	15	13	9	13	13	6	12	13	10	113
6	10	16	15	13	16	10	12	9	17	9	127
7	8	8	12	10	11	12	16	16	11	11	115
8	14	9	17	10	14	13	16	11	10	14	128
9	11	7	14	19	11	14	12	13	10	13	124
Totals	104	126	138	120	114	111	126	117	121	123	1200

Table 18.1. Number of occurrences of first and second digits of 1200 random numbers generated by the RND function of BASIC on a microcomputer.

(a) Test of column totals: The observed totals for the 10 columns were tested against the expected value of 120. The calculated χ^2 was 6.57, with a probability of about 0.70, with 9 d.f. (degrees of freedom).

(b) Test of row totals: The observed totals for the 10 rows were tested against the expected value of 120. The calculated value of χ^2 was 3.62, with a probability of about 0.93, 9 d.f.

(c) Digit frequency: The observed total for each of the ten digits was tested against the expected total of 240; the number of 1's, for example, $123 + 126 = 249$. The calculated χ^2 was 7.86, with a probability of about 0.55 for 9 d.f.

(d) Odd and even digits: The numbers of odd and even digits, 1209 and 1191, were compared with their expected numbers of 1200 each. The calculated value of χ^2 was 0.135, with a probability of about 0.7 for 1 d.f.

(e) Pairs and non-pairs: There were 122 pairs of numbers (00, 11, 22, ...) and 1078 non-pairs. When compared with their expected numbers of 120 and 1080, the χ^2 value was 0.038, 1 d.f., with a probability of about 0.85.

(f) Overall number: The overall numbers of occurrences of each of the 100 types of number pairs was tested against the expected value of 12 in each case. The calculated χ^2 value was 85.67, with 99 d.f., and a probability of about 0.8.

(g) Overall contingency test: This is a test for the unexpected joint occurrence or absence of number pairs, with the expected number based on the observed occurrence of individual numbers. For example, the table indicates that 8 occurred as the first digit 128 times, and that 2 occurred as the second digit 138 times. Based on this, we would expect 82 to occur (128 x 138)/1200, or 14.72 times in this specific series of 1200 numbers. The digit pair 82 occurred 17 times. Similar expected and observed values may be calculated and compared for all 100 number pairs. For the series of number in Table 18.1, a contingency χ^2 of 74.10 was calculated. This value has a probability of about 0.7, with 81 d.f.

A final test involves a comparison of the values given in Table 18.1 with the values expected from a random sampling. The basis for the comparison is the expected mean number of 12. If the computer's random number generator is working properly, we can anticipate some variability around the mean number of 12 for each pair. That is, we anticipate that some pairs will occur 11 times, some 10 times, some 13 times, etc. We would be quite surprised to find that all the numbers in the table were 12; in fact, we would suspect our random number generator to be defective. The distribution of values is expected to follow the Poisson distribution, because we are dealing with discrete occurrences of a rare random event. That is, in the case of Table 18.1, the occurrence of any number pair, "72" for example, should be a random event happening with a chance of 0.01.

Assuming the process is random, the probability of any number N occurring in Table 18.1 may be found with

$$P_N = \frac{m^N e^{-m}}{N!} \tag{18.2}$$

where m is the mean number, in this case 12. If we wished to find the probability of obtaining a value of 6 in Table 18.1, the equation would be solved with

$$P_6 = \frac{12^6 e^{-12}}{6!} \cong 0.02548$$

In Table 18.1, we would expect the value 6 to appear 2 or 3 times.

Beginning with 0 and solving Equation 18.2 for successive values of N will give the expected frequencies of various values appearing in Table 18.1. This equation is easily solved with BASIC because of the following relationship:

$$P_N = \frac{m}{N} P_{N-1}$$

Assuming that the value of m is not so large that rounding error by the computer becomes significant, a series of Poisson probabilities for values of $0, 1, 2, \ldots$, may be generated by the following short BASIC program:

```
10 REM    PROGRAM FOR GENERATING THE FIRST 21 VALUES
20 REM    FOR A POISSON PROBABILITY SERIES
30 REM    P...  PROBABILITY OF OCCURRENCE OF VALUE N
40 REM    M...  MEAN VALUE OF N
50 REM    N...  POISSON VARIABLE
60 REM
100 LET M = 5 :           REM MEAN VALUE IS DEFINED HERE
110 LET P = EXP(-M) :     REM PROBABILITY OF N=0
120 PRINT "N", "PROBABILITY OF N"
130 PRINT "0", P
140 FOR N = 1 TO 20
150    LET P = P * M/N
160    PRINT N, P
170 NEXT N
180 END
```

If probabilities are multiplied by the total number of occurrences, the expected frequencies will be given. The expected and observed frequencies of occurrence may be tested with the usual goodness-of-fit test, and the calculated χ^2 value compared with a tabular value. Frequencies for values lying away from the mean may have to be added together to increase the size of the expected value. The chi-square test requires that expected values be sufficiently large to avoid bias, with some statisticians recommending a minimum of 5; see your favorite statistics book for more details.

The Poisson analysis of the values of Table 18.1 is summarized in Table 18.2. The calculated χ^2 value is 12.90 with 11 d.f., and a probability of about 0.3.

Value	Poisson Expected Probability	Poisson Expected Frequencies	Expected Frequencies (Grouped)	Observed Frequencies (Table 18.1)
0	0.6144×10^{-5}	0.00		
1	0.7373×10^{-4}	0.01		
2	0.4424×10^{-3}	0.04		
3	0.1770×10^{-2}	0.18	4.58	4
4	0.5309×10^{-2}	0.53		
5	0.1274×10^{-1}	1.27		
6	0.2548×10^{-1}	2.55		
7	0.4368×10^{-1}	4.37	4.37	4
8	0.6552×10^{-1}	6.55	6.55	2
9	0.8736×10^{-1}	8.74	8.74	10
10	0.1048	10.48	10.48	15
11	0.1144	11.44	11.44	16
12	0.1144	11.44	11.44	6
13	0.1056	10.56	10.56	13
14	0.9049×10^{-1}	9.05	9.05	7
15	0.7239×10^{-1}	7.24	7.24	7
16	0.5429×10^{-1}	5.43	5.43	8
17	0.3832×10^{-1}	3.83	3.83	4
18	0.2555×10^{-1}	2.56		
19	0.1614×10^{-1}	1.61	6.30	4
20	0.9682×10^{-2}	0.97		
>20	0.1160×10^{-1}	1.16		
Total	1.000	100.01	100.01	100

Table 18.2. Frequency analysis of values from Table 18.1 compared with expectation of randomness based on Poisson distribution.

This Poisson test may find some types of nonrandom behavior of the random number generator that the previous tests will not find. However, it will not detect other types. For example, suppose the values less than 8 in Table 18.1 were all clustered in the upper left corner of the table, and all the values greater than 14 were in the lower right. The Poisson test would not detect this evidently nonrandom arrangement.

Overall, the set of two-digit random numbers of Table 18.1 passes the tests for randomness given here.

Exercise 18-7: Set up a BASIC program to generate and sort 1000 random numbers as in Table 18.1. This can be accomplished readily using a tallying technique which increments by 1 the elements of a 10 by 10 matrix having subscripts corresponding to the first and second digits of the random number. Write your program to calculate the χ^2 value for the overall goodness-of-fit test comparing all the values of your table with the expected value of 10 (99 d.f.).

Exercise 18-8: Modify the program of Exercise 18-7 to calculate the χ^2 value for the following goodness-of-fit tests: row totals, column totals, digit frequency, odd and even frequency, and pairs and non-pairs.

Exercise 18-9: Modify the program of Exercise 18-7 to perform the contingency test (81 d.f.) described in Section 18.6 above.

Exercise 18-10: Modify the program of Exercise 18-7 to calculate a value of χ^2 for the goodness-of-fit test of randomness based on the Poisson expectation of values in your 10 by 10 table.

Conclusion

This chapter has introduced you to the basic concept of employing random numbers generated by the computer to simulate some simple random processes. The models are themselves not of great biological significance because they all deal with simple systems which are already well understood. Their main purpose in this chapter has been to demonstrate the function of the Monte Carlo technique, and to provide numbers for testing the randomness of the BASIC random number generation process. The models presented in the next two chapters will have more utility for students interested in stochastic models of biological processes.

CHAPTER 19

MODELING
OF SAMPLING PROCESSES

Of the many activities carried out as biological research, the one most subject to random variation is the process of sampling. By definition, samples are taken in order to obtain information about an entire population of potential measurements of some variable of interest. In cases where the population is infinite, it is not possible to sample the whole population. If the population is finite, a complete sample is possible theoretically, but it usually is not practical. Therefore, one or more samples are taken so that estimates can be made of population parameters such as the mean, the median, the range or the standard deviation. An estimate of a population parameter is called a statistic. Although statistics will vary among samples taken from a given population, the parameters are constant. Since one seldom has data for an entire population, it is usually impossible to compare sample statistics with population parameters in order to obtain information about their validity. Such a comparison may be readily accomplished, however, by modeling the sampling process using Monte Carlo methods. Much of what we know about various probability distributions such as the chi-square distribution or the normal distribution has resulted from long tedious hours of investigating the nature of random numbers, largely without the aid of the high-speed computer.

The objective of this chapter is to introduce some common methods for modeling a variety of biological sampling processes. Biological variables are of two distinct types, continuous and discrete. A continuous variable is one which has an infinite number of values within its range. Examples of continuous variables include weights, lengths, volumes, rates, concentrations of chemical constituents, and lengths of time periods. Discrete variables may be expressed only as integer values. Examples include the number of organisms in a population, the number of births and deaths, and the number of leaves on a branch. In the previous chapter, most of the models were concerned with discrete events like coin flips, radioactive decay of atoms, and interaction of gametes. The sampling models of this chapter involve both types of variables.

19.1 Sampling a Community of Organisms

Many programs of ecological research require estimates of the number of organisms of different species in biological communities. These data may be used to prepare tabulations of relative composition, or to calculate any of the various indices of biological diversity. In either case, a pertinent question is how large a sample must be collected in order to have a reliable measure of the community composition. One of the problems with real communities is that variability between samples may be the result of a non-uniform environment. Every environment is composed of a mixture of microhabitats which may not be evident to the person doing the sampling. Another complication is that communities are not unlimited in size, so that it is possible to sample a real community out of existence. Simulated sampling provides a means of evaluating the effect of sample size on estimates of population parameters for a model community which is homogeneous and unlimited in size. A simple discrete model of random sampling of a community is described here, and serves to introduce some general techniques of using uniform random numbers to sample events or items having various probabilities.

Assume we have a community with a known fraction of organisms of five different types say A, B, C, D and E. For this simple example, we will assume the fraction of each of the five types is the same, namely 0.2. The fraction of a given type in the community is also the probability that any organism drawn at random from the community is a representative of that type. Therefore, we can simulate the collection of a sample organism from the community by selecting a random number between 0 and 1, and then comparing the random number with the cumulative probabilities of each type. For example, if a uniform random number R is less than 0.2, then we shall assume it represents the collection of an organism of Type A. If R falls between 0.2 and 0.4, then we shall assume it represents collecting an organism of Type B. If R is between 0.4 and 0.6 then it will represent collecting an organism of Type C, and so forth. In this way a single uniform random number can be made to represent one of the 5 types of organisms.

By repeatedly selecting a random number and finding the organism that it represents, a simulated sample of organisms may be collected. The program for sampling may be written to include a tally of the numbers of organisms of each type in the simulated sample. When the sampling process is completed, various derived sample statistics may be calculated and compared with the known population parameters.

Exercise 19-1: Use the technique described above to simulate the sampling of a hypothetical biological community composed of 8 species present in the following proportions:

Species A: 0.495   Species B: 0.255   Species C: 0.124   Species D: 0.061

Species E: 0.032   Species F: 0.016   Species G: 0.009   Species H: 0.008

Simulate the collection of 4 samples. Each sample should consist of 200 organisms. Write your program so that the output consists of a list for each species showing the number collected in each sample, the relative frequency of the species in each sample (frequency = number collected/200), and an estimate of frequency from the average of the four samples.

19.2 Sample Size and an Index of Community Diversity

Various indices of species diversity have been used often to compare one biological community with another. Most of these indices are attempts to provide a single parameter to characterize both the number of species and the relative abundance of each species in a community. The Shannon-Weaver diversity index, which has been used frequently, is based on the equation

$$H = -\sum_{i=1}^{s} p_i \ln p_i \qquad (19.1)$$

where H is the estimate of community diversity, p_i is the proportion of the total number of individuals making up the ith species, and s is the total number of species. Although the value of a Shannon-Weaver index is not directly dependent upon sample size, its accuracy is improved by increasing sample size. Pielou (1977) provides a thorough, valuable discussion of the theory and application of indices of diversity.

A question that frequently arises when calculating diversity is: How large a sample must be taken to obtain an accurate measure of diversity? This question may be examined by simulating the sampling process on a community that previously has been characterized carefully.

Exercise 19-2: Use the sampling simulation to determine the effect of sample size on the variability of the diversity index of Equation 19.1. Write your program to collect samples of sizes 25, 50, 100, 200, and 400 individuals. Simulate the collection of samples from a community with a fixed, known composition. For each sample size, collect 6 samples. Calculate the diversity index of each sample. Print out your results as a table or graph to show the reduction in variability of estimated H as sample size increases. Find the parameter value of H for the population as a whole to be used for comparison. The algae listed below should be used as the community to be sampled, with the numbers of cells serving as parameter values for your simulation.

No.	Taxon	number ml^{-1}	Frequency
1	*Cyclotella stelligera*	554	0.3506
2	*Fragilaria crotonensis*	265	0.1677
3	*Tabellaria fenestrata*	197	0.1247
4	*Melosira granulata*	179	0.1133
5	*Cyclotella michiganiana*	115	0.0728
6	*Dinobryon divergens*	107	0.0677
7	*Cyclotella meneghiniana*	85	0.0538
8	Unidentified flagellates	70	0.0443
9	*Anabaena flos-aquae*	8	0.0051
	Totals	1580	1.000

(These data show dominant taxa of algae in Lake Michigan near Grand Haven, 25 July 1969, from Schelske et al. (1971).)

19.3 Spatial Distribution of Organisms

The distribution of organisms in space may be simulated by generating uniform random numbers to define the coordinates of location for each individual. The technique can be used to locate them in one-dimensional space (e.g. on a transect through a community), two-dimensional space (e.g. trees in a forest) or three-dimensional space (e.g. plankton in a portion of ocean). Locating organisms in this way assumes that each is completely independent of the others, neither attracted nor repelled by them.

In practice, the two-dimensional simulation may proceed by generating a random number to set an organism's location on the x-axis, and then generating another number to set the location on the y-axis. The location of each organism may be depicted graphically, for example on the video or plotter output of the computer. The simulation data may be analyzed by dividing the "space" into a matrix of discrete units, and finding how many organisms are located in each unit. (An almost identical procedure was carried out in Section 18.6, for the test of the first two digits of a set of random numbers. The difference is that in Section 18.6, the initial "location" was not continuous and was determined by a single random number rather than two.)

The location of individuals generated by the random numbers should be random. Conventionally, their locations are tested for randomness using the Poisson test outlined in Section 18.6.

Exercise 19-3: Write a computer program to simulate the random placement of 1000 individuals in a two-dimensional space, measuring 10 by 10 units. Use the graphics screen to observe the locations of the simulated organisms. As the positions of the organisms are being calculated, your program should convert the values of the x- and y-axis locations to integers which may be summed and stored in a 10 by 10 matrix. After all of the simulated organisms are located on the screen and stored in the matrix, your program should find the number of spaces having $0, 1, 2, 3, \ldots, 25$ organisms, and store these numbers in an array. This array is a frequency distribution which should be tested against the expected frequency distribution obtained from the Poisson distribution (Equation 18.2) with a mean of 10. The procedures shown in Section 18.6 apply directly to this exercise.

The real challenge of this exercise is to write the simulation and the analysis into the same program.

19.4 Random Sampling from the Poisson Distribution

Because the Poisson distribution describes the random location of individual events or objects, is used quite frequently in a variety of simulation studies. Several techniques have been developed to sample from the distribution. For example, Wiebe (1971) used a technique of finding and summing Poisson probabilities, beginning with P_0, until the sum of the probabilities was greater than some uniform random number generated previously for comparison.

A more efficient technique uses an algorithm given by Knuth (1981), given in the flowchart of Figure 19.1.

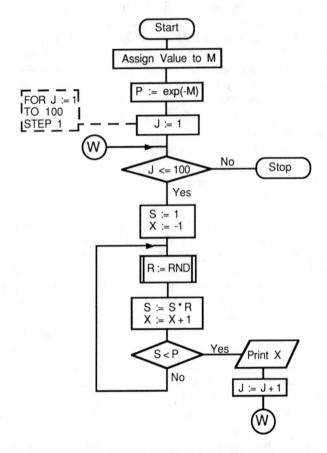

Figure 19.1. Flowchart for a BASIC program to generate 100 numbers drawn randomly from a Poisson distribution. M is the mean of the distribution being sampled, X is the Poisson variate.

With lengthy simulations a third method, the table-lookup technique, is slightly more rapid than either of the methods above, and may save considerable time. The number of calculations is minimized. The trick is to set up an array of 1000 or more subscripted numbers, and then assign digits $0, 1, 2, \ldots$, to the array with the number of 0's assigned in proportion to the Poisson probability of 0, the number of 1's proportional to the probability of 1's, etc. A 3-digit random number can be used to examine the content of the corresponding element of the array, and produce a Poisson variate. The method is extremely rapid and useful for lengthy simulations (Keen and Nassar 1981).

Exercise 19-3: Write a program to simulate the random germination of radish seeds in a garden. Assume a mean germination time of 7.0 days for 1000 seeds, all of which germinate. Use the Poisson number generator of Figure 19.1 to find the day of germination for each seed. Graph the resulting simulation data as a histogram, showing numbers of seeds germinating for the sequence of days after planting.

19.5 Random Sampling from a Normally-Distributed Population

The normal or Gaussian distribution is a continuous, symmetrical distribution, which may be fully described by a mean and a standard deviation. This sampling distribution is used frequently in stochastic biological simulation to obtain measurements such as height, weight, and concentrations of chemical components. These and similar measurements can be expected to be distributed normally in many situations.

There are numerous techniques that serve to generate random, normal variates with a computer's random number generator. One of the most efficient is the Box-Muller algorithm (Swartzman and Kaluzny 1987), given in the flowchart of Figure 19.2. The process generates two independent normally distributed random numbers, drawn from the "standard" normal distribution with a mean of zero and a standard deviation of one.

A number drawn from a standard normal distribution may be transformed to a variable from any other normal distribution using the following equation:

$$N_R = N_s\sigma + \mu \qquad (19.3)$$

where N_s is the number from the standard normal distribution, σ is the standard deviation and μ is the mean of the new normal distribution.

Exercise 19-5: Use the procedure diagrammed in Figure 19.2 to generate a sample of 2000 normally distributed random numbers, having a mean of 50 and a standard deviation of 10. Write your program so that it calculates and prints out the mean and standard deviation of the sample; this will provide a method of checking on the validity of the method. Round each of the generated numbers to the nearest whole integer, and plot the frequency of the integers with a histogram.

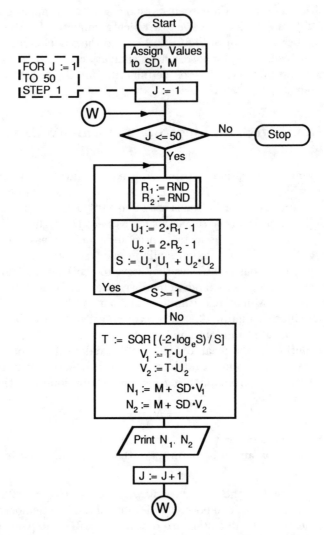

Figure 19.2. Flowchart for generating 100 numbers drawn as a random sample from a normally-distributed population. The population has a mean M and a standard deviation SD.

19.6 Random Sampling from the Exponential Distribution

The exponential distribution is a continuous distribution in which the origin is the start of the frequency curve. This is the frequency distribution which resulted in the exponential radioactive decay curve discussed in Section 18.3. Random distributions of this type characterize processes which have a certain probability of failure. For example, it may be used to describe the lifetimes of organisms having a constant probability of death, i.e. the Type B survival shown in Figure 9.1.

One way to generate this distribution is to set a probability of death or failure or decay, and then generate a series of uniform random numbers, checking each number against the probability until one random number is less than the probability. The count of the random numbers generated is then equal to the lifetime being simulated. This was the technique discussed in Section 18.3.

As an alternative approach, the following equation may be used to convert a single uniform random number between 0 and 1 into an exponentially distributed random number:

$$N_e = -\mu \log(R) \qquad (19.4)$$

where N_e is the exponentially distributed random number, μ is the mean of the exponential distribution, and R is the uniform random number.

Exercise 19-6: Assume a certain species of ornamental carp follow an exponential survival curve as adults, and that the mean survival in a particular Japanese temple pond is 14.3 years. Use Equation 19.4 to simulate the lifetimes of 200 adult carp placed in this pool in 1910. Plot number of carp remaining in the pool as a function of time since 1910. According to your simulation, would any of them still be alive?

19.7 Simulation of Simple Mark and Recapture Methods

Mark and recapture methods are often used to estimate parameters of animal populations, including size, birth and death rates, migration rates, etc. They are particularly useful where other, direct census methods cannot be used, particularly in the case of fish, or migratory birds. We will introduce the fundamental ideas behind the method and outline a simulation based on these. You should be aware that there has been much recent research on these methods, and that the models discussed here are obsolete; however, they are useful starting points for understanding current techniques. An introduction to some of the recent methods can be found in Otis et al. (1978), Seber (1982), and Pollock (1981).

The mark and recapture method of estimating populations of organisms involves taking a moderately large sample, marking the sampled organisms in some way, and returning them to the original group. At some subsequent time, a second sample is collected and inspected for the presence of marked individuals. The Peterson equation for estimating the size of the population is based on a simple proportion:

$$P_p = \frac{SM}{R} \qquad (19.5)$$

where P_p is the Peterson estimate of population size, S is the number of organisms in the sample, M is the number of marked individuals in the population. R is the number of "recaptures", or marked organisms that appeared in the second sample. It is possible to repeat the determination with several samples; on the occasion of each sampling, the unmarked individuals in the sample are marked and returned to the population, increasing the value of M.

This estimating procedure may be simulated by using Monte Carlo techniques. The simulation begins after some marked organisms have been released into the population. The total population size, the number of marked organisms, and the sample size are the necessary inputs. The sample size is assumed to be fixed for a series of samples, although the simulation is easily modified to accept a randomly varying sample size. The simulation proceeds with the establishment of the ratio between the number of marked organisms in the sample and the total population. The collection of a sample is simulated by generating a series of uniform random numbers between 0 and 1; each number is checked in turn against the ratio. If the number is less than the ratio, the organism is considered to be marked, and if greater it is unmarked. During this process, the ratio of marked to unmarked organisms must be adjusted to reflect changes in the ratio due to sampling removal. The process of checking random numbers is repeated a number of times equivalent to the sample size. At the end of the simulated sample collection, the size of the population may be estimated with the Peterson equation above. The sampling procedure may be repeated, after the unmarked organisms are passed to the marked category, and the sample returned to the population. The flowchart for this simulation is shown in Figure 19.3.

In addition to the Peterson estimate, two other equations are useful. The first is the Schnabel estimator:

$$P_S = \frac{\Sigma(SM)}{\Sigma R} \qquad (19.6)$$

The second is the Schumacher-Eschmeyer estimator:

$$P_{SE} = \frac{\Sigma\left(M^2 S\right)}{\Sigma(MR)} \qquad (19.7)$$

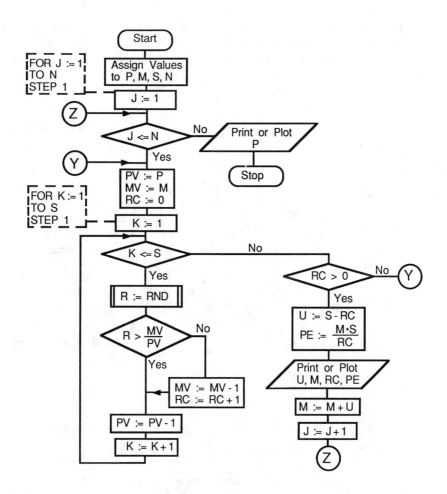

Figure 19.3. Flowchart for stochastic simulation of a mark-and-recapture process using the Peterson index.

P is the population size; a constant during a simulation.

M is the number marked when simulation begins; a constant.

N is number of samples to be collected; a constant.

S is number of organisms to be collected; a constant.

PV is population size; a variable used in sampling procedure.

MV is number of marked organisms in the population; a variable.

U is the number of unmarked organisms in a sample.

PE is the estimated size of the population.

RC is number of marked organisms recaptured in a sample.

Both of these estimates have been proposed to obtain better estimates where a series of samples has been taken from the population. The estimates of these methods become increasingly accurate as the number of samples increases.

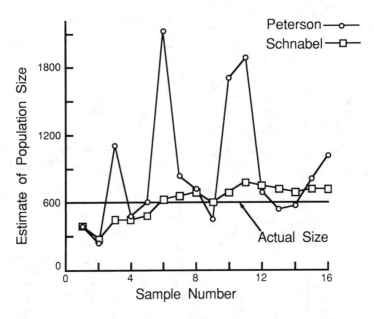

Figure 19.4. Graph of data from mark-and-recapture simulation, based on flowchart of Figure 19.3.

Exercise 19-7: Use the information given above and the flowchart to write a program for the mark and recapture simulation. Set up your simulation with a population size of more than 1000 individuals, with about 5 percent of the individuals already marked. Use a sample size of between 20 and 50. Repeat the sampling a minimum of 10 times. Plot the population estimates against sample number as in Figure 19.4. Also calculate and plot either the Schumacher-Eschmeyer or Schnabel estimate for the same series of samples. Then answer the following question: What assumptions are implied in this simulated estimating procedure which are probably rarely true of real populations?

Conclusion

This chapter has presented a number of Monte Carlo methods for working with different aspects of the sampling process. The techniques that you have learned here for generating specific sampling distributions will be used in subsequent chapters.

CHAPTER 20

RANDOM WALKS
AND RELATED
STOCHASTIC PROCESSES

There are many processes in biology which involve discrete changes in some variable that occur as a result of random fluctuations. These stochastic processes result in the condition of the variable being definable only with a probability statement. In a random walk, the direction of movement is given with a statement of probability, so that the movement between any two points in a given number of movements is stated as a probability. The random walk and similar stochastic processes are characterized by a "lack of memory"; that is, the direction taken toward future states or locations depends only on the current state or location, and not on previous situations. In this chapter we will discuss several biological processes that can be effectively simulated by the random walk process. Additional information on the theory of these processes may be found in Olinick (1978), Feller (1968), Berg (1983), and other books on probability theory.

20.1 The Classic Random Walk Model

The random walk model is sometimes called a "drunkard's walk". Both names are derived from theoretical problems suggested by an inebriated person who has an equal probability of staggering either one step to the left or one step to the right, and perhaps either backwards or forwards. Once in the new position, the same probability applies to the next step. A variety of important biological models are based on the random walk.

A typical random walk pattern is obtained if we keep track of the total number of heads and tails that occur as we flip a coin repeatedly. At each flip, the coin has the same probability of coming up heads or tails. In any series of flips, sometimes there are more heads than tails, and sometimes more tails than heads. In a typical run, the lead (difference between total numbers of heads and tails) may change repeatedly. What is surprising is that in a very long series of flips, the lead may depart a rather long distance from the point of equality of numbers of heads and tails (Figure 20.1). This occurs because no matter how far from equality the lead may be, it has just as great a chance of moving further away from equality as moving towards it. At any time, there is only a 50 percent chance of moving toward equality.

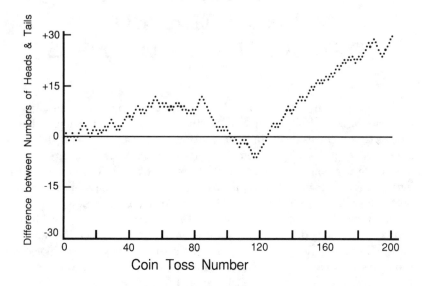

Figure 20.1. An example of random walk results from a coin-toss simulation.

An interesting experiment which can be conducted using a coin toss simulation involves counting the number of times the difference between heads and tails equals zero during a long series of tosses. Another interesting statistic to obtain during such an experiment is the toss number of the last crossover (change in lead) before the end of the series. This number can be expressed as the ratio of the last crossover, K, to the total number of tosses, N. It can be shown that the last crossover is just as likely to come in the first half of the series as in the last half. Thus, the average value of K/N for many runs would be close to 0.5, as expected. Where it might occur on any given run is not predictable.

Coin flipping experiments may seem rather unrelated to biological processes, but the principle illustrated here has many applications to biologically significant processes such as diffusion, spatial movement of organisms, changes in gene frequencies, population growth, and most situations which involve a probabilistic outcome. Feller (1968) discusses further some details about the classic random walk simulation.

Exercise 20-1: Write and implement a program for the classical coin flipping simulation described above. Set it up to simulate 10 series of 1000 coin tosses per series. Plot the data for the first two series relative to a zero line as shown in Figure 20.1. Have the computer keep track of the total number of crossovers, and the toss number of the last crossover. Print out these results in a short table, along with a calculation of the average number of the last crossover. (Note that a crossover is not defined merely by a zero difference between cumulative heads and tails, but by a shift from a head-to-tail lead or vice versa.)

20.2 Diffusion

The motion of particles in a liquid or gas medium is caused by thermal energy. The average instantaneous velocity of a particle or molecule can be quite high. For example, a molecule with a molecular weight of 1000 has an average velocity of 5000 cm sec^{-1} at 30°C. Such a molecule would obviously go places rapidly, except that it collides many times a second with the molecules of the liquid in which it is suspended. Because of these random collisions, the particle meanders about in the medium in a random walk. If a group of such molecules were followed from some initial identical starting point, they would all wander about, slowly getting farther apart, and away from the initial location. In a simple diffusion process like this, the average distance the particles move from the starting point is zero. That is, as many particles go left as go right, and up as down. Most importantly from the viewpoint of a study of random diffusion, the average absolute distance moved from the original location is a function of the square root of the time since departure.

Exercise 20-2: Use the Monte Carlo method to simulate the random movement of a particle in two-dimensional space, that is, Brownian movement or a random walk. One of the simplest ways to approach this is to generate a random number between 0 and 1, and subtract 0.5 from it to obtain a number between +0.5 and -0.5. Let this number equal the distance a particle moves on the x-axis from its present position. Then generate a second random number and subtract 0.5 to obtain the movement on the y-axis. Adding the two generated numbers to the particle's location will give the coordinates for its next location. Doing this repeatedly will produce the consecutive coordinates of the particle's location. Connecting the points with straight lines will provide a typical two-dimensional random walk pattern. Plot out the random movement of a particle beginning at a central location on a graph of coordinates. There is some bias in the simulation because distance moved each time is drawn from a uniform distribution between -0.5 and +0.5, rather than a normal distribution.

Exercise 20-3: Modify the preceding simulation to plot the final positions of 100 particles after each has moved randomly 50 times from a common point source. This is an approximate simulation of the diffusion process. Calculate the average distance moved by the particles from the source, using the formula for the straight-line distance from a point:

$$d = (x^2 + y^2)^{1/2}$$

20.3 A Direct Approach to Population Growth as a Stochastic Process

In this section we will outline a direct approach for examining the population effects of randomizing the processes of birth and death. The translation from the usual deterministic approaches used in Chapters 7 through 9 is made easier if the time interval is short enough so that the per capita rates of birth and death are both less than one. With per capita rates less than one, the birth rate may also be considered the probability that a given individual will give birth to a new individual during one interval of time. In the same way, the death rate can be thought of as the probability that a given individual will die during the time interval. The population is assumed to reproduce continuously, rather than seasonally.

We will begin with the simplifying assumption that the probabilities of birth and death are independent of each other, and are independent of age and sex. Thus, we are working under many of the same assumptions made for the models of homogeneous populations in Chapter 7. Based on these assumptions, the probability P_+ that any given individual will cause the population to increase in size by one individual is given by the equation

$$P_+ = b(1 - d) \tag{20.1}$$

where b is the birth rate and d is the death rate. This also may be viewed as the probability that an individual will cause a birth, but not a compensating death.

The probability that a given individual will cause the population to decrease by one, P_-, is defined by the equation

$$P_- = d(1 - b) \tag{20.2}$$

This also may be considered to be the probability of dying without producing a compensating birth.

The probability that an individual will cause no change in the size of the population, P_o, is

$$P_o = 1 - P_+ - P_- \tag{20.3}$$

After substitution this becomes

$$P_o = db + (1 - b)(1 - d) \tag{20.4}$$

Thus, the population will remain unchanged in size as a result of individuals that (1) neither die nor give birth during a time interval, or (2) both give birth and die.

To simulate the population activities of one individual over one unit of time, a uniform random number between 0 and 1 is generated. The random number is compared with P_+, P_-, and P_o (as in the sampling exercises of Chapter 19), to decide whether the individual causes an increase in population size. This process is repeated for each individual in the population. The number of increases minus the number of decreases is added to the previous population size to give the population size for the start of the next time interval. Applying this procedure over several intervals of time produces a simple Monte Carlo simulation of population growth.

Exercise 20-4: Write a program for the simple stochastic simulation of population growth described above. Set birth rate $b = 0.30$ and death rate $d = 0.27$. Output should consist of log population size plotted over time. Beginning with a population size of 10, allow the simulation to run for 150 intervals of time in order to observe the behavior of the model. The population size may drop to zero; if this extinction occurs, restart your simulation. The simulation data plotted on a logarithmic scale should approximate a line with a slope of $b - d$.

Exercise 20-5: Modify the simulation of Exercise 20-4 above so that the simulation of growth may be repeated several times. Write your program so that the size of each population is recorded at times 5, 10, 20, 40, and 80 units. The numbers for any population that drops to extinction should be discarded. After you have obtained data for 10 populations, have the computer calculate the mean and the variance of population size of the ten populations at each of the above time units. Note how the variance of population size increases with time.

Use the following equation from Poole (1974) to calculate the predicted value for the variance at each of these time intervals:

$$\text{Var}\,(N_t) = N_0 \frac{b+d}{b-d}\left(e^{t(b-d)} - 1\right)\left(e^{t(b-d)}\right) \tag{20.5}$$

How closely do the simulated variances approach those predicted by Equation 20.5?

20.4 A Probabilistic Model for Growth Based on Variance Estimates

Another approach employed in the development of stochastic models of population growth uses calculated values for variance to set up the size of random population fluctuations. The technique as described by Poole (1974) makes use of variance calculations for pure birth and death process which are discussed in detail by Goel and Richter-Dyn (1974).

The expected or most probable estimate of population size in moving from time t to time $t+1$ is

$$E\,(N_{t+1}) = N_t e^{(b-d)} \tag{20.6}$$

where b and d are defined as before. This expected value is the mean that would result from a very large number of trials using random numbers as in the section above. It is also the value given by the equation for exponential growth (Equation 1.6). The variance of this estimate is obtained from Equation 20.5 above, setting $t = 1$:

$$\text{Var}\,(N_{t+1}) = N_t \frac{b+d}{b-d}\left(e^{(b-d)} - 1\right)\left(e^{(b-d)}\right) \tag{20.7}$$

The standard deviation for the estimate of N_{t+1} is

$$S_N = \sqrt{\text{Var}\,(N_{t+1})} \tag{20.8}$$

Poole (1974) outlines the following technique for developing a Monte Carlo simulation of population growth based on calculations of variance:

(1) Calculate the expected value for N_{t+1} using the values of N_t, b and d.

(2) Calculate the variance of the estimate using Equation 20.6, and the standard deviation using Equation 20.8.

(3) Use the Box-Muller algorithm (Figure 19.2) to generate a random variate, R_n, from a standard normal distribution having a mean of zero and a standard deviation of one.

(4) Use the random normal variate from (3), the standard deviation from (2), and the expected value from (1) to produce a value for the next population size with the following equation:

$$N_{t+1} = \text{E}\,(N_{t+1}) + R_n S_N \qquad (20.9)$$

Exercise 20-6: Use the technique outlined above to develop a stochastic simulation of population growth. Begin with a population of size 10, and set $b = 0.30$ and $d = 0.27$. Run the simulation over a period of 150 time units, plotting log of population size, as in Exercise 20-4. You will have to set a trap for population sizes less than zero, since these are possible with this method. Note that all the terms to the right of N_t in Equation 20.7 are constants, so that calculations in your program may be simplified by using an equation of the form

$$S_N = \sqrt{K_N N_t} \qquad (20.10)$$

20.5 Genetic Drift

Existence of a large population was one of the requirements necessary for maintaining the Hardy-Weinberg genetic equilibrium discussed in Chapter 10. If populations are small, then significant changes from generation to generation can occur in the frequencies of alleles. These changes result in a random walk of gene frequencies known as genetic drift. Drift can be so extreme that it results in the fixation of one allele and elimination of other alleles for a given locus. For a simple A/a model system, fixation can be defined formally as the condition in which the frequency q of gene a becomes 1 or 0. In smaller populations the random changes will be larger, and fixation will occur more frequently than in larger populations.

These random walks of gene frequency may be simulated using uniform random numbers to determine the genetic contribution from the parents of each offspring. The simulation may be simplified by the following conditions:

1) The simulation will be limited to a consideration of a single locus involving only two alleles. We will follow Chapter 10 in labelling them A and a, with frequencies p and q.

2) We will also follow Chapter 10 in labelling the three possible genotypes as AA, Aa, and aa. Their frequencies in each generation will
have the corresponding labels D, H, and R.

3) There is no selection against either allele operating in the population.

4) Generations will be discrete, so that the genotypes of a given generation will be determined only from the genotypes of the previous
generation.

5) The size N of the population is constant, with the same number
of offspring being produced each generation.

6) Gene and genotype frequencies will be the same for males and
females.

7) Mating is random, so that the probability of a given gene coming
from either parent is equal to the frequency of that gene in the
population.

The simulation is set up to determine the genotype of each offspring by
sampling from the population of available genes just as in the simulation
of monohybrid crosses of Section 18.4. The gene donated by the first
parent of a given offspring is determined by generating a uniform random
number between 0 and 1. If the number is less than p, the first parent will
contribute one A allele to the offspring. Otherwise, it will contribute the
a allele. The process is repeated for the gene from the second parent. As
the genotypes are determined, they may be summed in the same way as in
Exercise 18-5. After all N offspring have been produced, the genotypes are
used to calculate the values of p and q for the next generation, following
the normal procedures of Chapter 10 (Figure 20.2).

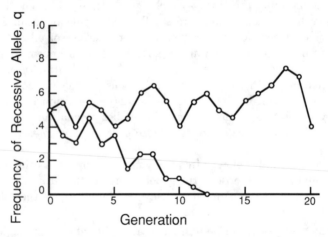

Figure 20.2. Sample results from a simulation of genetic drift in
a small population.

Exercise 20-7: Use the information given above to develop a Monte Carlo simulation of genetic drift. Write the program so that it considers populations with size N of 10, 20 and 40. For each population size, make five simulation runs, plotting the value of q, recessive gene frequency, against generation number. Begin each run with $N/2$ individuals with the AA genotype, and $N/2$ individuals with the aa genotype; there will be zero individuals initially with the Aa genotype.

Your program should keep track of the number of generations required for fixation for each simulation run, and determine the average number of time intervals to fixation for each population size. Genetic theory predicts mean time to fixation for any locus to be about $4N$ generations. If you have some combination of a fast computer, a compiled version of BASIC, and a lot of spare time, you might try a more elaborate simulation, making more than 5 runs per population size N, and also attempting some larger values of N.

20.6 Stochastic Simulation of Weather

Chapters 11 and 12 described some models of weather that are important as input to many ecological and physiological models. The deterministic models presented in those chapters ignored variability in weather patterns. The day-to-day variation characteristic of weather can be more important for some models than the overall annual patterns. Weather variations can be introduced easily into these models using stochastic techniques.

As a simple model of stochastic variation in weather, we will consider the effect of cloud cover on light intensity. For a hypothetical location, assume that most storm fronts arrive every five days. About 10 percent of the storm fronts will arrive four days after the arrival of the previous front. About 20 percent of the storms arrive 6 days after the arrival of the previous front. We can assume that 6 days is the maximum interval between arrivals. The storm front brings with it some cloud cover of varying density and duration. Assume that the cloud cover will last up to three days, with duration equally distributed among the possibilities of 1, 2, or 3 days. For any given day with cloud cover, assume that the cover will absorb between 35 and 85 percent of the light that would otherwise strike the ground; any percentage in this range will have an equal chance of occurring.

With the selection of three random numbers to determine the arrival times of fronts, duration of cloud cover, and percentage light absorption, a workable simulation of light variability can be produced.

Exercise 20-8: Using the assumptions and values in the discussion above, write a program to simulate the variability in light intensity for a location at 45°N latitude during the months of June, July and August (Julian days 152 through 243). To model maximum light intensity use Equation 11.1, assuming that the vernal equinox falls on day 79. Your output should consist of a plot of daily light intensity over the three months.

20.7 Simulation of Natural Selection by Random Processes

Because of the obvious ways that the form and function of organisms are adapted to their environments, it appears that they have been "designed". Critics of natural selection as an evolutionary process repeatedly have stated that random, probabilistic processes such as mutation and selection cannot lead to a designed organism. Common criticisms include such statements as "A functioning eye appearing by chance mechanisms has the same probability as coming up with a dictionary after dynamiting a print shop." Dawkins (1986) outlined an instructive simulation based on Monte Carlo sampling that demonstrates a method for random and chance events to produce a "designed" endpoint. The model is a sort of reverse random walk.

The simulation proceeds in two stages. The first stage is to mimic a print-shop explosion as a method for producing a phrase in English. While there are many ways to do this, the following method also is modified easily to make up the second stage:

(1) Assign each letter of the alphabet to one element of a BASIC string array with subscripts numbered 0 through 25; for example, Z$(0)="A", Z$(1)="B", ..., Z$(25)="Z". (This assignment may be done with READ and DATA statements.) A single space and any other punctuation needed in your target phrase (see next step) may be assigned as elements to subscripts numbered 26, 27, etc.

(2) Select a "target phrase" consisting of a meaningful set of a few words. Assign the numerical value of each letter of this phrase in sequence to a subscripted array. This array will consist of numbers. A phrase of about 20 to 30 letters and spaces is convenient. Suppose that you pick the phrase "BASIC BIOLOGICAL MODELING IS FUN". The subscripted array of numbers that corresponds to this phrase would be set up, for example, as $T(1)=1$, $T(2)=0$, $T(3)=18$, ..., $T(32)=13$. This array can be set up for example with a READ-DATA sequence. (It might also be set up with a FOR-NEXT loop using BASIC's string-handling ability.)

(3) With a short random sampling routine, select a series (32 in this example) of uniform random numbers from the range 0 to 26. Assign these as elements to another subscripted array. For example, B(1)=17, B(2)=3, B(3)=21, etc., where 17, 3, and 21 are the first three of the 32 random numbers.

(4) Print out the English equivalent of this random trial, by printing the element of the string array from Step (1) whose subscript corresponds to the element of the array in Step (3). In the example here, the computer would print out Z$(17), Z$(3), Z$(21), etc., or "PDV..."

(5) Compare the sequence of random numbers in the subscripted array with the sequence in the array for the target phrase. To do this, the program would check whether B(1)=T(1), B(2)=T(2), ..., B(32)=T(32).

(6) If all the elements of the two arrays do not match, then the program should proceed by going back to step (3). If the phrases do match letter for letter, then the program should stop.

Each of the steps (3) through (5) may be accomplished with a short FOR-NEXT loop. These may be included in a larger loop that includes steps (3) to (6).

It is helpful if a counter for this larger loop is included. After the counter reaches several hundred, the modeler usually gets tired of looking at random combinations of letters. Rarely will the target phrase be reached by chance selection of letters; this is not surprising. There is only a 1/27 chance of getting the correct first letter with a random draw, a $1/(27^2)$ chance of the first two letters occurring correctly, and a $1/(27^{32})$ chance of all 32 letters appearing in correct order.

Dawkins pointed out that natural selection does not work by completely random trials, and suggested that the steps above be modified to include a simulation of reproduction. The second stage of the simulation includes reproduction and should proceed as follows:

Steps (1) through (5) are carried out as above.

(6) If the number sequences of the target phrase and the randomly selected phrase do not match, then the random sequence "reproduces" itself.

(7) Reproduction is carried out by producing copies of the array as "offspring", but with a small probability of error in the process. This introduction of error is a simulation of mutation processes.

(8) The number sequence for each of the "offspring" copies is compared to the sequence of the target phrase, with a score attached to each offspring indicating the degree of resemblance to the target.

(9) If the resemblance between at least one offspring and the target sequence is not complete, then the offspring with the highest score

is selected for further reproduction. This step simulates the process of natural selection. The simulation proceeds with step (7) after printing out the letter equivalent of the favored offspring, and the value of a loop counter for the (7)-(9) generation loop.

Thus, the initial random set of letters is altered randomly each generation, but the elements that resemble the target phrase are retained from one generation to the next. Depending upon the number of offspring per generation and the procedure used for mutation, the random phrase will be altered more or less rapidly and come to resemble the target phrase, simulating the action of natural selection. The use of a target phrase does not resemble the action of natural selection, which has no apparent long-term goal. In this simulation it serves merely as an arbitrary measure of fitness for each generation.

Exercise 20-9: Write and implement the two BASIC programs for the simulation outlined above. In the second program, limit repro-duction to 10 "offspring" per "generation". Each of the offspring should differ from the "parent" by one randomly selected letter at one random location. With only 10 offspring, it is possible that all of them might be less like the target phrase than the parent; this can increase considerably the time to obtain a copy of the target phrase. The rate of change may be increased by ensuring that one of the offspring exactly resembles the parent. It can be increased even more by limiting mutation to those elements of the array that do not resemble the target phrase. In this latter case, which you should use in your simulation, a target phrase of 25 letters should be reached in approximately 60 generations. Your output should consist of a printout of several random letter sets from the first pro-gram. For the second program, produce a printout of the selected offspring in each generation leading to the target phrase.

Conclusion

The ideas associated with random walks are important across a range of biological fields from genetics and microbiology through ecology. In some of these fields, the simulations that are based on random walks become quite complex and may require considerable time to accomplish even on large, fast computers. The fundamental concept of departure from deterministic expectation needs to be kept in mind constantly by persons performing research and evaluating data and results.

CHAPTER 21
MARKOV CHAIN SIMULATIONS
IN BIOLOGY

A Markov process is a general type of stochastic process. The random walks described in the previous chapter are one type of Markov process. In this chapter we will consider Markov chains. They are defined much like random walks, but the methods of simulation and programming are based on a different approach. Formally, Markov chains are stochastic processes that have a finite number of discrete states. The probability of being in any state at a given time depends only on the state in the previous time interval, and the probabilities of moving from that state to other states. Markov processes are named for the 19-century Russian mathematician who first described them. They have been applied across most fields of biological study.

21.1 A Simple Markov Chain

A simple example may help in introducing and understanding Markov chains. Suppose you are studying an individual predator with three species of prey, for example a single lion that eats zebra, gnu, and kudu. We will call these prey types A, B, and C. (At this point, you could produce a lion predation simulation using Monte Carlo sampling of the three prey based on their abundance, as in Chapter 19. This would not be a Markov process, however.) If you spent a long period recording the sequence of prey that the lion eats, you would have data that look like this: BAACBBCACAABAA.... These data can be used as the basis for modeling lion predation with a Markov process. One looks at the current prey type (B, for example) and calculates the probability that the next prey item will be an A, another B, or a C. These three probabilities must sum to 1.

If you assume that the probability of eating a B (or C or A) after an A does not depend on what preceded the A, then the predation process can be considered Markovian. That is, the next state depends only on the current state and the probabilities of moving to the next state. A diagram that shows the different combinations of current and next prey types may be constructed as in Figure 21.1.

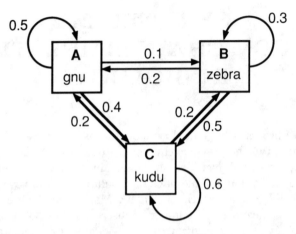

Figure 21.1. Transition diagram for a Markov chain simulation of a lion preying on three different types of prey.

The process of taking a sequence of prey can be simulated in a straight-forward manner. The sequence is started with any one of the prey types. Based on the probabilities of moving from the current state to the next state, selecting a uniform random number will determine which one of the possible states follows the current one. From the new state, a random number is selected to find the next state, and so on. A wide range of biological processes has been simulated using the basic Markov chain model.

Exercise 21-1: Write and implement a program to simulate predator-prey interactions, based on the diagram of Figure 21.1. Your output should consist of a series of 1000 letters showing the sequence of prey items taken. In addition, print out the tally of the total numbers of each prey type selected by the predator.

21.2 An Enzyme–Substrate Interaction

In a simple, idealized enzyme-substrate system (see Section 2.2), the enzyme may exist either in the uncombined form or in the combined enzyme-substrate complex. The enzyme may be thought of as having two possible Markov states (Bartholomay 1964). The enzyme will have stated probabilities of moving from one form to the other, as diagrammed in Figure 21.2. As shown, there are five transition probabilities associated with the system:

   a) P_S is the probability that the enzyme will interact with any one of S substrate molecules to form ES during one time interval.
   b) P_2 is the probability that the enzyme will remain as a free enzyme molecule during one time interval.
   c) P_3 is the probability that the enzyme-substrate compound will become uncombined to yield free substrate and free enzyme during one time interval.
   d) P_4 is the probability that the enzyme-substrate compound will remain combined during one time interval.
   e) P_5 is the probability that the enzyme-substrate combination will break down to yield product and free enzyme during one time interval.

In this system, $P_S + P_2$ must be unity, and $P_3 + P_4 + P_5$ must likewise equal unity.

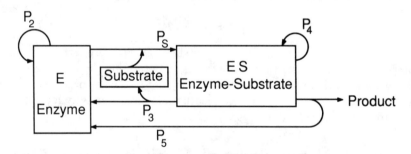

Figure 21.2. Diagram for a Markov chain simulation of an enzyme-substrate system, showing various transition probabilities.

The probabilities P_3, P_4, and P_5 are all constant. P_S and P_2 will vary with the concentration of substrate. In this system, substrate concentration is considered to be the number of substrate molecules that can interact with the enzyme molecule. This number is important because the response of enzyme to varying substrate concentration is the primary behavior of interest here.

With the path from ES to E split into two, this system involves a variation on the Markov process introduced above in Section 21.1. Along one path the breakdown of the enzyme-substrate compound yields a product, and along the other it does not. Although the biochemical difference is significant, this division creates few difficulties for analyzing the process.

Probabilities P_S and P_2 are constant for the duration of any given simulation, but vary with the number of substrate molecules, S. For a single unit of time, P_1 is the probability that a single enzyme molecule will react with a single substrate molecule. So, $1 - P_1$ is the probability that the molecule will not react. If the molecule is given n opportunities to react, the probability that it will not react is $(1 - P_1)^n$. Therefore, $1 - (1 - P_1)^n$ is the probability that the enzyme will react at least once in n opportunities. The probability of an enzyme molecule reacting with any one of S substrate molecules is

$$P_S = 1 - (1 - P_1)^S \tag{21.1}$$

As stated before,

$$P_2 = 1 - P_S \tag{21.2}$$

These two equations describe the change in probability of forming the enzyme-substrate combination as a function of S, substrate concentration. In a simulation of this process, the substrate concentration is assumed to be constant over the duration of the experiment. This assumption is like that of experimental measurements of initial reaction velocity.

Note that

$$(1 - P_1)^S \approx \exp(-P_1 S) \tag{21.3}$$

As with other stochastic simulations, this simulation of the enzyme-substrate interaction uses uniform random numbers to determine the next state of the system. In this case, the number of product molecules produced in a given span of time would provide a measure of enzyme activity for different substrate concentrations.

Exercise 21-2: Write and implement a program for the enzyme simulation described above. Use the following transition probabilities:

$$P_1 = 0.0002 \qquad P_3 = 0.01 \qquad P_4 = 0.95$$

Set S initially to a value of 2, and find the number of product molecules produced in 1000 units of time. Then repeat the simulation with S increased each time by a factor of 2; that is, $2, 4, 8, \ldots, 1024$. A flowchart for this simulation is found in Figure 21.3. As output from your program, plot the number of product molecules as a function of the number of substrate molecules.

In addition to the graphical output, have your program produce a listing of the number of product molecules at the different densities. Use these simulation data with the CURFIT program of Chapter 3 (or a similar program) to find how closely the hyperbolic

equation actually fits the simulation data. The CURFIT program
will produce best-fit estimates of K_m and V_{max} for the Michaelis-
Menten equation. Try to determine the relationship between these
values and the probabilities used in the simulation.

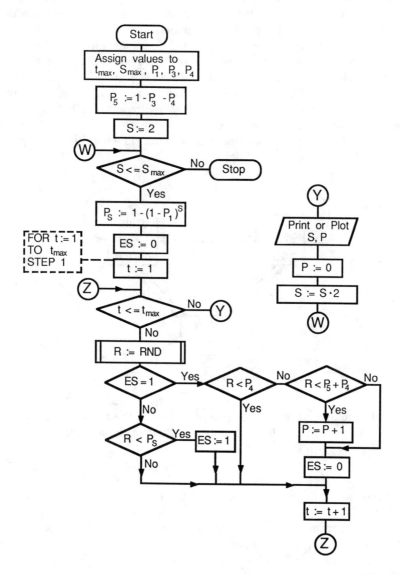

Figure 21.3. Flowchart for a Markov chain simulation of an
enzyme-substrate system.

21.3 Using a Transition Matrix to Model Markov Processes

One of the more useful ways of representing Markov processes is with a transition matrix, based on the transition diagram. (The transition diagrams are sometimes termed digraphs.) Figure 21.4 is a transition diagram that shows a simple Markov model of secondary succession in an old field of temperate regions; see Horn (1975) for more information. State 1 of the diagram represents grasses and other "pioneer" species that are early invaders of an abandoned cultivated field. State 2 represents rapidly growing tree species such as aspen, which require full sunlight. State 3 represents hardwood tree species such as red maple which can grow in partial shade. The diagram shows some hypothetical probabilities for changes among the different states, based on a time unit of several years.

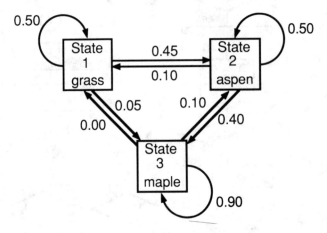

Figure 21.4. Transition diagram (digraph) with values for probability of transition between different states of old-field succession for any given site.

These probabilities may be assigned as elements of a transition matrix. Information about future states of the system may be obtained by multiplying the transition matrix by a state vector. To use the method with the probabilities given in Figure 21.4, we first set up the transition matrix for the nine probabilities involved:

$$\boldsymbol{P} = \begin{bmatrix} P_{11} & P_{12} & P_{13} \\ P_{21} & P_{22} & P_{23} \\ P_{31} & P_{32} & P_{33} \end{bmatrix} \tag{21.4}$$

Here, for example, P_{21} refers to the probability of a transition to State 2 from State 1. If the actual probability values from Figure 21.4 are assigned to the matrix, then

$$P = \begin{bmatrix} 0.50 & 0.10 & 0.00 \\ 0.45 & 0.50 & 0.10 \\ 0.05 & 0.40 & 0.90 \end{bmatrix}$$

Notice that the probabilities in each column sum to unity. This occurs because they represent all the possible transitions from a given state.

The column vector that represents the different states is

$$N = \begin{bmatrix} N_1 \\ N_2 \\ N_3 \end{bmatrix} \tag{21.5}$$

The values assigned to the column vector can vary with the experiment. In this case, it may represent the numbers of different plots of the old field that are in the three states at any given time. Alternatively, it might represent the probability of being in each of the three states at a given time. In this latter case, these proportions must sum to unity.

The state of the system after one time interval is obtained by multiplying the transition matrix by the state vector. If the original vector contained the number of plots of each state at time t, then the product vector would contain the number of plots of each type at time $t + 1$. If the original vector contained the proportions of states at time t, then the product vector would contain the proportion of each state at time $t + 1$.

If the transition matrix is multiplied repeatedly by successive state vectors, the state vector should give the ultimate equilibrium value of the system. In this case, it would show the numbers or proportion of each of the three states in the "climax community".

Exercise 21-3: Use the method of matrix multiplication that is described above to write a program for a simple simulation of secondary succession. Use the probabilities given in Figure 21.4. Begin your simulation with the proportion of State 1 = 1; the initial proportions of the other states will be zero. Allow the simulation to proceed for about 20 time units. Plot the proportions of each state over time.

21.4 A Markov Predation Model

An extended Markov chain may be used as the basis for a model of the dynamics of predator-prey interactions. In this instance, the interactions will be those of a wolf pack and a moose herd in an isolated area; the classic example is the wolf predation on the moose herd of Isle Royale National Park (Peterson 1984). The model is based on the different states of a roving, preying wolf pack.

The pack is assumed to operate as a unit, and to exist in one of four activity modes. At any time the pack will be hunting for moose, testing a moose (i.e. chasing, harassing and trying to kill it), eating a moose, or resting after eating. A variety of other activities will occur of course, but we are concerned here only with wolf-moose interactions. We assume that wolves exist only on moose meat.

The meeting of a pack and a moose is assumed to occur at random. With a moose herd of constant size, the probability of transition between hunting and testing is a constant. The outcome of a testing episode will have probabilities that vary with the age of the moose. Healthy young adult moose have a high probability of escaping or holding the pack at bay, while very young moose have a much lower probability of surviving an attack. Likewise, older moose have a lesser probability of escaping an attack, because they frequently have an arthritic condition or other debilitating problems. The moose are assumed to be met randomly while the pack is hunting. The chance of encountering a moose of a given age is assumed to be proportional to the relative densities of the different age classes. When the pack kills a moose, the length of time spent eating the moose will vary with the size of the moose, which depends on the age of the moose. The amount of wolf food that a mature moose provides can be four times that of a calf. The pack is assumed to rest for a constant period after eating each kill.

Given these assumptions, let us set up a general plan for a Markov-like simulation of the wolf pack. Some of the states, such as eating and resting, are essentially a series with transition probabilities of one. The simulation can make short-cuts through these serial identical states. Like the previous simulations of this chapter, the focus of the process is the current state of the pack, and the probabilities of moving to a different state over a unit of time.

Assume that the simulation begins with the wolf pack in hunting state. A random number is generated and compared with the probability of finding a moose. If the number is greater than the finding probability, it is assumed that no moose was found during the time interval. The elapsed time counter is increased by one unit, and the pack remains in hunting state. If a moose is found, then the pack moves to a testing state, and time is increased by one unit. There are three testing states depending on the age of the encountered moose, so a second random number is generated to determine which of the three moose age classes has been encountered. This random number is compared to the proportions of each age class in a sampling procedure like that of Section 19.1. Once the age class of the moose is established, a third random number is generated to be compared with the probability of a kill within that age class. If this random number is less than the probability of a kill, the pack will move to a state of eating moose. If no kill occurs, the wolf pack will revert to the hunting state, and the moose is presumed to go its merry way. If the moose is killed, the pack will stay in the eating state for a fixed time, depending upon the dead moose's age. The age class also determines how much food the pack gets to eat. After eating, the pack is assume to spend one day in a resting state, and then to return to the hunting state with a probability of 1.

The result of this model is a scenario describing the activities of a wolf pack at various times. Over a lengthy simulation, several estimates could be made about the variability of the time spent in various activities. The primary objective of this simulation is to provide a means of gaining new insights into an important biological process.

Exercise 21-4: Use the information above as the basis for writing a program to simulate a wolf-moose interaction. The probabilities given below apply over 0.1 day; this should be used as the unit of time increment in your simulation. The moose population is divided into three age classes with the following characteristics:

	Age Class I	Age Class II	Age Class III
Age Range	0-1 yrs	2-5 yrs	6-20 yrs
Average Edible Weight	100 kg	300 kg	400 kg
Fraction of Population	0.32	0.44	0.24
Probability of death at testing	0.62	0.02	0.34
Time to eat a moose	0.5 days	1.5 days	2.0 days

Assume that 0.7 is the probability of the pack in the hunting state finding a moose. Testing requires 0.1 day. An entire day is spent resting after each moose is eaten.

Allow your simulation to run for about 1000 tenth-days, and print out the following simulation results: amount of time spent in each activity mode (hunting, testing, eating, resting), total number of moose killed, and the average amount of moose meat eaten by the pack per day.

21.5 A Simulation of Cattle Reproduction

This simulation is based on a supposition that a dairy farmer wants to maximize the profitability of his dairy herd by reducing the number of bulls that he keeps. The farmer also wants to avoid any significant impact on the overall reproduction of the herd. A simulation may be performed based on a Markov chain of reproductive probabilities, using different numbers of cows per bull. The object of the simulation is to determine the maximum number of cows that may be effectively served by one bull.

The simulation is based on the following assumptions about the reproductive physiology of the cows:

1) The estrous cycle of each cow is independent of the other cows. That is, no cow has any effect on the time of ovulation of any other cow.

2) The mean length of the estrous cycle of cows is 28 days with the following distribution:

26% have a cycle of 28 days

68% fall in the range of 27 to 29 days

95% fall in the range of 26 to 30 days

99% fall in the range of 25 to 31 days

100% fall in the range of 24 to 32 days.

3) Cows may be successfully fertilized only on the day that they ovulate. If they are not successfully fertilized, they return to the beginning of their cycle.

4) A bull has a 90 percent probability of successfully fertilizing the first cow he mates with during a given day. However, because of reduced numbers of sperm, the probability of successful fertilization drops to about 33 percent for the second cow during a given day, and to about 11 percent for the third cow. The general rule is that the probability of a successful fertilization is one-third that of the previous attempt on any given day.

Based on these assumptions the simulation of dairy herd reproduction would proceed in the following way.

An array of subscripted numbers is set up to record the length of each cow's estrous cycle. A series of uniform random numbers is generated to determine cycle lengths based on the probabilities in assumption (2) above. Thus, if a given random number is less than 0.26, the cow is assumed to have a 28-day cycle, and this number is assigned to the array of cycle lengths. If the random number is greater than 0.26 but is less than 0.68, the cow is assumed to have a cycle of either 27 or 29 days. An additional random number is generated and compared with 0.5 to decide between 27 and 29 days. This cycle length is then assigned to the array, and so on. With this procedure, all the cows are assigned a length for their estrous cycle. These values are constant for a given simulation.

A second array is set up to indicate the location of each cow in its estrous cycle. The cycle length for each cow is determined by the value in the first array; for example, 29 days. Then a uniform random number between 1 and 29 is selected, and assigned to the second array. Thus, the two arrays represent the cycle lengths for each cow, and the location of each cow in her cycle.

After these two arrays are set up, the simulation progresses by time increments of one day. On each day, the simulation program proceeds stepwise through the second array, and adds one day to the cycle location. This value is compared against the value for cycle length in the first array. If the time in the estrous cycle does not equal the cycle length, then the simulation simply moves on to the next cow in the array.

If the value in the location cycle does equal the length of estrous, then ovulation is assumed to have occurred. Mating is simulated by generating a random number and comparing it with the probability of successful fertilization. Regardless of the success or failure of the mating, the probability of success is divided by three in case there is another attempt at mating during the same day. When a successful mating does occur, a pregnancy counter is incremented by one, and the value of the cow's cycle location is set to a large negative number, for example -100000. This effectively removes the cow from the mating pool for the duration of the simulation. Thus, the simulation program checks each cow each day and keeps track of matings and pregnancies.

Exercise 21-5: Write a program for the simulation of dairy herd reproduction that is described above. Run your simulation with some number of cows for 100 simulation days, repeating the simulation 10 times to find the average number of cows that are successfully fertilized. Use the program to find the approximate number of cows a single bull may serve and still achieve a 90 percent fertilization success over a 100-day period.

Conclusion

The models based in Markov processes that you have looked at in this chapter are a small sample of a great variety of such models that may be useful in biology. Some will appear in succeeding chapters. Like most stochastic models, they are in some ways weaker than the deterministic models for the same phenomena. They do not make specific predictions about the outcome of the modeled processes. However, they do make predictions about the range of expected outcomes, and hence may be quite powerful in this respect. Certainly they provide more insight into the function of many biological systems which are characteristically stochastic.

PART FOUR
SUPPLEMENTARY MODELS

Three chapters make up this fourth and final segment. A variety of interesting models are presented in cellular biology and biochemistry, in developmental biology, and in epidemiology. These fields have been covered partially or not at all in preceding sections. No new simulation techniques are introduced, so the models explained in this section are in that sense supplementary to the preceding chapters. However, if your biological interests lie in these fields, the material will surely have an important primary significance.

Although the models require no new techniques for computer simulations, they are generally "advanced", with most having numerous equations. A few of the models have stochastic components. The BASIC programs required for some simulation exercises will be longer that any programs written for previous exercises. However, most of the simulations are modifications of previous simulations, so the actual writing of BASIC programs should not be very difficult.

Simulations based on these "long" programs may require protracted computing times. Other exercises specify lengthy periods of simulated time with very many iterations of various Euler loops for numerical integration. These time-consuming simulations may provoke interest in working with compiled versions of BASIC, or with faster computers.

These chapters significantly widen the book's coverage of different disciplines of biology. Despite this, major gaps remain, which must be expected in a workable introductory text. Thus, if you should work through all 24 chapters of the book, you will not be exposed to even a majority of the important models in all the fields of biology. However, you should be able to write programs for computer simulation of any adequately quantitative model.

CHAPTER 22

MODELS
OF CELLULAR FUNCTION

In this chapter we will consider simulations based on a number of different models that have as a central theme the function of individual cells. The models describe a variety of different cellular events, from reactions of multienzyme systems through mechanisms of control of cell growth, nutrient uptake, and division. Several models are elaborations of simpler models you have encountered in previous chapters. The models of this chapter lie in the following fields: (a) cellular biochemistry and enzyme control; (b) cell growth and its control by nutrient uptake; (c) ribosome-protein control of cell division; and (d) control of protein synthesis by gene repression.

22.1 Dynamics of Multienzyme Systems

Biochemists traditionally have followed a reductionist approach in their laboratory procedures, in which living materials are broken down and reactions are studied individually. This approach is very powerful and has produced large sets of data about the kinetics of individual enzyme-catalyzed reactions. These data have been used to formulate a number of conceptual models describing the functional relationships between components of cellular biochemical systems. Glycolysis, Kreb's cycle, the pentose phosphate cycle, and various biosynthetic pathways are all familiar conceptual models of multienzyme systems. Each is supported by much evidence derived from many independent experiments.

Synthetic experimental methods in which enzymes and other components of cells are combined to form metabolic systems in the laboratory have not met with much success. However, models of chemical kinetics can allow biochemists to simulate the behavior of multienzyme systems. These simulations can help with interpreting the results of experiments and with planning new ones. Simulation can also aid in determining the relative importance of different factors in relationship to observed system behavior. This is important because many multienzyme systems are nonlinear, and behave counter-intuitively.

Multienzyme systems may involve more than 50 chemical reactants, often transported between several cell components and involved in reactions catalyzed by 30 or more different enzymes. Each enzyme involved may show different kinetic behavior and may respond to feedback control. Clearly, such systems offer a considerable challenge to simulation programmers. In the following five sections of this chapter we will examine two different approaches to the problems of modeling multienzyme systems, and look at several interesting biochemical systems as examples of these approaches.

22.2 A Mass Action Model of Glycolysis

One of the basic methods for simulating multienzyme systems is the direct mass action approach. Enzymes are considered to react with substrates to form complexes which in turn break down to form products. The elements of this approach were described previously in connection with the Chance-Cleland model in Section 6.4. This approach is particularly useful because it may be adapted to almost any type of enzymatic reaction. It can be used for reversible as well as non-reversible reactions. A principal disadvantage is an assumption that reaction rates are fixed, regardless of substrate concentration. This assumption limits the range of substrate concentrations over which the model is likely to be realistic. Another disadvantage is the considerable amount of computing time required for some simulations.

The mass action models for multienzymes use the direct approach of solving each chemical equation. Most of the early simulations used this approach and included the enzymes as chemical reactants in the system. One such early attempt was a study of the control of glucose metabolism, a classic simulation implemented by Chance et al. (1960) on the UNIVAC I computer. Despite some primitive simulation techniques, this investigation allowed comprehensive analysis of a multienzyme system so that simulation results could be compared with the dynamics of intact cells.

The glycolysis model presented here is based on this early simulation. It has been simplified greatly to make it usable with BASIC on most personal computers. The model described below is a truncated version of the original model and covers only the anaerobic aspects of glycolysis. For a description of glycolysis, you should review a biochemistry textbook such as Zubay (1988).

The chemical equations used in the model of anaerobic glycolysis proceed in the following sequence:

1. $GLU + HEX \xrightarrow{k_1} CHEX$

2. $CHEX + A3P \xrightarrow{k_2} HEX + H6P + A2P$

3. $H6P + PFK \xrightarrow{k_3} CPFK$

4. $CPFK + A3P \xrightarrow{k_4} PFK + FPP + A2P$

5. $FPP \xrightarrow{k_5} DHAP + GAP$

6. $DHAP + HNAD \xrightarrow{k_6} AGP + NAD$

7. $AGP + NAD \xrightarrow{k_7} DHAP + HNAD$

7A. $DHAP \xrightarrow{k_A} GAP$

7B. $GAP \xrightarrow{k_B} DHAP$

8. $GAP + CGPDH \xrightarrow{k_8} CX + HNAD$

9. $CX + PI \xrightarrow{k_9} DPGA + GPDH$

10. $NAD + GPDH \xrightarrow{k_0} CGPDH$

11. $DPGA + A2P \xrightarrow{k_C} A3P + P3GA$

12. $P3GA + A2P \xrightarrow{k_D} A3P + PYR$

13. $PYR + HNAD \xrightarrow{k_E} LAC + NAD$

14. $LAC + NAD \xrightarrow{k_F} PYR + HNAD$

20+. $A3P \xrightarrow{k_G} A2P + PI$

The numbers given to these equations refer to the numbers used by Chance et al. (1960). Reactions 7A and 7B have been added to the original set of equations. Reaction 20+ summarizes the results of ATP utilization by the system outside of anaerobic glycolysis. k_1 through k_G are the rate constants for the reactions. The names of the reactants in these equations have been chosen to conform generally with standard biochemical usage. However, some of the standard names have been modified so they can be used as the names of variables with most dialects of BASIC, to simplify writing programs for simulation. The full names of the reactants in the equations are given in Table 22.1.

A2P	Adenosine diphosphate	GPDH	Glyceraldehyde 3-phosphate
A3P	Adenosine triphosphate		dehydrogenase
AGP	α-Glycerophosphate	H6P	Hexose 6-Phosphate
CGPDH	Glyceraldehyde 3-phosphate	HEX	Hexokinase
	dehydrogenase-NAD complex	HNAD	Reduced nicotinamide
CHEX	Hexokinase-complex		adenine dinucleotide
CPFK	Phosphofructokinase complex	LAC	Lactic acid
CX	Glyceraldehyde 3-phosphate	NAD	Nicotinamide adenine
	dehydrogenase-acyl complex		dinucleotide
DHAP	Dihydroxyacetone phosphate	P3GA	3-Phospho glyceric acid
DPGA	Diphospho glyceric acid	PFK	Phosphofructokinase
FPP	Fructose 1,6-diphosphate	PI	Inorganic Phosphate
GAP	Glyceraldehyde 3-phosphate	PYR	Pyruvic acid
GLU	Glucose		

Table 22.1. Names of reactants involved in a model of anaerobic glycolysis, with corresponding abbreviations usable as variable names for a simulation written in BASIC.

The glycolysis model described here has employed most of the simplifying assumptions used by Chance et al. (1960). To reduce the number of equations involved, some enzymes are assumed to be operating well below V_{max} so that simple first-order reactions may be employed, e.g. the reactions numbered 5, 7A and 7B. Hexose isomerase and enolase are assumed to be sufficiently rapid that they can be combined with other reactions. Altogether, such simplifications have reduced the number of equations by about a third. Some of the reactions involved in glycolysis are reversible and are represented in this series by pairs of reactions, for example numbers 6 and 7.

A simulation based on this model may be programmed using the familiar two-stage simple Euler integration. Calculations are made somewhat clearer if the first stage is divided into two parts as in Section 6.2. Part 1 of the first stage involves calculating flux rates for the different reactions given above. For example, the flux rates for the first two reactions would be found with:

$$F_1 = k_1[GLU][HEX]$$

$$F_2 = k_2[CHEX][A3P]$$

The remainder of the flux rates F_3 through F_G are found similarly. In part 2 of the first stage Euler solution, the changes in the different reactants are estimated over time Δt; for example:

$$\Delta[GLU] = (-F1)\Delta t$$

$$\Delta[NAD] = (+F_6 - F_7 - F_0 + F_E - F_F)\Delta t$$

The second stage of the Euler solution then follows the usual procedure, for example:

$$[GLU] \leftarrow [GLU] + \Delta[GLU]$$

The objective of the glycolysis simulation described here is to demonstrate the direct mass action approach in a system that is familiar to most biologists. The model provides a relatively realistic simulation of anaerobic glycolysis in that conversion of glucose to pyruvate and lactate is accompanied by an increase in ATP ($A3P$) with negligible effect on NAD and NADH.

This simulation is relatively cumbersome on most personal computers with a BASIC interpreter. A lot of computer time is required to show a significant conversion of glucose to pyruvate. A small increase in computation speed may be obtained by algebraically combining Part 2 of the first Euler stage with the second stage. However, trying to increase the speed of the simulation by increasing the size of Δt will result in instability.

Exercise 22-1: Write and implement the simulation of anaerobic glycolysis that is described above. Use the following rate constants which are based on those of Chance et al. (1960):

$$k_1 = 3 \times 10^{-5} \mu\text{mole}^{-1}\mu\text{sec}^{-1} \qquad k_2 = 1 \times 10^{-4} \mu\text{mole}^{-1}\text{sec}^{-1}$$

$$k_3 = 4 \times 10^{-4} \mu\text{mole}^{-1}\mu\text{sec}^{-1} \qquad k_4 = 4 \times 10^{-4} \mu\text{mole}^{-1}\mu\text{sec}^{-1}$$

$$k_5 = 1 \times 10^{-3} \mu\text{sec}^{-1} \qquad k_6 = 2 \times 10^{-5} \mu\text{mole}^{-1}\mu\text{sec}^{-1}$$

$$k_7 = 8 \times 10^{-7} \mu\text{mole}^{-1}\mu\text{sec}^{-1} \qquad k_A = 1 \times 10^{-2} \mu\text{sec}^{-1}$$

$$k_B = 1 \times 10^{-2} \mu\text{sec}^{-1} \qquad k_8 = 6 \times 10^{-3} \mu\text{mole}^{-1}\mu\text{sec}^{-1}$$

$$k_9 = 4 \times 10^{-6} \mu\text{mole}^{-1}\mu\text{sec}^{-1} \qquad k_0 = 6 \times 10^{-5} \mu\text{mole}^{-1}\mu\text{sec}^{-1}$$

$$k_C = 1 \times 10^{-4} \mu\text{mole}^{-1}\mu\text{sec}^{-1} \qquad k_D = 5 \times 10^{-5} \mu\text{mole}^{-1}\mu\text{sec}^{-1}$$

$$k_E = 5 \times 10^{-6} \mu\text{mole}^{-1}\mu\text{sec}^{-1} \qquad k_F = 1 \times 10^{-7} \mu\text{mole}^{-1}\mu\text{sec}^{-1}$$

$$k_G = 1 \times 10^{-3} \mu\text{sec}^{-1}$$

As initial concentrations of reactants, use the following values, all given as μmoles liter^{-1}:

$GLU = 3000$	$H6P = 0$	$FPP = 0$
$DHAP = 0$	$AGP = 0$	$GAP = 0$
$DPGA = 0$	$P3GA = 0$	$PYR = 1000$
$LAC = 1000$	$NAD = 100$	$HNAD = 100$
$A3P = 500$	$A2P = 100$	$HEX = 10$
$PFK = 10$	$GPDH = 0$	$PI = 4000$
$CHEX = 0$	$CPFK = 0$	$CGPDH = 50$
$CX = 0$		

The major long-term changes in the system will be in the concentrations of glucose (GLU), lactate (LAC), and ATP ($A3P$). As the output from your simulation, plot concentrations of these three variables over 4 msec (4000 μsec). Use a Δt value of 1 μsec.

22.3 Generalized Mass Action Simulations for Single Substrate Reactions

This section will describe a general approach for simulating a series of enzyme-catalyzed reactions. The basic reaction has already been described in Section 6.3 and is represented as follows:

$$S + E \underset{k_2}{\overset{k_1}{\rightleftarrows}} ES \underset{k_4}{\overset{k_3}{\rightleftarrows}} E + P$$

where S is substrate, E is the enzyme, ES is the enzyme-substrate complex, and P is product. The rate constants for the four reactions are k_1

through k_4 . The fluxes in a single reaction of this type are as follows:

$$F_1 = k_1[S][E] \tag{22.1}$$

$$F_2 = k_2[ES] \tag{22.2}$$

$$F_3 = k_3[ES] \tag{22.3}$$

$$F_4 = k_4[E][P] \tag{22.4}$$

This model may be programmed for simulation using the two-stage simple Euler procedure used in Chapter 6 and also used for the simulation of the preceding section. Thus, Equations 22.1-22.4 define fluxes for the first part of stage one. The second part of stage one involves finding the changes in each of the reactants. For example, the change in $[S]$ is

$$\Delta[S] = (\Delta F_1 + F_2)\Delta t \tag{22.5}$$

Stage two is the update procedure for the reactants. $[S]$ is updated, for example, with

$$[S] = [S] + \Delta[S]$$

This approach was followed in Exercise 6-4 to simulate the reaction involving a single enzyme. With slight modifications, the technique may be generalized for simulating a sequence of single substrate reactions of the following type:

$$S_1 \longrightarrow S_2 \longrightarrow S_3 \longrightarrow \cdots S_i \cdots \longrightarrow S_{n-1} \longrightarrow S_n$$

Because the same procedures of Euler integration are involved in each reaction of the sequence, subscripted variables may be used in the simulation. The BASIC statements required for calculating the flux rates of Equations 22.1 through 22.4 would be written as follows:

```
10 REM EULER STAGE 1 - PART ONE
20 FOR I = 1 TO (N-1)
30    F1(I) = K1(I) * S(I) * E(I)
40    F2(I) = K2(I) * ES(I)
50    F3(I) = K3(I) * ES(I)
60    F4(I) = K4(I) * E(I) * S(I+1)
70 NEXT I
```

Notice that the product of the reaction involving S_i becomes S_{i+1}. The remainder of the Euler procedure may be given with a second FOR-NEXT loop:

```
110 REM EULER STAGE 1-PART TWO + STAGE 2
120 FOR I = 1 TO (N-1)
130    S(I) = S(I) + (-F1(I) + F2(I)) * DT
140    E(I) = E(I) + (-F1(I) + F2(I) + F3(I) - F4(I)) * DT
150    ES(I) = ES(I) + (F1(I) - F2(I) - F3(I) + F4(I)) * DT
160    S(I+1) = S(I+1) + (F3(I) - F4(I)) * DT
170 NEXT I
```

It is important to notice that the statements 130 through 160 are an algebraic combination of the first Euler stage (part 2) and of the second stage. This procedure decreases the number of statements and the computation time at the expense of some clarity.

Exercise 22-2: Use the subscripted mass action technique described above to simulate a hypothetical reaction pathway consisting of five intermediates connected by four reactions catalyzed by four different single enzymes:

$$S_1 \xrightarrow{\ E_1\ } S_2 \xrightarrow{\ E_2\ } S_3 \xrightarrow{\ E_3\ } S_4 \xrightarrow{\ E_4\ } S_5$$

The rate constants should be given the following values:

$i = 1:$	$k_1 = 0.01$	$k_2 = 0.02$	$k_3 = 0.02$	$k_4 = 0.001$
$i = 2:$	$k_1 = 0.001$	$k_2 = 0.02$	$k_3 = 0.02$	$k_4 = 0.01$
$i = 3:$	$k_1 = 0.01$	$k_2 = 0.002$	$k_3 = 0.02$	$k_4 = 0.001$
$i = 4:$	$k_1 = 0.01$	$k_2 = 0.02$	$k_3 = 0.02$	$k_4 = 0.01$

Set initial concentrations of substrates as follows:

$$S_1 = 1000 \qquad S_2 = 160 \qquad S_3 = 50 \qquad S_4 = 80 \qquad S_5 = 0$$

The total concentrations of all four enzymes will be 10 units. (All units in this simulation are arbitrary.) At the start of the simulation the substrates will be almost at saturation levels. Thus, initial amounts of the enzyme-substrate complex forms will be relatively high:

$$ES_1 = 9.96 \qquad ES_2 = 9.50 \qquad ES_3 = 9.70 \qquad ES_4 = 9.60$$

The free enzymes will therefore have the following initial concentrations:

$$E_1 = 0.04 \qquad E_2 = 0.50 \qquad E_3 = 0.30 \qquad E_4 = 0.40$$

Use $\Delta t = 0.1$ time unit. Run the simulation for at least 500 time units. As output, plot the substrate concentrations through time.

The program may be modified in a variety of ways to produce interesting simulations. As one example, the concentrations of S_1 and S_5 may be held constant, and the other intermediates allowed to come to equilibrium. As another example, all the values of k_4 can be set to zero to simulate the irreversible condition in which no product can revert to substrate.

22.4 The Rate Law Approach for Sequences of Single Substrate Reactions

The mass action approach described in the previous two sections produces simulations that require a lot of computing time, even when run on large fast computers using compiled languages such as FORTRAN or Pascal. In addition, the nonlinear mass action equations and the size of the coefficients can produce large errors or system instability with simple techniques of numerical integration. These problems have caused modelers to look for other techniques of simulating cellular chemistry.

The rate law approach described by Rhodes et al. (1968) seems to offer an acceptable alternative to the mass action approach. Initial trials by Rhodes found the rate law technique to be about 100 times faster in computation than a comparable simulation based on the mass action approach. Essentially identical simulation data were generated by both methods.

The rate law technique employs rate expressions like the Michaelis-Menten equation to determine the flux between a given substrate and its product. The Michaelis-Menten equation is the specific rate law for the non-reversible single substrate reaction. For the reversible single substrate reaction, the rate law is described by Savageau (1976) as follows:

$$V_{\text{net}} = \frac{\frac{V_s}{K_s}[S] - \frac{V_p}{K_p}[P]}{1 + \frac{[S]}{K_s} + \frac{[P]}{K_p}} \tag{22.6}$$

where V_{net} forward reaction velocity, or flux, for an enzyme catalyzed reaction with a single substrate. $[S]$ and $[P]$ are concentrations of substrate and product, V_s and V_p are maximum velocities for the forward and reverse reactions, and K_s and K_p are Michaelis-Menten constants for the forward and reverse reactions. Note that if $[P]$ is zero, this equation reverts to the standard form of the Michaelis-Menten equation. Equation 22.6 appears to be an adequate model for most single substrate reactions.

The four constants involved in Equation 22.6 are related to the equilibrium constant for the reaction by the Haldane relationship:

$$K_{eq} = \frac{V_s K_p}{V_p K_s} \tag{22.7}$$

Thus, to be internally consistent the four constants usually provided as input for this model are K_{eq}, K_s, V_p, and V_s. The value of K_p is found with

$$K_p = \frac{K_{eq} V_p K_s}{V_p} \tag{22.8}$$

The ratios V_s/K_s and V_p/K_p in Equation 22.6 are made up of constants, so that it is possible to set these equal to new terms V_F and V_R respectively. This will both simplify the equation and make it run rather more rapidly in a computer simulation.

For simulating sequential reactions, BASIC programs based on the rate law can be much shorter than programs based on the mass action approach. This can be illustrated with a simulation based on a sequence of substrates as in Section 22.3 above. Again, subscripts can be used for all variables and constants of the model. The program should begin with the input of values for the constants K_{eq}, K_s, V_p, and V_s, followed by calculation of the constants K_p, V_F, and V_R. Initial values of the substrate concentrations S_i then should be assigned. The BASIC statements for part one of the first stage of simple Euler integration would be

```
210 REM EULER STAGE 1 - PART ONE
220 FOR I = 1 TO (N-1)
230    F(I) = (VF(I)*S(I) - VR(I)*S(I+1)) / (1 + S(I)/KS(I)
       + S(I+1)/KP(I))
240 NEXT I
```

As in the BASIC statements in the preceding section, the product of the reaction involving S_i becomes S_{i+1}. Similarly, the Euler procedure is completed with a second FOR-NEXT loop:

```
310 REM EULER STAGE 1-PART TWO + STAGE 2
320 FOR I = 1 TO (N-1)
330    S(I) = S(I) - F(I) * DT
340    S(I+1) = S(I+1) + F(I) * DT
350 NEXT I
```

Here again, statements 330 and 340 are an algebraic combination of part two of the first Euler stage and of the second stage.

The advantage of the rate law approach is obvious when you compare the number of BASIC statements required for numerical integration in this simulation with the number needed for the mass action approach. Only a single flux equation and two update equations can accomplish the procedure that required four flux equations and four update equations in the previous section. In terms of computer-time requirements, stability of the system is more important than program brevity, because it allows larger values for the Euler time increment Δt.

Exercise 22-3: Using the rate law approach, write and implement a BASIC program to simulate the hypothetical reaction sequence that was given in Exercise 22-2. Simulation results almost identical with those of Exercise 22-2 should result from use of the following constants:

$$i = 1: \quad K_{eq} = 10.0 \quad K_s = 0.2 \quad V_s = 0.2 \quad V_p = 0.2$$
$$i = 2: \quad K_{eq} = 0.1 \quad K_s = 2.0 \quad V_s = 0.2 \quad V_p = 0.2$$
$$i = 3: \quad K_{eq} = 100.0 \quad K_s = 0.02 \quad V_s = 0.2 \quad V_p = 0.02$$
$$i = 4: \quad K_{eq} = 1.0 \quad K_s = 0.2 \quad V_s = 0.2 \quad V_p = 0.2$$

Set initial values of all five substrate concentrations to those given in Exercise 22-2. Use a Δt value of 1.

22.5 The Yates-Pardee Model of Feedback Control

Many metabolic pathways are controlled by an end product metabolite which feeds back to inhibit one of the initial reactions in a sequence. This simple form of control, found to exist in a variety of biosynthetic pathways, was first described by Yates and Pardee (1956).

The Yates-Pardee control process may be represented by the following sequence, based on the model of sequential reactions in Sections 22.3 and 22.4 above:

$$S_0 \xrightarrow{E_0} S_1 \xrightarrow{E_1} S_2 \xrightarrow{E_2} S_3 \xrightarrow{E_3} S_4 \xrightarrow{E_4} S_5$$

S_0 through S_5 are hypothetical substrates and products which form an unbranched metabolic pathway. E_0 through E_4 are enzymes which catalyze the reactions involved in the sequence. S_4 is a product which feeds back to inhibit enzyme E_0 allosterically. This model is similar to one investigated in detail by Walter (1974).

Assuming that only reactions with single substrates are involved, we may simulate this process by employing the rate law approach described in Section 22.4. The rate law (Equation 22.6) would be an adequate model for the reactions catalyzed by E_1 through E_4. The reaction catalyzed by enzyme E_0 requires an equation based on the concerted model for allosteric control developed by Monod et al. (1965). This model is described in Section 2.7. Assuming no allosteric activation in the sequence we are considering, the forward reaction velocity may be found with the following equation for allosteric inhibition:

$$F_0 = V_0 \frac{\alpha(1+\alpha)^3}{L(1+\beta)^4 + (1+\alpha)^4} \qquad (22.9)$$

In this equation, V_0 is the maximum velocity, and L is the equilibrium constant for conversion of enzyme from the active form to the inactive form. α is $[S_0]/K_0$, the ratio of the substrate concentration to its dissociation constant. β is $[S_4]/K_i$, the ratio of inhibitor concentration to its dissociation constant. (See Section 2.7 for further details and notation.)

A program for a simulation based on this model will be similar to programs for simulating the sequential reaction of the preceding section. Equation 22.9 must be included to determine the flux from S_0 to S_1.

Exercise 22-4: Write and implement a BASIC program to produce a simulation of the Yates-Pardee model of control of sequential reactions outlined above. Use the following values for the constants of the four reactions that follow Equation 22.6:

$$K_{eq} = 10 \qquad V_s = 10 \qquad V_p = 1 \qquad K_s = 20$$

For the controlled reaction, use the following constants:

$$K_0 = 10 \qquad K_i = 5 \qquad L = 200 \qquad V_0 = 10$$

Begin the simulation with substrate concentrations S_1 through S_5 set at 100. S_0 should be set at 600. Write your program so that concentrations of S_0 and S_5 are held constant at their initial values, to allow the system to come to a steady state. Plot the concentrations of S_1 and S_4 through time, until the system approaches steady state. The result should be an oscillating system, resulting from a lag in feedback.

22.6 Allosteric Control of Phosphofructokinase

A model of glycolysis was presented in Section 22.2 above. The key enzyme that controls glycolysis is considered to be phosphofructokinase, the enzyme that catalyzes the conversion of fructose 6-phosphate to fructose diphosphate. The key position of the enzyme at the head of the energy producing section of the pathway permits the whole glycolysis pathway to be controlled from this point. Phosphofructokinase activity is inhibited by adenosine triphosphate (ATP), and activated by adenosine monophosphate (AMP). This results in a slowing of the rate of glycolysis when an adequate supply of ATP energy is available from other sources such as the Kreb's cycle, and an acceleration of glycolysis when ATP concentrations decline and AMP concentrations increase.

Oscillations of various metabolites associated with glycolysis have been observed experimentally in a variety of tissues and microbes. Richter and Betz (1976) developed a model to determine whether lags in the feedback pathways of phosphofructokinase could account for the magnitude and frequency of the observed oscillations.

The model of Richter and Betz was based on a sequence of reactions, shown in the following diagram:

$$
\begin{array}{ccccccc}
 & A3P & & A2P & & & \\
V_0 & + & V_p & + & V_r & & V_c \\
\xrightarrow{} & F6P & \xrightarrow{} & FPP & \xrightarrow{} & A3P & \xrightarrow{} & \text{consumption} \\
 & \Big\downarrow V_s & & & & & \\
 & \text{storage} & & & & &
\end{array}
$$

The chemical notation of the diagram follows that given in Table 22.1. In this model, $F6P$ enters the pathway at a constant rate, V_o. Phosphofructokinase catalyzes the reaction of $F6P$ with $A3P$, resulting in a formation of FPP and $A2P$. The rate of this reaction, V_p, is determined by a rate law expression based on the Monod model. The product, FPP, enters a sequence of reactions with velocity V_r; ultimately the sequence will yield four $A3P$ and consume four $A2P$. The energy-requiring reactions of the cell will consume $A3P$ at a rate described by V_c. Excess $F6P$ is assumed to be stored at a rate determined by V_s.

The velocity for the reaction catalyzed by phosphofructokinase is modeled with the following equation:

$$V_p = V_m \cdot \frac{\alpha(1+\alpha)^3}{\frac{L(1+\beta)^4}{(1+\gamma)^4} + (1+\alpha)^4} \cdot \frac{[A3P]}{(K_m + [A3P])} \tag{22.10}$$

The first half of the right side of this equation is based on the Monod

model for allosteric control discussed in Section 2.7. The second half is
a saturation term for the second substrate of the reaction, $A3P$. V_m is
the maximum velocity of the phosphofructokinase reaction, and L is the
dissociation constant for the enzyme. α is $[F6P]/K_f$, where K_f is the
dissociation constant for $F6P$. β is $[A3P]/K_i$, where K_i is the dissociation
constant for the allosteric inhibitor. γ is $[A1P]/K_a$, where K_a is the
dissociation constant for the allosteric activator, and $A1P$ is the notation
for AMP.

The velocity for the sequential reactions involved in using FPP to make
$A3P$ is modeled with the following equation:

$$V_r = V_n \cdot \frac{[FPP]}{(K_n + [FPP])} \cdot \frac{[A2P]}{(K_r + [A2P])} \tag{22.11}$$

V_n is the maximum velocity for the reaction, K_n and K_r are half-saturation
constants. The rate of consumption of $A3P$ is modeled by a first-order
decay equation:

$$V_c = k_c[A3P] \tag{22.12}$$

with k_c representing the rate constant for consumption. The storage rate
of $F6P$ is described by the equation

$$V_s = k_s[A3P][F6P] \tag{22.13}$$

where k_s is the rate constant for the storage reaction.

The adenosine nucleotides follow this reaction:

$$A3P + A1P \underset{V_2}{\overset{V_1}{\rightleftharpoons}} 2\, A2P$$

The velocities of the forward and reverse reactions, V_1 and V_2, are determined by the following second-order equations:

$$V_1 = k_1[A3P][A1P] \tag{22.14}$$

$$V_2 = k_2[A2P][A2P] \tag{22.15}$$

This model assumes the total concentration of the three nucleotides is a
constant, N_o. Based on this assumption, the concentration of $A1P$ may
be found at any time with the equation

$$[A1P] = N_o - [A3P] - [A2P] \tag{22.16}$$

Looking at Equations 22.10 through 22.16, you can see that this model
involves aspects of both the mass action approach and the rate law approach. The principal equations, 22.10 and 22.11, use the rate law approach.

In setting up a simulation of phosphofructokinase control of glycolysis based on this model, the second part of stage one of a simple Euler integration will consist of the following equations for changes in concentrations of reactants:

$$\Delta[F6P] = (V_o - V_p - V_s)\,\Delta t \qquad (22.17)$$

$$\Delta[FPP] = (V_p - V_r)\,\Delta t \qquad (22.18)$$

$$\Delta[A3P] = (4V_r - V_p - V_s - V_c - V_1 + V_2)\,\Delta t \qquad (22.19)$$

$$\Delta[A2P] = (-4V_r + V_p + V_s + V_c + 2V_1 - 2V_2)\,\Delta t \qquad (22.20)$$

Exercise 22-5: Write and implement a BASIC program to simulate the control of glycolysis following the Richter and Betz model above. The following constants may be used in setting up your simulation:

$k_1 = 100$ ml μmole^{-1} min^{-1} $K_a = 0.01$ μmole ml^{-1}

$k_2 = 50$ ml μmole^{-1} min^{-1} $K_f = 0.03$ μmole ml^{-1}

$k_s = 6$ ml μmole^{-1} min^{-1} $K_m = 0.01$ μmole ml^{-1}

$V_o = 24$ μmole ml^{-1} min^{-1} $K_n = 1.0$ μmole ml^{-1}

$V_m = 33$ μmole ml^{-1} min^{-1} $K_r = 0.3$ μmole ml^{-1}

$V_n = 20$ μmole ml^{-1} min^{-1} $K_i = 0.05$ μmole ml^{-1}

$k_c = 1$ min^{-1} $N_o = 3.3$ μmole ml^{-1}

$L = 250$

Begin your simulation with $[A3P] = 2.0$ μmole ml^{-1} and $[A2P] = 1.0$ μmole ml^{-1}. Use simple Euler integration with a Δt of 0.001 min. Plot the concentrations of $A3P$, $A1P$ and FPP over a simulated two-minute period to observe glycolytic oscillations. After a period of initial instability, the system should show regular oscillations with a frequency of about 4.4 cycles per minute.

22.7 Control of Cell Growth by Nutrient Uptake

The growth of cells is dependent on the nutrients available for transport into the cell and rates of use in the cell. Clearly, the processes of nutrient uptake and growth involve many pathways of the types described in the preceding sections of this chapter. We will consider here and in the next section some models for the control of cell growth based on the rate of nutrient uptake and the processing of nutrients in the cell. Instead of considering only the nutrient concentrations outside of the cell, they also take into account the internal nutrient concentrations. These multistage nutrient models represent an increase in complexity from the models considered in Chapter 8, just as the multienzyme models in the preceding sections were more complex than the models in Chapter 6. Although the models below may be applied to essentially any cell, for simplicity we will consider the experimental conditions of microbial cells growing in a culture medium, as in Chapter 8.

Caperon (1968) and Droop (1968) recognized that the growth of cells was a function of the internal nutrient concentration rather than the external concentration. These authors developed models for internal pools of nutrients, representing unassimilated nutrients available for biosynthesis. Under conditions of fluctuating nutrients, these models predicted growth rates much closer to experimental results than simple models like those of Chapter 8. Cells are observed to continue growing for a period after they are transferred from optimal nutrient concentrations to sub-optimal. Conversely, there is a lag in cell growth when cells are transferred from sub-optimal to optimal nutrient concentrations. There appears to be good justification for modeling growth response to nutrient concentrations as a process with at least two stages.

In the models considered here, the first stage involves nutrient uptake to form an internal pool of nutrients. The second stage involves assimilation of internal nutrients and their conversion to new cell biomass. Nutrient uptake is generally carried on by mediated transport processes such as those described in Section 14.6. Nutrient uptake by the cell may therefore be described by a hyperbolic equation of the Michaelis-Menten type:

$$\nu = V_m \frac{N}{(K_n + N)} \qquad (22.21)$$

where ν is the rate of nutrient uptake per unit of biomass, V_m is the maximum rate of nutrient uptake per unit of biomass, N is the external nutrient concentration, and K_n is the external nutrient concentration at which the rate of uptake is half of the maximum value.

Assuming the cells are growing in a batch culture, Equation 22.21 may be used also to describe change in nutrient concentration of the culture medium. If cell biomass concentration is represented by B, then change of external nutrient concentration is

$$\frac{dN}{dt} = -\nu B \qquad (22.22)$$

Likewise, with q representing internal nutrient concentration of a cell, change in internal concentration is found with the equation

$$\frac{dq}{dt} = \nu - \mu q \tag{22.23}$$

The units of internal nutrient concentration, q, are grams of nutrient per gram of cell biomass, or μmoles of nutrient per cell, or some other similar measure. Droop (1973) named q the "cell quota". As described in Chapter 8, μ is a specific growth rate with units of time^{-1}.

To find values of μ, the Monod growth model of Chapter 8 is modified so that μ depends on the internal nutrient concentration q. In addition, there is defined a threshold of minimum internal nutrient concentration, q_o, which represents the level of internal nutrient needed for growth. The value of μ for any internal concentration is found with

$$\mu = \mu_m \frac{(q - q_o)}{K_q + (q - q_o)} \tag{22.24}$$

where μ_m is the maximum growth rate, K_q is the half- saturation constant for growth, and $(q - q_o)$ is the internal nutrient concentration available for growth. The term in parentheses defines the nutrient pool, or unassimilated nutrient. Thus, q_o might be considered to be the assimilated nutrient that is not available for stimulating further cell growth. The value of μ is constrained to be zero or greater. While cells will lose biomass if q drops below q_o, this loss is considered to be due to respiration, cell leakage, and mortality. The cell functions of respiration, leakage, etc., are combined in the model, and proceed at a constant rate R, as described in Section 8.1. Figure 22.1 shows the general pattern of growth rate vs. internal nutrient concentration that has been determined by experiment, for example Caperon and Myer (1972).

In simulating cell growth based on the model of two-stage nutrient uptake, the changes in biomass and nutrient concentration for simple Euler integration procedures may be found with the following equations:

$$\Delta N = (RqB - \nu B)\,\Delta t \tag{22.25}$$

$$\Delta q = (\nu - \mu q)\,\Delta t \tag{22.26}$$

$$\Delta B = (\mu B - RB)\,\Delta t \tag{22.27}$$

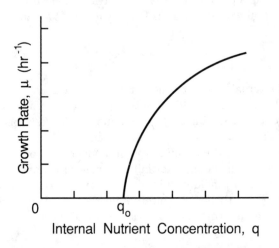

Figure 22.1. Generalized graph showing cell growth rate as a function of internal concentration of nutrients.

Exercise 22-6: Write and implement a BASIC program for simulating microbial growth in a batch culture based on the model given above. Assume the following values for the necessary constants:

$$q_o = 0.02 \text{ mg}N \text{ mg}B^{-1} \qquad K_n = 0.05 \text{ mg l}^{-1}$$

$$K_q = 0.03 \text{ mg}N \text{ mg}B^{-1} \qquad R = 0.01 \text{ hr}^{-1}$$

$$V_m = 0.03 \text{ mg}N \text{ mg}B^{-1}\text{hr}^{-1} \qquad \mu_m = 0.1 \text{ hr}^{-1}$$

Begin your simulation with these values for the variables:

$$N = 0.5 \text{ mg l}^{-1} \qquad B = 0.01 \text{ mg l}^{-1} \qquad q = q_o$$

Plot concentrations of external nutrients, internal nutrients, and cell biomass for 240 simulated hours of growth. Use a Δt value of 0.1 hr for the simple Euler integration. Note particularly the relationship of growth to concentrations of free nutrient and internal nutrient.

22.8 Cellular Control of Nutrient Uptake by Internal Nutrient Pool

Rhee (1973) showed that the rate of uptake of phosphate by the alga Scenedesmus is controlled by the cellular concentration of labile polyphosphates. Thus, the size of the internal nutrient pool appears capable of causing inhibition of uptake of phosphate from the external environment. To model this inhibition, Rhee (1973) proposed an equation of this form:

$$\nu = V_m \cdot \frac{N}{(K_n + N)} \cdot \frac{K_i}{(i + K_i)} \tag{22.28}$$

where i is the concentration of inhibitor, in this case polyphosphate, and K_i is inhibitor concentration which causes the maximum uptake rate to be reduced by one-half. DiToro (1980) introduced a generalized equation to model the inhibition of uptake by internal nutrient concentration:

$$\nu = V_m \cdot \frac{N}{(K_n + N)} \cdot \frac{K_i}{(q - q_o + K_i)} \tag{22.29}$$

The control described by this model may account for the variability in the apparent values for V_m and K_n noted by Rhee (1978) and others.

Exercise 22-7: Use the modified model for rate of nutrient uptake, Equation 22.29, to produce a simulation of cellular growth. Use the approach described in Exercise 22-6, with the same values for coefficients and initial concentrations. Include control of nutrient uptake with K_i from this model set at 0.08.

The multistage models for nutrient limitation are intended to account for observations including the lag phase of cell growth in batch culture, luxury consumption of nutrients, and multiple limiting nutrients. Development of models of growth response to nutrients remains an active area of research. These models are a common area of interest for cellular biologists, for cell physiologists, for ecologists concerned with algal and microbial growth, and for bioengineers studying industrial fermentation.

22.9 A Model of Cellular Growth

Models of cell division are an important component of the study of cellular function. They lie between the models of enzyme function and nutrient uptake studied above, and the models of development of organisms to be considered in the next chapter.

Shuler et al. (1979) presented an elaborate model for growth and division of bacteria. Their model involved 14 differential equations for describing changes in concentrations of precursors, protein, nucleic acids, enzymes, carbohydrates, and waste products. The model attempts to consider the effect of cell geometry on growth, and predicts temporal events during the cycles of DNA synthesis and cell division.

A much simpler model for control of cell division is presented here, modified from the description of Alberghina (1977, 1978). The cycle of cell growth is divided into two interdependent systems. One system serves to model the synthesis of ribosomes and proteins, which are the structural components of cell growth. The second system models DNA replication and controls the start of cell division. The model is diagrammed in Figure 22.2.

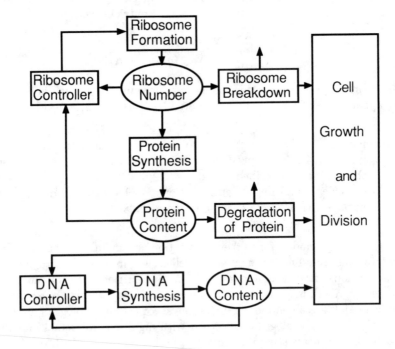

Figure 22.2. Diagram of a conceptual model of some components of the cycle of cell growth and division.

In the first system, the synthesis of proteins and the formation of ribosomes are tightly linked. Protein synthesis depends upon the presence of ribosomes, and a cellular reference ratio of ribosomes to protein is assumed to exist. For mammalian cells, the changes of protein and ribosome content can be modeled with two equations. The first describes the change in numbers of ribosomes in a cell. The number of ribosomes is represented by R. P, the protein content, is expressed as numbers of amino acids polymerized as protein. Then

$$\frac{dR}{dt} = k_1[\rho P - R]_+ - k_2 R \qquad (22.30)$$

The constant ρ represents the cellular reference level of the number of ribosomes that should be present per unit of protein. The bracketed term indicates that ribosomes will be formed at a rate proportional to the difference between number of ribosomes that should be present given the current protein level, and the numbers actually present. (The + subscript indicates that the expression in brackets must be positive or zero; see Section 13.5.) k_1 is the rate constant for formation of ribosomes and k_2 is the rate constant for breakdown of ribosomes. Both have units of time^{-1}.

Change of protein content of a cell is found with

$$\frac{dP}{dt} = k_3[R]_+ - k_4 P \qquad (22.31)$$

k_3 is an activity constant for ribosomes, giving the number of amino acids polymerized per unit time per ribosome. k_4 is the rate constant for protein degradation.

Cell division is controlled by the second system, a simple model of DNA replication. The replication process is assumed to begin only after protein content P exceeds some threshold level P_D. DNA is assumed to replicate at a constant rate, which implies saturation of the synthetic mechanism with the required molecular components. Given this situation,

$$\frac{dD}{dt} = k_5 \qquad (22.32)$$

where D is the cell's DNA content expressed as genome units. In the usual diploid, non-replicating condition of mammalian cells, the number of genome units is unity ($D = 1$). The replication constant k_5 has units of D time^{-1}.

The DNA model system must be constrained so that DNA replication will proceed only until replication is complete and DNA content has doubled, that is, when $D = 2$. The DNA system thus operates when $P > P_D$, and stops when $D \geq 2$. Generally, the time required for replication is $1/k_5$. When replication ends, a refractory period τ_D begins. The cells prepares for division during this period, and then divides at the end of τ_D. The cell content of ribosomes, protein, and DNA is assumed to be halved by the actual division of the cell.

Exercise 22-8: Write a program to simulate cell growth and division using the model described above. Alberghina (1978) provides the following appropriate constants for mammalian cells, based on data from mouse fibroblasts:

$$k_1 = 1 \text{ min}^{-1}$$
$$k_2 = 0 \text{ min}^{-1}$$
$$k_3 = 650 \text{ ribosome}^{-1} \text{ min}^{-1}$$
$$k_4 = 0.333 \times 10^{-3} \text{ min}^{-1}$$
$$k_5 = 2.5 \times 10^{-3} D \text{ min}^{-1}$$
$$\tau_D = 200 \text{ min}$$
$$P_D = 6.25 \times 10^{12} \text{ amino acids}$$
$$\rho = 1.7 \times 10^{-6} \text{ ribosomes (amino acid)}^{-1}$$

Begin your simulation with $R = 8.5 \times 10^6$, $P = 5 \times 10^{12}$, and $D = 1$. Use simple two-stage Euler integration, finding changes in P, R, and D as the first stage, and updating their values in the second. Use a Δt value of 1. Allow the simulation to proceed for 2000 minutes, which should be sufficient time to permit two cycles of cell division. Plot concentrations of P, D, and R.

22.10 Control of Protein Synthesis by Gene Repression

A simple model for the control of protein synthesis by gene repression was proposed by Maynard Smith (1968). His model is based on the conceptual model of Jacob and Monod (1961), which is diagrammed in Figure 22.3. The model describes how the production of ribose nucleic acid (mRNA), needed for the synthesis of an enzyme, is controlled by the concentration of the product. According to the Jacob and Monod hypothesis, the structural gene produces mRNA only when the operator is not interacting with active repressor. The active repressor results from the interaction of the repressor protein with the corepressor molecules. Thus, when the product (corepressor) is in sufficiently high concentration, it interacts with repressor protein to produce active repressor, which "shuts off" the operator and thus stops mRNA production by the structural gene. In bacteria, there is only a single gene responsible for each type of enzyme protein. Additional information about the Jacob and Monod model may be found in most texts of cell biology or biochemistry.

The symbolism for the following model development is found in Figure 22.3. It may be assumed that the process of repression and derepression of the operator involves two reversible reactions.

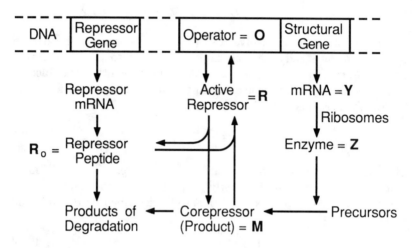

Figure 22.3. Diagram of the conceptual model of gene repression of Jacob and Monod (1961).

The first concerns the interaction of repressor protein with corepressor (product) to form active repressor as follows:

$$M + R_0 \rightleftharpoons R$$

M is the product or corepressor, R_o is the inactive repressor protein, and R is active repressor.

This reaction may be represented by the following equilibrium expression

$$K = \frac{[R]}{[R_o][M]} \tag{22.33}$$

with K the equilibrium constant. The equation can be rearranged:

$$[R] = K[R_o][M] \tag{22.34}$$

Assuming repressor protein R_o is in steady-state between production and degradation, its concentration would be constant. The value for $[R]$ then becomes a direct function of $[M]$:

$$[R] = b[M] \qquad (22.35)$$

where b represents the combined constants of Equation 22.34.

A second reaction describes the interaction of active repressor with the operator to form the repressor-operator complex, OR:

$$R + O \rightleftharpoons OR$$

The reversible reaction involves two processes, repression and derepression. By using steady-state assumptions similar to those employed in Chapter 2, especially in Section 2.3, it may be shown that the fraction of time that the structural gene exists in the unrepressed condition, ϕ, may be described by this equation:

$$\phi = \frac{a}{a + b[M]} \qquad (22.36)$$

where a is a constant. M and b are defined above. The rate of mRNA synthesis is assumed to be directly proportional to the fraction of time that the gene is unrepressed, and thus to be described by the differential equation

$$\left(\frac{d[Y]}{dt} \right)_s = \frac{c}{a + b[M]} \qquad (22.37)$$

where the constant c is the product of a and the proportionality constant for RNA production.

There is a continuous breakdown or decay of mRNA which is assumed to be described by another differential equation for first-order decay:

$$\left(\frac{d[Y]}{dt} \right)_d = -k[Y] \qquad (22.38)$$

where k is the rate constant for breakdown. The overall equation for net change in mRNA is given by the summation of these two equations:

$$\frac{d[Y]}{dt} = \frac{c}{a + b[M]} - k[Y] \qquad (22.39)$$

The rate of enzyme formation and breakdown can be described similarly by this equation:

$$\frac{d[Z]}{dt} = n[Y] - f[Z] \qquad (22.40)$$

where n is a rate constant for mRNA-dependent synthesis of enzyme, with units of enzyme mRNA^{-1} time^{-1}. f is a rate constant for first-order decay or breakdown of the enzyme.

If the enzyme is assumed to be saturated with the precursor P required for the formation of corepressor-product M, then the rate of formation of M is a direct function of enzyme concentration. The rate of breakdown or conversion of M to something else is assumed to be directly dependent on its concentration. The differential equation for change in the concentration of corepressor-product therefore resembles that of the enzyme:

$$\frac{d[M]}{dt} = g[Z] - h[M] \qquad (22.41)$$

where g is a rate constant for enzymatic synthesis of M and h is a rate constant for first-order decay of M.

Exercise 22-9: Write and implement a program to simulate the control of protein synthesis following the Jacob and Monod model, based on Equations 22.39, 22.40, and 22.41. Simple two-stage Euler numerical integration is adequate. Begin your simulation with $[Y] = [Z] = [M] = 0$. Set $\Delta t = 1$. Use the following values for constants in your simulation:

$$a = 5 \qquad b = 0.00025 \qquad c = 1 \qquad f = 0.1$$

$$g = 100 \qquad h = 0.1 \qquad k = 0.1 \qquad n = 2$$

After some initial fluctuations, the system should stabilize at concentrations of $[Y] = 1$, $[Z] = 20$, and $[M] = 20000$.

22.11 A Stochastic Simulation of the Operon Model

A deterministic version of the Jacob and Monod model was given in the previous section. The rate of mRNA transcription from a structural gene was assumed to be a function of the fraction of time the operon was in the unrepressed state. In bacteria there is only a single DNA molecule, and presumably only a single operon for a given structural gene. At any given time, mRNA transcription will be either fully on or fully off. Thus, the system provides an interesting example of statistical dynamics, and can be described with a model based on the probability of the state of the operator. Such a model will be of the general type discussed in Chapter 21.

The operator controlling the operon may be considered to exist in either of two possible states. The various probabilities associated with the transition between the states are shown in Figure 22.4. In this system, O_r is the operator in the repressed condition, and O_u is the operator in

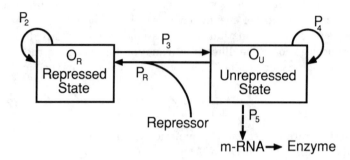

Figure 22.4. Diagram of gene repression system, showing function of probabilities of transition.

the unrepressed condition. The probability of converting from O_u to O_r during any time interval Δt is designated as P_r. This probability is a function of repressor concentration, and follows the equation

$$P_r = 1 - (1 - P_1)^R \qquad (22.42)$$

in which R is the number of repressor molecules present. P_1 is the probability of interaction between O_u and a single repressor molecule, and P_r is the probability of interaction between O_u and at least one repressor molecule. P_1 is a constant for the system, while P_r varies with R.

For any single time unit Δt the probability of remaining in the repressed condition is P_2, the probability of reverting to the unrepressed condition is P_3, and the probability of remaining in the unrepressed condition is P_4. Note that

$$P_3 = 1 - P_2 \qquad (22.43)$$

and that

$$P_4 = 1 - P_r \qquad (22.44)$$

During any time interval Δt that the operon is in the unrepressed condition, the probability of an mRNA molecule being transcribed is P_5. This probability is constant, but results in the transcription of a single mRNA molecule only when the operator is in the unrepressed state.

The mRNA is assumed to be degraded following first-order kinetics, as in Equation 22.40 above. The approximate expression for the decay over an interval of time Δt is:

$$Y_{t+\Delta t} = [Y_t - kY_t]_+ \qquad (22.45)$$

where Y_t represents the amount of mRNA present at the beginning of the time interval Δt, and k is the decay rate constant. The rate of synthesis of enzyme is assumed to be a direct function of the amount of mRNA present, as in Equation 22.41. The approximate equation describing this relationship over an interval Δt is

$$Z_{t+\Delta t} = [Z_t + nY_t - fZ_t]_+ \qquad (22.46)$$

Z_t represents the concentration of enzyme at time t. As in Equation 22.41, n is a rate constant for mRNA-dependent enzyme synthesis, and f is the constant for first-order degradation of the enzyme. (The + subscript in the two equations indicates that the values of Y and Z must be positive or zero, but never less than zero; see Section 13.5.)

Exercise 22-10: Write and implement a program for the stochastic model of the operon described above. Use the following probability values:

$$P_1 = 0.001 \qquad P_2 = 0.99 \qquad P_5 = 0.5$$

For k, n, and f use the values given in Exercise 22-9. Set $\Delta t = 1$. Vary the number of repressor molecules as follows: 0, 1, 2, 4, 8, 15, 30, 60 125, 250, and 500. For each of these numbers, run the simulation for 1000 time units. Over the 1000 units, find the fraction of time spent in the unrepressed state, the average number of mRNA molecules present, and the average enzyme concentration. Begin each simulation run with Y and Z set to zero. Repeat the simulation for a total of three runs with each repressor number, in order to obtain an estimate of the variability caused by the randomness of the simulation. Your output should be in a tabular format, with columns for number of repressor molecules, fraction of time spent unrepressed, average amount of mRNA present, and average concentration of enzyme.

Conclusion

The models in this chapter have the common theme of function and growth of cells. The models selected are interesting representatives, but do not really begin to cover the diversity of models that have been developed for the subject material. Most of the exercises produce simulations that strain the speed limits of conventional BASIC running on most personal computers. The problem is not that the computers cannot perform the simulations; rather, it simply becomes inconvenient to interact with the personal computer when asking "what if..." questions based on results of previous simulations.

CHAPTER 23

MODELS OF DEVELOPMENT AND MORPHOGENESIS

A single cell, the fertilized egg, grows and proliferates to become an adult during the development of multicellular organisms. Biologists always have been fascinated by the great complexity produced by such apparent simplicity. Research and experimentation on development has proceeded along several rather different paths. There have been attempts to find the factors responsible for morphogenesis, which is the placement, pattern and shape of organs during development. At the cellular level biologists have studied the interactions among neighboring cells that produce patterns of differentiation. At the molecular level they have been concerned with the chemical products of differentiated vs. precursor cells, and of course with the activity of particular genes and their products.

Mathematical models for developmental processes have been published infrequently compared with models for other biological disciplines. There are probably a variety of reasons for this. Partly there is a traditionally strong research emphasis on description of development. Another reason is that developmental processes are so complex that simulations based on simple mathematical models rarely produce interesting insights. However, the published models of development are extremely diverse, covering a variety of developmental stages and an assortment of developmental processes for a number of plant and animal species (Ransom 1981, Malacinski and Bryant 1984).

Most models of development may be put into either of two general groups. Both are based on the elaboration of complex patterns from initial simplicity. The first group includes the models of cell sorting. In laboratory experiments, cells have been observed to sort themselves according to type. Models of this process are characterized principally by simulated movement of cells in two- or three-dimensional arrays. Some of the complex patterns of cell aggregation of developing organisms have been produced with these models (Sampson 1984, Malacinski and Bryant 1984, Meinhardt 1982).

The second group includes models of cellular proliferation. Cell division is modeled in one-, two- or three-dimensions, following some fixed set of "genetic" rules for growth and differentiation. Usually a biochemical gradient is involved as part of the growth process. Patterns that can be observed in development of multicellular organisms have been produced successfully using models of this second group.

In this chapter we will examine several models that may be used to simulate cell-to-cell interactions during development, and some others that can produce morphological patterns using simple, reasonable mechanisms.

23.1   A Model of Filament Growth

A simple model of growth in one dimension will be considered first because it illustrates some simulation techniques that are needed with other models. The system being modeled may be thought of as an unbranched filament of algal cells. Each cell in the row of cells can grow, and each will divide when it reaches a sufficiently large size. We are interested in modeling the number of cells in the filament and the size of the individual cells.

During a small unit of time, each cell is assumed to grow by a small fixed amount. At the end of each time unit, every cell in the filament is checked to find whether it has reached the size at which cell division is necessary. If it has become large enough, then the cell is divided into two smaller daughter cells that take its place in the filament. The total number of cells in the filament will increase by one.

A program based on this model for simulating growth will start with some number of cells, n, in the filament. A subscripted variable, for example C_i, can represent cell size. The simulation will start with cell sizes C_1, C_2, C_3, ... , C_n being assigned initial values. The simulation begins by increasing each C_i a small amount to represent cell growth. Then each C_i is checked to find whether it has exceeded C_{max}, the maximum cell size. If some cell, for example C_4, has exceeded this limit, the cell is divided into two cells. The filament length n will be increased by 1, and all C_i with i greater than 4 will be "pushed" into C_{i+1} . After all the cells have been checked, the growth process will resume. This model of growth is diagrammed in Figure 23.1.

Exercise 23-1: Write and implement a BASIC program to simulate growth of an algal filament based on the model explained above. Begin the simulation with a single cell ($n = 1$), having a size $C_i = 10$ units. Each cell in the filament will grow by one size unit for each unit of time. The cells should divide when they reach a size of 16 units or greater. When cell division occurs, assume that one

Before Division

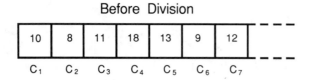

After Division

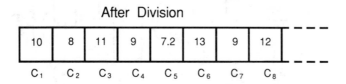

Figure 23.1. Diagram of growth process of algal model. Values of cell content C_i are shown inside cells. Cell 4 divides, and the daughter cells receive 50% and 40% of its contents. Cells originally numbered 5 and above are "pushed" up one number, and filament length increases by one.

daughter cell will receive 0.5 of the contents of the mother cell, and the other will receive 0.4 of the mother cell content; the missing 0.1 is assumed lost to the increased respiration of division. The output of your program should be a histogram showing the cell number on the x-axis and the size of the cell on the y-axis. Produce a set of histograms showing the results of growth after times of 12, 24, 36 and 48 units. A flowchart showing one method of programming this type of growth is given in Figure 23.2.

23.2 Chemical Gradients in Morphogenesis

Growth processes like the one described above are needed by most models of development. However, a principal objective of developmental models is simulating differentiation, so that model cells will develop differently after they are produced by a common mechanism. These models necessarily include a mechanism for changing cell form or function as the cells proliferate and grow. In most developmental models, part of the necessary mechanism is the presence of chemical gradients across developing cells. If the models also include chemical thresholds that switch different modes of development, then cells can develop differently at different chemical concentrations. Such models imitate the ability of genes to respond in a qualitative on-off manner to chemical repressors and derepressors. Many models of development are a combination of diffusion models from Chapter 14 and control models from Chapter 17.

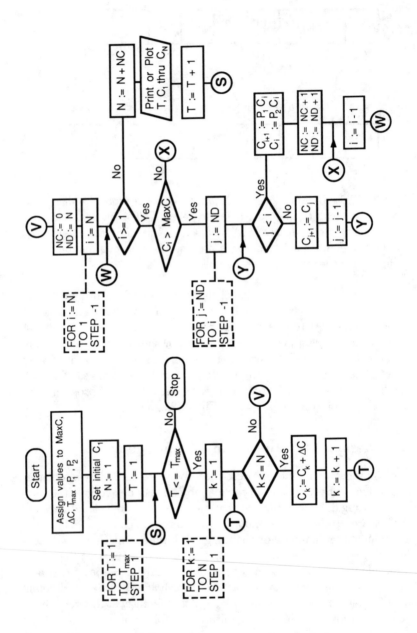

Figure 23.2. (opposite) Flowchart for simulation of growth of an algal filament. Filament begins with a single cell.

C_n is size of cell n.

$MaxC$ is maximum cell size before division; a constant.

ΔC is cell growth per unit of time; a constant.

P_n is proportion of mother cell going to daughter cell n; a constant.

N is number of cells in the filament.

T is a counter for the time loop.

NC is a counter for number of new cells in one time unit.

ND is a counter for increased length of filament in one time unit.

k is the counter for cell growth loop.

i is the counter for cell division loop.

j is the loop counter for "pushing" loop.

The models which follow in this chapter assume the existence of chemical gradients. The three basic mechanisms used to establish gradients are the diffusion of a substance between adjoining cells, the production of the substance, and the loss of the substance by degradation, deactivation, absorption or diffusive loss to environment. You have studied all of these mechanisms in previous chapters, so the following discussion is brief.

Production of a substance usually is assumed to be confined to small areas of the developing system. If substance is produced at a constant rate at site or cell i, the production will follow a familiar equation:

$$\left(\frac{dS_i}{dt}\right)_{\mathrm{P}} = k_P \tag{23.1}$$

where S_i is the amount or concentration of substance, and k_P is the production constant with units of S time^{-1}. In many models, the production equation may be modified:

$$\left(\frac{dS_i}{dt}\right)_{\mathrm{P}} = k_P Q_i \tag{23.2}$$

where Q_i may be an on-off "switch" with a value of either 1 or 0 depending upon some external or internal control or threshold (see Chapter 17). Q_i may also indicate density of some producer mechanisms in cell i, in which case k_P takes on units of S time^{-1} Q^{-1}. An example of this latter case is the control of protein synthesis by number of ribosomes, described in the previous chapter (Equation 22.31).

Loss of substance from cell i is modeled frequently with the equation for first-order decay:

$$\left(\frac{dS_i}{dt}\right)_{\text{L}} = k_L S_i \qquad (23.3)$$

where k_L is the rate constant for loss, including degradation, absorption, etc. k_L is usually a negative value.

Diffusion between adjacent cells has been described in Section 14.2. An algebraic combination of the flow equations explained there gives a convenient expression for the rate of change of substance due to diffusion:

$$\left(\frac{dS_i}{dt}\right)_{\text{D}} = k_D \left(S_{i-1} + S_{i+1} - 2S_i\right) \qquad (23.4)$$

where S_i is concentration of the diffusing substance in cell i of a series of cells (as in Figure 23.1). S_{i-1} and S_{i+1} are the concentrations in adjacent cells. k_D is the diffusion constant. Because most development models assume cell volume to be constant, S may indicate either concentration or amount.

Depending upon the requirements of a model, these expressions may be combined to find the net change in S_i:

$$\frac{dS_i}{dt} = \left(\frac{dS_i}{dt}\right)_{\text{D}} + \left(\frac{dS_i}{dt}\right)_{\text{L}} + \left(\frac{dS_i}{dt}\right)_{\text{P}} \qquad (23.5)$$

Equations such as this, called diffusion-reaction equations, may be solved with precision suitable for most models using the usual two-stage simple Euler integration technique.

23.3 Differentiation of Algal Filaments

A simple model of response to a gradient may be used to simulate growth of an algal filament having two different kinds of cells. This type of filament growth is observed for some blue-green algae (cyanobacteria). In species of *Nostoc*, *Anabaena*, and *Aphanizomenon*, the filament consists of the usual vegetative cells, with special cells called heterocysts occurring at intervals along the filament. The pattern of occurrence of heterocysts differs among species. Here we will follow the pattern for *Anabaena*, in which heterocysts occur at regular intervals along the filament, usually separated by 8 or 9 vegetative cells. The vegetative cells can divide regularly (as in Section 23.1 above), but heterocysts do not divide. When a filament lengthens so that 16 or 17 cells lie between a pair of heterocysts, a vegetative cell in the center of the length develops as a heterocyst.

The growth of a filament with two cell types may be modeled by assuming that heterocysts produce some chemical that inhibits the formation of heterocysts. This inhibitor diffuses along the filament away from each heterocyst (Baker and Herman 1972). As it moves from cell to cell, inhibitor may be lost by absorption or deactivation. When the inhibitor level reaches a sufficiently low level in a cell, a daughter cell may begin to develop as a heterocyst, and to produce inhibitor also. In *Anabaena*, the rates of production, diffusion, and loss are such that a distance of about 8 cells is required to decrease the concentration of inhibitor to a level allowing development of a heterocyst.

An elementary model of heterocyst formation can be based on some "genetic rules" which are biologically reasonable. The terminology is based on a filament that is n cells long, numbered consecutively 1, 2, 3, ... , i, ... , n, with concentration of inhibitor in cell i indicated by H_i.

1) Individual cells do not grow. The filament as a whole grows lengthwise with a random process. At regular time intervals, a single cell randomly chosen from the whole filament divides. In this simple case the cell, including its inhibitor concentration H_i, is merely duplicated as H_i and H_{i+1}.

2) Heterocysts do not divide.

3) Diffusion of inhibitor between cells follows Equation 23.4 above. The extra diffusive loss to the environment by the end cells with their larger free surface is modeled by setting up dummy cells 0 and $n + 1$, with $H_0 = H_1/2$ and $H_{n+1} = H_n/2$.

4) Heterocysts produce inhibitor at a constant rate, following Equation 23.2 above. A heterocyst is distinguished from a vegetative cell for modeling purposes solely by having the on-off switch for inhibitor turned on ($Q_i = 1$). For the vegetative cells the switch is off ($Q_i = 0$).

5) The switch is turned on in a cell when inhibitor concentration H_i drops below a threshold level H_P. No mechanism exists for turning off a switch.

6) Degradation of inhibitor in each cell including heterocysts follows a simple first-order decay function (Equation 23.3).

7) The processes of diffusion, production and degradation are rapid relative to the division process.

Exercise 23-2: Write a program that follows the rules above to simulate the growth of an algal filament with heterocysts. The flow of your program should be approximately as follows: (1) A cell division interval is initiated with diffusion, production, and degradation of inhibitor occurring over a period of T units of time. A check for levels of inhibitor below H_P is performed at each time unit; production of inhibitor is switched on when $H_i < H_P$. (2) After the lapse of T time units, a cell is chosen randomly to divide, and the pushing routine from Exercise 23.1 is then applied. (3) After division, the next interval of cell division begins.

Begin your simulation with a filament of 4 cells, with cell number 1 being a heterocyst. Start with inhibitor concentrations for cells 1 through 4 of 0.667, 0.333, 0.167, and 0.083 respectively. Set H_P = 0.010, T = 25 time units, and Δt = 1. Let $k_D = k_p = 0.1$ and $k_L = -0.05$. Allow your simulation to proceed for at least 40 intervals of cell division. Output may take several formats. An interesting output that displays growth of the filament is shown in Figure 23.3, and can be produced with the circle and block plotting features of the GRAPH program.

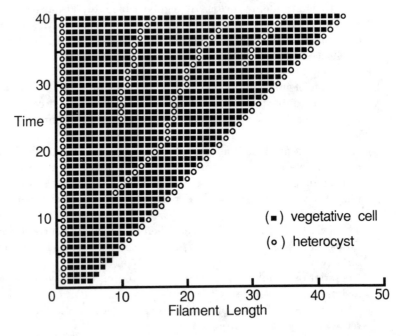

Figure 23.3. Output of a simulation for growth of an algal filament with heterocysts.

23.4 A Model of Pattern Formation in Plant Growth

As plants grow they produce leaves to absorb light and gases for photosynthesis. Leaves develop on plant stems in various patterns. In general the leaves appear positioned to optimize light absorption. The most common pattern in broad-leafed plants is a spiral arrangement, with single leaves growing in a spiral around the axis of the stem. A simple, general model for simulating spiral growth relies on the diffusion of a "leaf inhibitor", a substance produced by a developing leaf. This inhibitor will prevent the growth of other leaves until it drops below a threshold level at some point on the stem. (The process is exactly like the model system of heterocyst inhibition described in the previous section.) Named the "morphogen field theory" by botanists, this model of leaf development is described here based on the explanation of Charles-Edwards et al. (1986).

A cross-section of the growing tip (apical meristem) of a plant stem that is producing leaves is shown in Figure 23.4a. We can imagine a new leaf, a "primordium", starting to grow at point A on the circumference of the stem, and beginning to produce the leaf inhibitor. The substance will begin to diffuse through the cells around the periphery of the stem; it cannot effectively diffuse through the woody center of the stem. As in the model of algal heterocysts, a diffusion gradient of inhibitor will be set up across the cells. This gradient will produce a minimum concentration on the side of the stem opposite the primordium (Figure 23.4b). We expect that another leaf primordium will begin to develop at this point. The new leaf may be located further up the stem, where the inhibitor has dropped below the threshold level because of diffusion along the stem length.

The second primordium will begin to produce inhibitor also. The combined production of inhibitor by the two primordia will produce a diffusion gradient with a peak at A ($0°$ or $360°$) and at B ($180°$). A third primordium is therefore likely to develop at one or the other gradient minima between A and B (Figure 23.4c). The location of the minimum C is not likely to be precisely between A and B, at $90°$ or $270°$. The location of the minimum will depend upon the relative rates of production of inhibitor by A and B.

If we allow the system to proceed further, and a new third primordium at C begins to produce inhibitor, the result will be a fourth primordium located at point D, etc. When this system of inhibition is repeated several times as a stem grows and produces leaves, the angle between successive leaf sites will take on a constant value. The value will determine the leaf arrangement of the plant. Thus, leaf arrangement depends on relative rates of production of inhibitor by successive leaf primordia.

If the rates of inhibitor production are such that the ratio of concentration at B and A is 2:1, the angle between B and C will be $137.507\ldots°$, the Fibonacci angle. In fact, a large majority of plants that have a spiral arrangement of leaves around stems have the successive leaves growing at angles of $137.5°$ from each other. Mathematicians always have been intrigued by the theoretical properties of this angle; see Dixon (1981) for

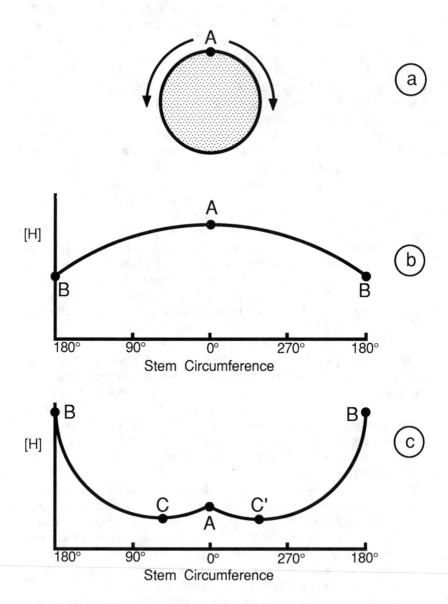

Figure 23.4. Diagram of morphogen field theory for leaf inhibitor. (a) Cross section of stem producing leaves. First leaf grows at A, producing inhibitor. Inhibitor diffuses around the stem away from A. (b) Concentration of inhibitor around stem periphery. Leaf is at A. Second leaf will grow at B, 180° from A. (c) Inhibitor concentration around stem, with leaves at A and B. Third leaf will begin to grow at C or C'.

example. From the practical viewpoint of the plant, the angle maximizes the exposure to sunlight of successive leaves on a stem, providing minimal self-shading.

This model diffusion mechanism can also apply to the arrangement of florets in the head of composite flowers. The florets grow in a spiral from the center of the flower head of plants like daisies (*Bellis*), sunflowers (*Helianthus*) and dandelions (*Taraxacum*). Florets are evolutionarily modified leaves, and are produced at an angle of about 137.5° from each other around the central area of the composite flower. The growing angle of 137.5° results not only in the primary spiral of florets as they are produced, but also in secondary spirals that run clockwise and counterclockwise in apparent designs.

Exercise 23-3: Generate a spiral of dots at successive angles of 137.50776°, to simulate the positions of florets growing in the head of a sunflower. A point may be generated around the origin of an *x-y* graph with

$$X = R \cos \theta$$

$$Y = R \sin \theta$$

where R is the radius from the origin and θ is the angle. (Most BASIC trigonometric functions use angular measurement of radians, which will require conversion of the 137.50776° angle.) "Grow" your flower beginning with $R = 0$ and $\theta = 0$. Increase R by one unit and θ by 137.50776° for each successive floret. Plot 1000 florets as single points on an *x-y* graph. The secondary spirals created by this angle of growth should be apparent.

Note that you may have to adjust the minimum and maximum values of x and y of your graph axes so that your flower will appear circular rather than oval on your video screen or printer.

23.5 A General Inhibitor-Activator Model

Biochemical gradients are important in the development of all embryos, but it is difficult to model gradients in the spherical embryos that are typical of most vertebrate species. As a result, the experimentation and simulation with gradients in development have focused on organisms that have obvious lengthwise components in development. Hence, there has been considerable work on embryos of insects, which have early developmental events taking place on a longitudinal axis, and on organisms like the hydras, which are similarly organized. The vertebrate limb, particularly in amphibia with regenerating capacities, has served as an experimental and simulation model also.

Gierer and Meinhardt (1972) and Meinhardt (1977) based a general model of pattern formation on three assumptions: (1) production and diffusion of an activating substance and (2) an inhibiting substance, and (3) a gradient of sources of activator and inhibitor. The processes involved with these are almost exactly like those of the preceding sections: production, diffusion, and loss including degradation, diffusion, etc. The following description is of the simplest and most general of several variations of their basic model.

The activator will activate the production of both inhibitor and activator. Inhibitor will inhibit only the production of activator. An allosteric behavior is assumed for the activator, so that two activator molecules are needed to cause the production of inhibitor and activator. A single inhibitor molecule can inhibit the production of activator.

Along the length of the developing organism, at any site i, the rate of change in activator concentration A_i is described with the following equation:

$$\frac{dA_i}{dt} = p_o a_i + p_A a_i \frac{A_i^2}{H_i} - m_A A_i + k_A \left(A_{i-1} + A_{1+i} - 2A_i\right) \qquad (23.6)$$

where a_i is the density of source and A_i is activator concentration at site i. A_{i-1} and A_{i+1} are concentrations of activator at adjacent sites $i-1$ and $i+1$. p_o is the constant for a small, basal rate of activator production. p_A, m_A, and k_A are rate constants for production, loss, and diffusion respectively. H_i is the concentration of inhibitor, and its rate of change is described with the equation

$$\frac{dH_i}{dt} = p_H h_i A_i^2 - m_H H_i + k_H \left(H_{i-1} + H_{i+1} - 2H_i\right) \qquad (23.7)$$

where h_i is density of the source of inhibitor, and p_H, m_H, and k_H are rate constants for production, loss, and diffusion of inhibitor.

Exercise 23-4: With a slight gradient of source density, the inhibitor-activator model can produce a steep gradient of activator. Write a BASIC program based on Equations 23.6 and 23.7 to simulate this phenomenon. Assume the developing organism is divided lengthwise into 20 segments. The densities of source of activator at each segment, a_i, should be set up with a gradient as follows:

$$a_1 = 0.60, \quad a_2 = 0.59, \quad a_3 = 0.58, \quad a_4 = 0.57, \ldots, \quad a_{20} = 0.41$$

For the simulation, assume that inhibitor densities h_i equal a_i for all segments. Use the following constants for the equations:

$$p_o = 0.0006 \quad p_A = 0.05 \qquad m_A = 0.0035 \quad k_A = 0.001$$

$$p_H = 0.025 \quad m_H = 0.0045 \quad k_H = 0.45$$

Set initial values of activator $A_i = 0.1$ in all segments, and values of inhibitor H_i similarly at 5.0. Use simple two-stage Euler integration with $\Delta t = 1$ to solve the equations. As output plot A_1 over time. The values of A_1 should stabilize at about 20.36 units after 4000 time units, following a sharp rise to about 30 units at 1000 time units. Write your program so that at the end of 4000 time units it will also produce a graph of A_i, H_i, and a_i over the values of i from 1 to 20. You may also wish to obtain a printed listing of the values of A_i and H_i for use in Exercise 23-5 below.

As in Section 23.3, using "dummy" sites may make the programming of diffusion easier. In this case, sites 0 and 21 are set up, with these two sites having concentrations of inhibitor and activator always equal to those of sites 1 and 20, respectively. This simulation requires a relatively large amount of computing time with most BASIC interpreters.

Sampson (1984) notes that Gierer and Meinhardt's (1972) activator-inhibitor model is in fact only an amplifier of a pre-existing gradient, and because of this it does not really describe the production of gradients during development. The criticism may not be very severe. The eggs of the fruit fly *Drosophila*, for example, are laid with existing chemical gradients produced by nurse cells of the ovary (Ingham 1988). The model of Gierer and Meinhardt is for relatively rapid processes of morphogen production and distribution. These processes are distinct from the slower, more permanent processes of cellular morphological differentiation, which are presumably caused by the morphogens.

The model was applied to the development of hydras, which are common small freshwater coelenterates (Gierer and Meinhardt 1972, Bode and Bode 1984). Hydras have remarkable regenerative capacities, including an ability to regenerate a head (tentacled region) from a small cut-out section of the middle of the body. In addition, the regeneration may be altered by grafting portions of other individuals onto developing pieces. The determination of the fate of the various pieces is rapid compared with actual growth and differentiation of cells.

The fundamental idea of the activator-inhibitor model is that the activator and inhibitor are produced and diffuse rapidly. These then determine the subsequent cellular differentiation, which will in turn slowly establish and modify the gradient of morphogen sources (Meinhardt 1984).

Exercise 23-5: A simple, common biology experiment with hydra involves the removal of the region with tentacles; this region will regenerate completely. This experiment can be simulated with Gierer and Meinhardt's model by assuming the gradient established in Exercise 23-4 is that of a hydra, with a tentacled region located at segments 1-4. If these sites are removed from the equilibrium situation of Exercise 23-4, a new steep gradient of activator will be produced, beginning at segment 5.

Write a program to simulate two simple hydra transplantation experiments. Using the equilibrium values of Exercise 23-4, remove the tentacled region, represented by segments 1-4. "Transplant" segments 5-8 from "another" hydra at equilibrium to the front of your hydra and allow the simulation to begin. That is, your modified hydra will have segment source densities of .56, .55, .54, .53, .56, .55, ... , .41. The value of A_i and H_i will be similarly duplicated for the first four segments. After the simulation runs for about 4000 time units, new equilibrium values of activator and inhibitor will be established, showing a high concentration of activator at the end of the animal that had the original segments 1-4.

For the second simulation experiment, begin as above by removing segments 1-4 from an equilibrium hydra. However, for this experiment replace them with segments 5-12 of another individual. (This will result in a hydra with a length of 24 segments.) Allow the simulation to proceed for 4000 time units. An equilibrium should develop with a peak concentration of activator in the middle of the body of the hydra. This simulates a concentration of activator that will leading to the production of a tentacled region in the central segments of the manipulated hydra. Such results have been observed experimentally, with the location of the tentacles depending on the length of the transplanted segment.

23.6 Morphogenetic Patterns

Wilby and Ede (1974) developed a model for a simple diffusing morphogen that relies on a switching function to develop a distinct pattern of the morphogen. Their model follows the fundamental diffusion-reaction model of Section 23.2. Wilby and Ede included a double switch to control production and loss of the morphogen.

The basic structure of their model is like that of the previous sections, with a series of connected sites or cells that function identically. The model requires an "initiator" region of a cell or cells at one end of the series, that is actively producing the morphogen at the start of the simulation. The morphogen flows through the series of cells following the usual diffusion process. Wilby and Ede postulated that there was in each cell a switching response to morphogen concentration. The production switch is off at low concentrations. As concentration increases and exceeds a threshold value, the production switch is turned on, and the site begins to produce morphogen. When concentrations increase further, past a second higher threshold, the productivity switch is turned off, and a switch for active degradation is turned on. The switches are thrown irreversibly. The mechanism is not out of line with known genetic mechanisms of development. The result of the model is a gradient of morphogen across the series of cells. The on-off switching will produce a series of peaks and valleys of concentrations.

As an elaboration of their model, Wilby and Ede set up a two-dimensional array of cells, and allowed diffusion in two dimensions. When the two-dimensional array was shaped like a developing vertebrate limb with a morphogen destroying boundary, patterns of morphogen concentration resembled that of cartilage location in the limb.

Exercise 23-6: Set up a simulation based on the Wilby and Ede modification of the model of Section 23.2. Use a series of 40 cells, each holding a concentration of morphogen M_i, for $i = 1$ to 40. Use the following for the initial concentrations:

$$M_1 = 1.5, \quad M_2 = 1.3, \quad M_3 = 1.1, \quad M_4 = 0.8,$$

$$M_5 = 0.5, \quad M_6 = 0, \dots, \quad M_{40} = 0$$

As a diffusion constant, set $k_D = 0.1$. Use a rate constant of 0.3 for production, with irreversible thresholds of 1.0 (switch on) and 5.0 (switch off). Use a degradation constant of 0.9, with an irreversible switch-on threshold of 5.0. Set the concentrations of dummy end cells (0 and 41) at 0.8 of the levels of the actual end cells (1 and 40). Write your program so that it can produce a separate graph at each interval of time showing the concentrations of morphogen plotted against number of cell in the series. Write your program to produce printouts of these graphs at times of 50, 100, 200 and 350 units of time.

Conclusion

The study of development and morphogenesis has made rapid progress in recent years, and significant progress has been made in the understanding of the immediate genetic control of molecular and cellular mechanisms of development. The current experimental emphasis on molecular genetics has produced an abundance of factual material that has not been successfully incorporated in simple models of development. The fundamental diffusion- reaction models explained in this chapter provide the basis for complex computer simulations. Given the relatively large amount of computing time needed for the simple models above, and the potential complexity of models of multiple gradients over many sites in three dimensions, it is apparent why supercomputers are needed for simulating even moderately complex developmental systems.

CHAPTER 24

MODELS OF EPIDEMICS

Simulation of the spread of infectious disease among individuals of a population is one of the really practical applications of computer simulation based on biological models. Models of epidemics have been of clear value as part of human and veterinary medical studies for several decades. They have been used less frequently in studying natural populations, usually because data for fitting the models are not available. Many of the models used in the study of epidemic disease also have been useful in sociological studies of the spread of rumors.

The vocabulary of epidemiology involves some restricted definitions for terms in general use. When a pathogen is transmitted to a new host, there may be a latent period between the time of infection and the beginning of the host's infectious period. The infectious period lasts as long as the host is capable of transmitting the disease to other individuals. The incubation period is the time between transmittal and the appearance of disease symptoms. The incubation period may be independent of the latent and infectious periods, depending upon the pathology of the disease. For some diseases there is an immune period after the infection has run its course. An immune individual cannot be infected, but may still be capable of transmitting the disease.

The biology of disease transmission results in several possible classes of the host organisms, which are important in modeling epidemics. A susceptible individual is not infected with the pathogen; the disease can be transmitted to susceptibles. A carrier is usually at some stage of infection, and can transmit the disease to susceptible individuals. During an epidemic, individuals may be removed from the infected group and thus be unable to infect susceptibles further. For example, diseased individuals in human or agricultural populations are frequently isolated or quarantined. Recovery and subsequent immunity also will remove individuals from the epidemic population, as will death caused by the disease. Vaccination for the pathogen will transfer an individual from the class of susceptibles to the removed class without passing through the class of infected carriers.

24.1 The General Epidemic Model

Most of the concepts used in the remainder of this chapter are illustrated in the classic general model for epidemics. The model was formulated by Kermack and McKendrick (1927, 1932; see also Lauwerier 1984). The biological characteristics of epidemic disease are simplified for the general epidemic model described here. Hosts can exist in only three classes: susceptibles, infected carriers, and removals. For simplicity, the latent period is assumed to be zero, so that all infected individuals are also carriers that infect susceptibles. The removed individuals have had the disease, but no longer participate in its spread because of immunity, isolation, or death.

The general model assumes a one-way movement of individuals through the successive classes of susceptibles, infected carriers, and removals. The population size (or density) is N, the number of individuals susceptible to the disease is S, the number of infected carriers is C, and the number of removals is R. The total population N at any time during the epidemic is the sum of the three host classes:

$$N = S + C + R \qquad (24.1)$$

Note that in the general model, deaths due to the epidemic disease do not decrease the value of N.

Like the mass action population models of Chapter 7, the rate of contact between carriers and susceptibles will be proportional to the product of their numbers. If the rate constant for disease transmission is k_1, then the rate at which susceptibles are infected and become carriers will be

$$\frac{dS}{dt} = -k_1 SC \qquad (24.2)$$

The change in numbers of carriers is determined by the rate at which susceptibles become carriers, and by the rate at which infected carriers are removed:

$$\frac{dC}{dt} = k_1 SC - k_2 C \qquad (24.3)$$

where k_2 is the coefficient for removal. The change in the number of removals is given by

$$\frac{dR}{dt} = k_2 C \qquad (24.4)$$

An epidemic simulation begins with a population N consisting mostly of susceptible individuals S, and a small number C of infected organisms. Removal numbers R are usually assumed to be zero initially.

Exercise 24-1: Write and implement a program to follow the course of a general epidemic involving S, C, and R, the three classes described above. Use coefficient values of 0.01 for k_1 and 0.02 for k_2. Set the total population size N to 20, and begin the simulation with a single carrier, so that $S = 19$, $C = 1$, and $R = 0$. Use simple Euler integration with $\Delta t = 0.1$. Plot the values of S, C, and R over 160 time units.

24.2 The Epidemic Curve

In addition to the decline in susceptibles and the corresponding increase in carriers, epidemiologists are frequently interested in the numbers of infected individuals that appear through time. Information about the new cases per unit of time will produce the "epidemic curve" when plotted through time. For example, new cases per week can be graphed over a period of several weeks. For some diseases, the epidemic curve may also be obtained by plotting deaths per unit time or hospitalizations per unit time. For simple epidemics the curve rises to a single peak and then symmetrically declines. An epidemic curve of new cases for simulations based on the general model can be obtained by plotting $-dS/dt$ from Equation 24.2 over time. The negative sign is needed because S always decreases during the course of the simple epidemic.

Exercise 24-2: Modify your program from Exercise 24-1 to produce a graph showing the epidemic curve for that exercise. Beginning from the start of the epidemic, plot $-dS/dt$ over time at intervals of $\Delta t = 0.1$. Use rate constants and initial values from Exercise 24-1. Estimate the area under the epidemic curve by accumulating the calculated values of $-dS$. The area under the curve should equal approximately the initial value of S.

24.3 An Analytical Model of Simple Epidemics

Bailey (1964) developed an elementary epidemic model based on further simplification of the simple general model given above. The biological characteristics of the epidemic disease are simplified by assuming only two classes of hosts: susceptibles and infected carriers. The latent period is zero, and the rate of removal also is assumed to be zero, so infected individuals remain carriers during the entire epidemic, with no death, recovery, or immunity. These assumptions reasonably fit some known diseases of plant and animal populations (Poole 1964). They also match approximately the characteristics of the common cold spreading through small, closed populations of humans, e.g. a school classroom.

As above, the population size is N, the number of individuals susceptible to the disease is S, and the number of infected carriers is C. Before the epidemic begins, all individuals are assumed to be susceptible. At any time during the epidemic,

$$N = S + C \qquad (24.5)$$

If the rate of disease transmission is k, then the rate at which susceptibles are infected and become carriers may be described as in Equation 24.2:

$$\frac{dS}{dt} = -kSC \qquad (24.6)$$

For this simple model, population size N is held constant and the equation may be written as

$$\frac{dS}{dt} = -kS(N - S) \qquad (24.7)$$

You may notice that this equation is very similar to the limited growth equation of Chapter 1 (Equation 1.16). It has a similar analytical solution:

$$S_t = \frac{S_0 N}{S_0 + (N - S_0)\, e^{kNt}} \qquad (24.8)$$

For any time after the epidemic begins at $t = 0$, this equation shows numbers of non-infected susceptibles (S_t) remaining in the population.

For simplicity, we can assume that a single individual becomes infected to start the epidemic. As the epidemic begins, C is 1 and S is $N - 1$. Equation 24.8 then simplifies to

$$S_t = \frac{(N - 1)N}{N - 1 + e^{kNt}} \qquad (24.9)$$

Exercise 24-3: Write and implement a BASIC program to show the decrease of susceptible individuals in a population during a simple epidemic that begins with a single infected individual. Use Equation 24.9 with $k = 0.01$, and plot the proportion of susceptibles (S_t/N) through time. Produce a graph that displays results for values of N of 10, 20, 40, 80, and 160. During an epidemic, the susceptible proportion of the population should decrease more rapidly for populations of larger size.

24.4   A Model for an Epidemic of Gonorrhea

The two-class model for general epidemics can provide the basis for a simulation of a gonorrhea epidemic in human populations. Gonorrhea is caused by the bacterial species *Neisseria gonorrhoeae*, which is spread almost exclusively by sexual contact. An infection can be painful and dangerous to an individual. The disease is the most prevalent communicable disease in the United States with about 21,000 new cases reported each week. These reported cases are estimated to be about a third of the actual cases (Grabowski 1983). The bacteria are readily transmitted, and many carriers show no external symptoms. Although previous infections offer no immunity to further infections, the disease can be treated successfully with antibiotics.

For gonorrhea epidemics, males and females must be modeled as two separate but interacting populations. Only the sexually promiscuous populations of males and females are considered in this simple model. The number of promiscuous males is indicated by N_M, and the number of females by N_F. Compared to the course of the disease the latent period is short, so both populations are divided only into susceptibles and carriers, as in the simple epidemic model above. Therefore,

$$N_F = S_F + C_F \qquad (24.10)$$

and

$$N_M = S_M + C_M \qquad (24.11)$$

where S_F and S_M indicate numbers of susceptibles and C_F and C_M indicate numbers of carriers, for females and males respectively.

The mass action model of homogeneous mixing and contact also is assumed for these populations. Hence, the rate of infection of susceptible males will be $k_1 C_F S_M$, and of susceptible females $k_3 C_M S_F$. k_1 and k_3 are the infection coefficients, with units of female^{-1} time^{-1} and male^{-1} time^{-1}. Homosexual transmission of gonorrhea is assumed to be insignificant.

In this model the infected carriers may seek treatment for the disease. Successful treatment will change a carrier to a susceptible. Treatment rates will be described with first-order rate constants, $k_2 C_M$ and $k_4 C_F$. Both of the coefficients have units of time^{-1}. In general, k_2 is larger than k_4 because infected males develop painful symptoms which cause them to seek treatment. Symptoms in females may be minor, so that they remain infective carriers for longer periods of time than do males.

The change in numbers of infected carriers of described by combining the expressions for gain and loss of carriers. For males the equation is

$$\frac{dC_M}{dt} = k_1 S_M C_F - k_2 C_M \qquad (24.12)$$

and for females it is

$$\frac{dC_F}{dt} = k_3 S_F C_M - k_4 C_F \qquad (24.13)$$

Exercise 24-4: Write a BASIC program based on Equations 24.10 through 24.13 to simulate the course of an epidemic of gonorrhea. Use comparatively large promiscuous groups of 900 males and 600 females. Based on information in Wallace (1972) and Braun (1983), the following are plausible daily rate coefficients:

$k_1 = 0.000032 \qquad k_2 = 0.2 \qquad k_3 = 0.00033 \qquad k_4 = 0.025$

Begin your simulation with one infected male carrying the disease and no infected females. Follow the numbers of infected males and females over a 12-year (4380 day) period. Simple Euler integration of the equations with $\Delta t = 1$ day will be adequate for this simulation. Your simulation should show an eventual steady-state of the numbers of infected males and females. The theory of this steady-state is discussed in Braun (1983) and Eisen (1988).

24.5 Thresholds of General Epidemics

While developing the general epidemic model of the preceding sections, Kermack and McKendrick (1927) demonstrated the "threshold theorem" of epidemiology. A simple statement of the theorem is that an epidemic will occur only if the number of susceptible individuals exceeds a certain value (Waltman 1974). The theorem can be shown in various ways. A simple method is to inspect Equation 24.3 above. If the change in the number of carriers is set to zero (i.e., the disease is not spreading), then the equation simplifies to

$$S = \frac{k_2}{k_1} \qquad (24.14)$$

Whenever the number of susceptibles S is greater than the ratio k_2/k_1, then number of infected carriers C will increase, because dC/dt will be positive. Whenever S is less than k_2/k_1, the number of carriers will decline. Interestingly, the rate of change of C does not depend directly on C.

The theorem agrees with the intuitive idea that the spread of a disease may be slowed or halted by increasing the rate of removal (e.g. quarantine), or by decreasing the rate of infection. Decreasing the number of susceptibles by vaccination, which takes them from susceptibles to removals directly, also accords with the theorem.

The threshold of k_2/k_1 is not a threshold value in the ordinary sense of the word. A value of S above the threshold does not necessarily indicate that a massive infection of susceptibles must occur. In essence, this threshold is analogous to the predator-prey or competition isocline values of Figures 7.4 and 7.5. Thus, if S is only slightly greater than k_2/k_1, C will increase only slightly until S drops to the threshold; C will then decrease toward zero. Figure 24.1 is a diagram of this property.

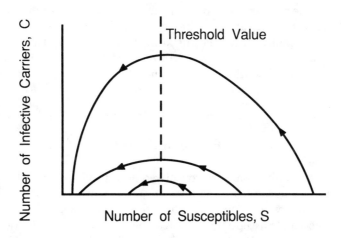

Figure 24.1. Phase plot of the number of susceptible and infected (=carrier) individuals for the general epidemic model. The arrowed trajectories show the time course of an epidemic for different starting values of susceptibles.

Exercise 24-5: Modify your program from Exercise 24-1 to produce a phase plot, S vs. C, of the course of an epidemic. Set $k_1 = 0.01$ and $k_2 = 0.12$. Plot the values of S vs. C for epidemics that begin with initial values of S of 12, 20, 40, and 80. Use initial values of $C = 1$ and $R = 0$ for each epidemic. The values of C should rise while S values are above the threshold level, and then should fall when S values drop below the threshold.

24.6 A Generalized Epidemic Model for a Non-Fatal Disease

The general model of the previous sections has been modified in numerous ways to provide more realism in simulation. Here we will consider the effects of adding newborn individuals to the pool of susceptibles, and of losing individuals from all epidemic classes through mortality except for deaths caused by the epidemic disease. As before, we will consider the three classes S, C, and R, which sum to N. In this model, N will not be a constant, but will be changed by both birth and death:

$$\frac{dN}{dt} = k_B N - k_D N \qquad (24.15)$$

where k_B is the rate constant for birth, and k_D the rate constant for death. This death rate is assumed to be the same for the different epidemic groups. The change in number of individuals spreading the disease as carriers will be determined by the rate of infection and recovery as in Equation 24.3, with additional losses due to death:

$$\frac{dC}{dt} = k_C CS - k_D C - k_R C \qquad (24.16)$$

where k_C, k_R and k_D are the rate constants for infection, removal, and death, as before.

All newborn individuals are considered susceptible, so the susceptible group will be increased by the addition of all newborn individuals. In addition to the loss of individuals through mortality, susceptibles will be lost to infection. Susceptibles may also be lost due to vaccination that transfers individuals directly from the susceptible to the removed group. The rate of change in the susceptibles is

$$\frac{dS}{dt} = k_B N - k_C CS - k_D S - k_V S \qquad (24.17)$$

where k_V is the rate constant for vaccination and the other constants are defined above. Increases in the removed group are due to the addition of immune individuals either through recovery or vaccination. Loss is due to mortality. The rate of change of the group is therefore

$$\frac{dR}{dt} = k_R C + k_V S - l_D R \qquad (24.18)$$

Exercise 24-6: Using the general model of this section, Horwitz and Montgomery (1974) simulated the spread of German measles (rubella) through the midwestern United States (Ohio, Indiana, Illinois, Michigan, and Wisconsin). They assumed a duration of two weeks for the infective-carrier stage, so they simplified their simulation by using two weeks as the time unit. This resulted in the following set of difference equations for modeling the changes in the different stages of measles:

$$\Delta N = k_B N - k_D N$$

$$\Delta C = k_C CS - k_D C - k_R C$$

$$\Delta S = k_B N - k_C CS - k_D S - k_V S$$

$$\Delta R = k_R C + k_V S - k_D R$$

The update procedures for the two-week periods are as usual: $N \leftarrow N + \Delta N$, etc.

Write and implement a BASIC program to simulate a measles epidemic using the following values derived from the study by Horwitz and Montgomery:

$$k_B = 7.28 \times 10^{-4} \quad k_D = 3.71 \times 10^{-4} \quad k_R = 1 \quad k_C = 1.12 \times 10^{-7}$$

Set $k_V = 0$. Begin your simulation with these initial values for the year 1966:

$$N = 38,480,000 \quad R = 30,000,000 \quad C = 416 \quad S = 8,479,584$$

Allow your simulation to proceed for an interval of 16 years (416 two-week periods). Plot values of C and S over this interval. You should observe three peaks of infection of measles.

After your simulation program is running successfully, modify it to include a vaccination program of 0.25 percent of the susceptible population each two weeks ($k_V = 0.0025$). Run your modified simulation. There should be an obvious impact of even this modest vaccination program on your results.

Exercise 24-7: Horwitz and Montgomery (1974) noted that the infection rate constant k_C is a function of time of the year, due to a variety of social and physiological factors. Their study used empirical coefficients to model seasonality, but a reasonable variability can be simulated with a sine curve. Modify your no-vaccination program from Exercise 27-6 by using a sine function for k_C. Let the average value of k_C be 1.12×10^{-7} with a range of $\pm .82 \times 10^{-8}$, a period of 26 two-week intervals, and a minimum at the thirteenth week of the year. As output, plot $\log_{10} C$ against time for 416 two-week periods (16 years). The log plot is needed in order to observe the seasonal fluctuations in the number of diseased individuals.

24.7 A Model of a Rabies Epidemic

The general epidemic model has been the basis of several simulations of rabies epidemics. Besides their significance in understanding the dynamics of a medically important disease, they provide an application of epidemic models to non-human populations (Bacon 1985). The following discussion is based on a model for the epidemic occurrence of rabies in populations of European foxes (Anderson et al. 1981). The model was designed to predict the dynamics of potential epidemics of rabies in the foxes of Great Britain, where populations were free of rabies.

Like the rubella model in the previous section, the model of rabies is based on the general epidemic model, with some modifications to bring the simulation closer to the realities of the particular disease. In modeling rabies in foxes, there are three classes of individuals in various stages of the disease: (a) uninfected, susceptible animals; (b) infected animals in which the disease is latent, so they are not yet carriers; and, (c) infected, rabid foxes which are capable of spreading the disease (Figure 24.2).

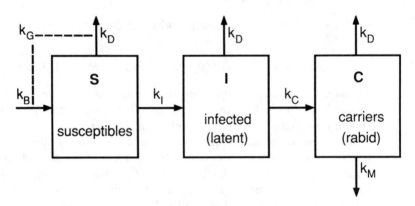

Figure 24.2. Compartmental diagram for the course of rabies in foxes. See text for terminology of the rate controls, k_X, which govern the flow among the three classes of individuals.

The disease is always fatal, so there is no immune or removed group. (Various plans for trapping and vaccination of wild foxes on a massive scale have been proposed seriously; these would create a removed group, of course.) In an infected fox, the latent period will vary from 2 to 12 weeks, and the animal behaves normally. However, infected animals are assumed not to reproduce. The rabid stage may last from 1 to 10 days; if the virus enters the spinal cord, the animal is paralyzed. If it enters the brain, the animal will display typical "furious rabies" symptoms. About equal numbers of infected foxes will develop either final stage.

The territoriality of foxes limits the upper densities of fox populations. To model this phenomenon, Anderson et al. (1981) included a logistic term in their equations for the increase in numbers of susceptible animals.

As before, N is the total population size or density. The densities of the separate disease classes are S (susceptibles), I (infected but not behaviorally rabid), and C (animals that can transmit the disease). At any time,

$$N = S + I + C \tag{24.19}$$

The rate of change in the number of susceptible foxes is described by

$$\frac{dS}{dt} = k_B S - k_D S - k_I C S - k_G S \frac{N}{K} \tag{24.20}$$

In this equation, k_I is the rate constant for the infection of susceptible foxes by rabid foxes; these susceptibles are transferred to the infected class. k_B is the birth rate constant, with only uninfected animals capable of producing offspring. k_D is the death rate constant, which is applicable to non-rabid deaths uniformly across all disease categories. In susceptibles, the birth and death rate constants may be combined as a single growth constant k_G:

$$k_G = k_B - k_D \tag{24.21}$$

k_G is equivalent to the population growth constant r of Chapter 7. K is the carrying capacity of the environment for fox population numbers. The logistic part of the Equation 24.20 becomes clearer, because

$$k_G S - k_G S \frac{N}{K}$$

is identical with the form of the logistic expression

$$rN \left(1 - \frac{N}{K} \right)$$

from Equation 7.3. The rabid foxes may be thought of as predators on the susceptibles. Thus, Equation 24.20 above is precisely equivalent to Equation 7.12, the Leslie-Gower model for changes in a prey population.

The change in the density of infected foxes in the latent stage of rabies is given by

$$\frac{dI}{dt} = k_I C S - k_C I - k_D I - k_G \frac{N}{K} \tag{24.22}$$

k_C is the rate constant for the per capita transition from the infected to the rabid state. The remainder of the terminology of the equation is discussed above.

The change of numbers in the rabid stage of the disease is given by

$$\frac{dC}{dt} = k_C I - k_D C - k_M C - k_G C \frac{N}{K} \qquad (24.23)$$

k_M is the rate constant for rabies-caused mortality of foxes in the final stage of rabies. The equations governing the three stages of the disease may be summed algebraically to provide an equation for the entire population of foxes:

$$\frac{dN}{dt} = k_G S - k_D N - k_M C - k_G N \frac{N}{K} \qquad (24.24)$$

Exercise 24-8: Write a program to simulate the course of a rabies epidemic in a population of European foxes, based on the model equations above. The following are rate constants derived from Anderson et al. (1981), with time units of days:

$$k_B = 0.00274 \qquad k_D = 0.00137 \qquad k_G = 0.00137$$

$$k_I = 0.21833 \qquad k_C = 0.033562 \qquad k_M = 0.200$$

Set carrying capacity K at 2.0 foxes km^{-2}. For this simulation, simple Euler integration will be adequate with $\Delta t = 1$. Begin your simulation with the population at the carrying capacity, with $N = 2$. Initiate the epidemic with $I = 0.1$ and $C = 0$. Follow the epidemic for a period of 40 years (14600 days). As output, plot N and C over the 40-year period. The results of your simulation should be a fairly realistic description of an epidemic, with large initial fluctuations of population size in response to a fatal disease. The epidemic should end with a small constant percentage of rabid individuals in the population.

24.8 Modeling the Spatial Spread of Disease

An important part of the study of epidemic disease is spread of disease from sites with an infected population to sites without the disease. The basis for most models of the spatial spread of disease is simple diffusion from a site with the disease to surrounding locations without the disease. In this section we will introduce a method for modeling diffusion across a landscape, and then apply it to the spread of rabies.

A workable method for modeling diffusion in two dimensions involves a simple expansion of the methods used in previous chapters for diffusion along a one-dimensional gradient (Watt 1968). We will model our diffusion landscape as a two-dimensional grid. Figure 24.3 shows a sample landscape with 100 sites, set up in a 10 × 10 grid. Diffusion of a "substance" (e.g. rabid foxes) among the sites is modeled by focusing on each site in turn, and finding the flow of substance between the focal site and each of the surrounding sites. The flow of substance will follow the simple diffusion process described in Chapter 14. There are some minor complexities involved in the two-dimensional model which require some discussion.

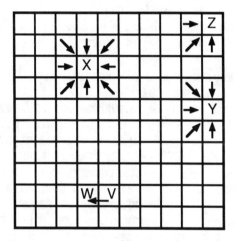

Figure 24.3. A grid of 100 sites for simulating diffusion between adjacent sites. Number of adjacent sites will vary with the location of the site (X, Y, and Z). Change in concentrations due to diffusion between two sites may be calculated when either is the center of focus. By convention, the change is calculated only when focusing on the gaining site (V and W); see text for details.

In this example, we will consider the surrounding sites to be only the eight sites located at the edges and at the corners of any grid site, as at location X of Figure 24.3. This convention has to be modified for some locations. The sites on the edges of the grid have only five surrounding sites (e.g. location Y), and the sites at the corners have only three surrounding sites (e.g. location Z). A computer program to simulate diffusion in this grid must account for these differences in surroundings.

To model diffusion among all of the sites, we will have to focus on each site of the grid in turn. When we focus on site X for example, we will consider the diffusion among X and the eight surrounding sites. This can be tricky, because diffusion can work to increase or decrease X. As a simple illustration, suppose that the concentrations of sites V and W are such that there is a net movement of substance from V to W. We could calculate the loss in V and gain in W when we are focusing on either site V or site W. We decide arbitrarily to calculate the change in concentration of two sites only when there is a net movement into the site of focus. Thus, in this example we would change the concentrations of V and W only when we focus on W.

Diffusion calculations may be performed with the usual two-stage Euler process. The changes in each grid site are calculated as the first stage, and then the concentrations are updated as the second stage. The following BASIC listing shows the calculation of diffusion among sites of a 10×10 grid for a single unit of the Euler increment Δt.

```
5 REM BASIC ROUTINE FOR 2-DIMENSIONAL DIFFUSION
6 :
10 REM    KW           = DIFFUSION COEFFICIENT
20 REM    W            = DIFFUSION BETWEEN SITES
30 REM    M            = SIZE OF DIFFUSION GRID
40 REM    DT           = EULER TIME INCREMENT
50 REM    C(I,J)       = CONCENTRATION AT SITE I,J
60 REM    DC(I,J)      = CHANGES IN CONCENTRATION AT SITE I,J
70 REM    I,J          = COUNTERS FOR FOCUSING ON SITES
80 REM    K,L          = COUNTERS FOR SURROUNDING SITES
90 REM    IL,IU,JL,JU  = LIMITS TO K & L FOR EDGE OR CORNER SITES
100 :
110 M = 10 :   REM 10 X 10 GRID
120 :
130 FOR I = 1 TO M
140    FOR J = 1 TO M
150       IL = I-1 :   IF IL<1 THEN IL=1
160       IU = I+1 :   IF IU>M THEN IU=M
170       JL = J-1 :   IF JL<1 THEN JL=1
180       KL = J+1 :   IF JU>M THEN JU=M
190       :
200       FOR K = IL TO IU
210          FOR L = JL TO JU
220             W = KW * ( C(K,L) - C(I,J) )
230             IF W<0 THEN W=0
240             DC(I,J) = DC(I,J) + W * DT
250             DC(K,L) = DC(K,L) - W * DT
260          NEXT L
270       NEXT K
280    NEXT J
```

```
290 NEXT I
300 :
310 FOR I = 1 TO M
320    FOR J = 1 TO M
330       C(I,J) = C(I,J) + DC(I,J)
340    NEXT J
350 NEXT I
```

Exercise 24-9: Based on the listing above, write a BASIC program to show diffusion in a 5×5 grid. Use a diffusion coefficient of 0.1, and set $\Delta t = 1$. Begin with all sites having concentrations of 0, except site (5,5) which should have an initial concentration of 25. Write your program to print the concentrations of each site at times of 0, 1, 2, 5, 20, and 100.

The above method for simulating diffusion provides the basis for the simulation of the spatial spread of disease. Realistic simulation of the spread of a disease like rabies may require an extremely fast supercomputer (Murray 1987). However, the spread of rabies through a very limited area can be simulated with BASIC, to demonstrate the elements that are used in more elaborate simulations.

In order to simulate the spatial dynamics of a rabies epidemic, the rabies simulation of Section 24.7 above is set up so that population growth and disease transfer among the different classes occur at each site of a grid. Thus, on a 5×5 grid, there will be 25 separate simulations proceeding simultaneously, each using the processes you worked with in Exercise 24-8. The spatial simulation is simplified by assuming that only rabid foxes move from site to site across the landscape.

To model the movement of rabid foxes, the model equation for the rabid foxes (stage C) will require modification. Net flow of rabid individuals between two adjacent sites i, j and k, l is given with

$$W = k_W \left(C_{kl} - C_{ij} \right) \tag{24.25}$$

Here, C_{ij} and C_{kl} are the densities of the rabid animals and k_W is the "wandering coefficient" for these animals. This equation is given as Line 220 in the BASIC listing of the diffusion routine.

The BASIC statements for the first-stage of the Euler integration of Equations 24.20, 24.22, 24.23 and 24.24 would be placed between Lines 140 and 150 of the BASIC diffusion routine.

Exercise 24-10: Modify your program from Exercise 24-8 to incorporate the spread of rabid foxes among adjacent sites on a 6×6

landscape grid. Although few data are available on the wandering of rabid foxes (Murray 1987), 0.01 is not unreasonable as a first estimate for k_W. Set up your simulation with the same constants as in Exercise 24-8. Begin with each site except (6,6) having a fox population at the carrying capacity of 2.0 individuals, all of which are susceptible; site (6,6) begins with 1.9 individuals. The simulated rabies epidemic should start at time zero with a density of 0.1 individuals of class I in site (6,6). All other classes at all the sites will have a density of 0. Output of your simulation may take several forms, but your program should print out at least the 6×6 matrix of the densities of the rabid C stage at appropriate time intervals. From your simulation it should be possible to observe the "epidemic wave" of rabies moving across the landscape from the site of the original infection.

Conclusion

The biomedical literature contains many examples of simulations of epidemics of various diseases. Most of these proceed from complex models of specific diseases, but they are based on the elements of epidemic models covered in this chapter. Because of their complexities, the more recent simulations have required computing techniques and hardware more advanced than BASIC operating on personal computers. Stochastic models have been important in developing the theory of epidemics (e.g. Bailey 1957, Bartlett 1960, Irwin 1964), but their role in simulation appears to be less than that of deterministic models. Recent work in chaotic dynamics of epidemics (e.g. Schaffer and Kot 1986) has resulted in major new directions for epidemiological modeling.

APPENDIX 1

THE MINIMUM
BASIC REQUIREMENT

We attempt to give enough information here about the BASIC language so that exercises in the book can be completed successfully. The coverage of BASIC is not intended to be comprehensive. Although users of the book are not assumed to have programming experience, there is no formal presentation of programming procedures. The elements of the language are learned easily in one or two sessions with a computer.

BASIC is the acronym for Beginner's All-purpose Symbolic Instruction Code, a computer language developed at Dartmouth College by Kemeny and Kurtz in 1964. Dozens of BASIC interpreters exist for large and small computers. We concentrate on the necessary elements that are common to most dialects. Procedures for entry into BASIC interpreters vary among machines, so users will have to obtain operating instructions for their particular installation. A reference manual for the user's particular dialect of BASIC will be a useful adjunct to this appendix. Many comprehensive instruction books have been published for BASIC in general and also for specific popular dialects. These books can be useful supplements for the following discussion.

A number of "structured" versions of BASIC have become available particularly for microcomputers. These offer numerous features which are attractive to programmers with some experience, but beginning programmers find them more difficult to learn. This appendix covers the elements of the standard BASICs, which apply generally to the newer versions also.

A1.1 Modes

BASIC operates in two modes or levels. BASIC starts in the immediate, command, direct, or calculator mode. The computer will respond immediately to instructions it receives. A list of some important commands for the direct mode are given at the end of this appendix.

The second mode is called the indirect, deferred, or programmed mode. Programs are typed into the computer in this mode, with a line number indicating to the computer that you are in this mode. Because the exercises in this book focus on writing programs, we will begin with this subject.

A1.2 BASIC Programs

A BASIC program consists of a series of instructions, also called commands or statements, which are arranged as lines beginning with numbers. The computer performs the commands in sequence, beginning with the lowest number. Line numbers must be integers, beginning with 0 or 1; there may be an upper limit, e.g. 65529. Line numbers need not be spaced one unit apart: 1, 2, 3, etc. It is a good idea to leave room in your numbering to allow you to add statements later. If you type in lines with numbers out of order, BASIC will arrange them from low to high.

A1.3 Numeric Constants

Numeric constants are static values that you use in your program. BASIC is flexible in working with numeric constants, shifting without distinction among integers and numbers with fixed or floating decimal points, and automatically handling the location of the decimal point. Constants may be signed or unsigned. They may be expressed in floating decimal form ("scientific notation"), so that -0.000511 becomes -5.11E-4 or some similar term depending upon the dialect; this is interpreted as -5.11×10^{-4}. Commas are never used in BASIC variables; three million is entered as 3000000 or 3E6; some dialects will permit spaces, as in 3 000 000.

A1.4 Numeric Variables

Variables in BASIC are names you choose to indicate values in a program. The value may or may not change with calculations in your program. If your program does not give a value to a variable, the computer gives it a value of zero. Rules for naming variables differ considerably among BASIC dialects; names acceptable to all consist of a letter of the alphabet, or a letter followed by a single digit, for example Q, Q5, P0, A1. Subscripted variables in BASIC are discussed below.

A1.5 Strings

A string constant is a series of alphanumeric characters (letters, numbers and some punctuation marks), enclosed between double quotation marks. (Some BASICs use single rather than double quotes; some use both.) The possible lengths of string constants may be limited, depending upon the version of BASIC. Some sample string constants are "NUMBER OF BACTERIA", "Plants per hectare", "3/01/75". Because double quotation marks denote the beginning and end of strings, they cannot be included easily within a string constant. String constants are named like numeric constants, but a dollar sign is appended to the name, for example A$, P8$, Z5$. BASIC is generally noted for its ease of handling string variables; however, this usually will not be important for biological simulations.

A1.6 Algebraic Expressions

A BASIC expression is any combination of variables or constants, perhaps combined with operators or parentheses. Expressions are always part of statements (see Section A1.7 below). An example of an expression might be:

$$U * (S \char`^ 2 / (S + K * (1 + (I/M))))$$

In more familiar algebraic notation this would be

$$U \cdot \frac{S^2}{S + K(1 + I/M)}$$

The arithmetic operators in BASIC (with alternate forms in brackets) are

Symbols	Meaning
^ [**]	Exponentiation
-	Negation
*	Multiplication
/	Division
-	Subtraction
+	Addition

Algebraic expressions are evaluated according to rules of precedence; most BASICs proceed in the following order:

1. Expressions enclosed in parentheses
2. Functions, built-in or user-defined (see Section A1.13 below)
3. Exponentiation
4. Negation
5. Multiplication and division
6. Addition and subtraction

An expression is evaluated from left to right, with operations in parentheses performed first. Then, scanning again from left to right, all functions are evaluated, then all exponentiation is performed, then all negation, and so on. Parentheses may be used freely in BASIC to indicate without question the proper order in which you wish your expression to be evaluated. Parentheses are always used in pairs, so that each expression must contain as many right parentheses as left.

A1.7 Assignment (LET) Statements

This is the fundamental statement in BASIC, and is used to assign values to variables. Examples include

```
150 LET Y = U * (S ^ 2 / (S + K * (1 + (I/M)))
160 P = 3.14159
170 Z = Z + 1
```

The computer evaluates the expression on the right side of the equals sign and assigns the resulting value to the variable on the left side. The previous value of the variable is lost. In most dialects of BASIC it is not necessary to include LET. The assignment statement is also used with string variables, for example

```
180 Y$ = "DENSITY OF KANGAROOS"
```

Note that the equal sign does not signify equality, but replacement of computer memory content. Some programmers consider this to be a BASIC flaw.

A1.8 Multiple Statements

Most dialects of BASIC allow multiple statements for a single line number. For example,

```
190 X = 75 :  P = X : D = 0
```

The colon is used to separate the statements in many BASICs. The technique is useful in several programming circumstances. Programs with fewer line numbers occupy less memory and run faster, an important factor with some machines having smaller memories.

A1.9 Remark (REM) Statements

The REM statement provides a way to insert comments into the program. An example of the standard form is

```
200 REM THIS IS A STANDARD FORM REMARK
```

The computer will ignore all information after the REM. Such statements are useful for naming programs, identifying their source, clarifying program steps, etc. Different dialects of BASIC may support additional methods of inserting remarks into programs.

A1.10 The PRINT Statement

The PRINT statement is the fundamental method for obtaining processed data from your program. Much of a programmer's time seems to be spent with PRINT statements attempting to get output to "look right". Statements have the form n PRINT e, where n is the line number, and e is a list of variables, constants or expressions. The statement can be used to print out messages and column headings. Example statements are:

```
210 PRINT Q2
220 PRINT X3/J2 + S
230 PRINT "NUMBER OF FLEAS PER DOG"
235 X$ = "ML OF BLOOD LOSS PER WEEK"
240 PRINT X$
250 PRINT
```

Line 250 causes a blank line to be printed.

Some features in various BASIC dialects for horizontal spacing of output include TAB, SPC, PRINT USING, etc. These tend to have different and equipment-specific performances. However, you can manage most programs for the exercises of this book with three techniques: the semicolon, the comma, and blank spaces between quotes.

If a semicolon separates two variables or expressions in a PRINT statement, no blank spaces are printed between the values of the variables. Some BASIC dialects print a leading blank as an unprinted plus-sign for positive numeric values, and some BASICs furnish a space after each number. Commas provide a method for tabulation of output. If a comma is put between variables, there is an automatic tab to the next built-in tab stop or printing zone before the second variable is printed. The stops are built-in at columns of variable width; usually they are about 15 characters apart. The 15-character spacing allows room for the maximum width of a floating point variable. The number of tab stops in a line will depend upon your printer, video screen, teletype, etc. The following provide some illustration of program lines and resulting output.

```
260 X = 5 :  Y = -66 :  Y$ = "Y-VALUE = "
270 PRINT X; Y
280 PRINT X, Y
290 PRINT X; " "; Y
300 PRINT Y$; Y
5-66
5                -66
5    -66
Y-VALUE = -66
```

A1.11 READ and DATA Statements

Numeric values may be assigned to variables in several ways in a BASIC program. Simple assignment statements are satisfactory and even preferable for a few variables, for example, A = 10 : B = 32, N1 = 100, etc. These are easy to read, to modify, and to correct if necessary. READ-DATA statements may be better suited for programs with many values to be assigned. The READ statement must work with DATA statements. A typical simple form might be

```
310 READ A, B, N1, N2, K1, C, Q$, D
320 DATA 10, 32, 1000, 3000, 2.5E5, 4.7E8, "OOPS!", 0.1
```

Most BASIC dialects permit much flexibility in use of READ and DATA statements. The values in line 320 could have been in seven different DATA statements. There may be more values in the DATA statement than variables in the READ statement; there can never be fewer. READ and DATA statements may be scattered through the program. The computer will treat all the values in a program's DATA statements as part of a common pool of data; READ statements will assign values beginning with the first value in the lowest numbered DATA statement. For clearer programming, DATA statements either should follow their associated READ statements directly, or should be placed together at the end of the program.

The RESTORE statement is used to reinitialize the DATA pool, so that the next READ statement will begin assigning the first DATA values in the program.

A1.12 The INPUT Statement

The INPUT statement permits the program to request that values be read from the keyboard during the execution of a program. Examples include

```
330 INPUT X
340 INPUT "WHAT IS THE VALUE OF X "; X
350 INPUT X,Y
360 INPUT M$
```

Upon encountering line 330, the program would halt with a question mark on the output device, and wait for the user to type in a value for X. INPUT statements with a preliminary message (Line 340) are permitted in most dialects of BASIC; the user may be notified of the nature of the expected input. Line 350 expects the user to type in two consecutive numeric values, separated by commas. Any alphanumeric input in response to Line 360 will be accepted.

A1.13 Mathematical Functions

Several mathematical functions are intrinsic to most dialects of BASIC. They may be used in any algebraic expression, and will produce a numeric value. In the following listing, X represents any numeric variable, constant, or expression. The list is alphabetical:

ABS(X)...finds the absolute value of X.

ATN(X)...finds the arctangent of X, the angle in radians whose tangent
is X.

COS(X)...finds the cosine of X, with X expressed in radians.

Note: if X is in degrees, use $COS(X*\pi/180)$

EXP(X)...finds the value of natural exponent of X, that is, e^X.

INT(X)...finds the integer part of X.

LOG(X)...finds the natural logarithm of X; that is, $\log_e X$.

Note: $\log_{10} X = (\log_e X)/(\log_e 10)$

RND(X)...finds a random number. BASIC dialects differ; see your manual.

SGN(X)...evaluated as 0 if X is 0, -1 if X is negative, and +1 if X is positive.

SIN(X)...finds the sine of X, with X expressed in radians. (See COS(X) note.)

SQR(X)...finds square root of X.

TAN(X)...finds the tangent of X, with X expressed in radians. (See COS(X) note.)

A1.14 FOR...NEXT Statements (Loops)

Computers excel at performing repeated tedious tasks. If you have a set of statements to be performed repeatedly, a FOR-NEXT loop is an efficient method for doing so. The FOR statement specifies the number of times that a certain segment of the program is to be repeated. The NEXT statement indicates that the computer is to loop back to the FOR statement to check whether the number of repetitions has been completed. Otherwise, control is passed to the statement following the NEXT statement. Examples of typical pairs of FOR-NEXT statements are

```
370 FOR Q = 1 TO 10
380 NEXT Q

390 FOR S = 0 TO 100 STEP 5
400 NEXT S

410 FOR R = 100 TO 0 STEP -2
420 NEXT R
```

The general form is: n FOR x = e_1 to e_2 STEP e_3. The line number is denoted by n, and e_1, e_2, and e_3 are numeric constants, variables, or expressions. x is a variable; some BASIC dialects require it to be unsubscripted. If STEP is not specified, it is assumed to be 1. STEP size may be less than 1, but this may cause difficulties in some BASIC dialects.

Each FOR statement must be paired with a NEXT statement. It is often desireable to nest FOR-NEXT loops, with one loop inside another. Associated FOR-NEXT statements must lie completely inside or completely outside other FOR...NEXT loops; overlapping loops can cause severe programming difficulties. Transferring in and out of FOR-NEXT loops with GO TO statements (see below) can cause a lot of programming misery. More subtle aspects of FOR-NEXT loops are discussed in Chapter 4.

A1.15 GOTO Statement

The GOTO statement is used to transfer control directly to another line in the program. It is used to jump back so that a routine can be repeated, or to jump ahead to bypass a part of a program. A typical example is

```
430 GOTO 3000
```

GOTO statements should be used sparingly and with careful planning. Transfer into and out of FOR-NEXT loops and subroutines with GOTO statements can produce many programming difficulties. The GOTO statement is frequently combined with an IF...THEN statement for branching in programs.

A1.16 IF...THEN Statements

IF...THEN statements allow statements to be executed only if certain logical conditions are met. The general form is

n   IF (expression) comparison (expression) THEN statement

Here, n is a line number, and (expression) may be an arithmetic expression, a variable, or a constant. The comparison between the expressions should be any one of those given in the following list. Different dialects have different symbols for some of these expressions.

Symbols	Meaning
=	is equal to
<> >< #	is not equal to
<	is less than
<= =<	is less than or equal to
>	is greater than
>= =>	is greater than or equal to

Here are some simple IF...THEN statements as examples:

```
440 IF N < 0 THEN N = 0
450 IF (N + P) <= (K + J) THEN GOTO 8000
460 IF N =< 30 THEN N = N * (N - N/L) : R = 20 :   GOTO 8200
```

If your dialect of BASIC permits statements like that in line 460, then note that all three operations will be executed only if the condition is met. If N > 30, then the computer moves on to the next line. No operations to the right of THEN will be executed.

Complex IF...THEN statements may be built up with multiple logical relations linked by logical operators. AND and OR are commonly used operators; others may be available in your BASIC dialect. Sample statements are

```
470 IF N > 20 OR N < 30 THEN GOTO 3000
480 IF X = 5 AND Y + X > N THEN Z = P + Q: GOTO 9700
```

Using the equal sign can be confusing because it does not signal variable replacement in an IF...THEN comparison. The equal sign also can be tricky, because some dialects demand absolute identity, while others behave as if 9.9999999999 were equal to 10.

A1.17   Subroutine: GOSUB and RETURN Statements

A subroutine is a sequence of statements used repeatedly or used at several parts of the same program. The GOSUB statement is used to transfer to a line number specified in the GOSUB statement. For example,

```
500 GOSUB 4000
```

This statement would transfer control of the program to statement 4000, which will be the first statement in the subroutine. The lines of instructions in the subroutine will be followed until a RETURN statement is reached. The RETURN causes control to be transferred to the statement immediately following the GOSUB.

Note that GOSUBs may be nested, with a subroutine called from within a subroutine. Most BASICs have a limit to the levels of nesting of subroutines. This is of limited concern for the exercises of this text.

A1.18 The STOP Statement

The STOP statement may be inserted into a program to halt its execution. Many BASICs respond to the statement with an indication of the line number of the STOP. The STOP (or END) statement should be inserted between the main program and any subroutines, to prevent unexpected entry into a subroutine after the main program has completed execution. Many versions of BASIC include the CONT command. When entered from the keyboard after a STOP statement (or END statement), the program will begin executing with the statement following the STOP.

A1.19 The END Statement

Usage of the END statement may vary somewhat between dialects of BASIC. With some, it must be the last statement in any program, and in this case its use elsewhere in the program may not be permitted. In other cases, it is a convenient alternative to the STOP statement to signal the end of the main program. In most dialects the program will simply cease execution at an END, with no indication of line number.

A1.20 Arrays and Subscripted Variables

Subscripted variables are numeric or string variables followed by an expression enclosed in parentheses. They are called subscripted because of the analogy to A_1, A_2, A_3, etc. For example,

A(1), A(2), A(3), ... are all variables in an array called A.

B(1), B(2), B(3), ... are all variables in an array called B.

Likewise, C(1,1), C(1,2), C(1,3), C(4,8)...are all variables in a two-dimensional array C.

In some BASIC dialects, the unsubscripted variables A, B, B1, C3, and C0 may all be used in the same program with the above subscripted variables A(I) and B(I). Some dialects permit non-integer subscripts, which may perform in unexpected ways. Subscripts are best kept as integers.

Many versions of BASIC automatically recognize arrays if they require not more than 10 elements (or 10 by 10 elements for two-dimensional arrays). For larger arrays, one must first save storage space with a DIM (=dimension) statement. These should be placed so that the computer only encounters the DIM statement once during the running of the program. The following statement would reserve storage for three lists, and two six-by-six arrays:

```
510 DIM A(25), B(100), I(1000), C3(6,6), D(6,6)
```

Some BASICs permit arrays to have more than two dimensions, so that it is possible to dimension an array with $F(2,8,3,100)$ for example. Arrays may be of almost any length. BASIC should notify you with an error message about lack of memory when your arrays become too large.

Values may be assigned to arrays in various ways. We can assign 24 values in sequence to an array Q with subscript J by varying J from 1 to 24:

```
520 DIM Q(24)
530 FOR J = 1 TO 24
540    READ Q(J)
550 NEXT J
560 DATA 12,19,3,8,7,0,2,7,4,0,5,1
570 DATA 8,18,10,3,3,0,0,0,21,98,43,1
```

One of the more useful features of subscripted variables is the ability to define one variable in terms of another. Consider array B which is to consist of the average of successive values in the Q array above:

```
580 DIM B(23)
590 FOR P = 1 TO 23
600    B(P) = (Q(P) + Q(P+1)) / 2
610 NEXT P
```

It is sometimes convenient to use subscripted arrays to store computed data and then to print it out at the end of calculations. Thus, we could obtain a printed list of the calculations above with the following lines:

```
620 PRINT "J", "Q(J)", "B(J)"
630 FOR J = 1 TO 23
640    PRINT J, Q(J), B(J)
650 NEXT J
660 PRINT ,Q(24)
```

Subscripted variables may be used in a variety of counting methods. Assume we have 1000 integers all between 1 and 16 which have been produced during a simulation. We need to find out how many 1s, 2s, 3s, etc., there are. The integers are stored in an array called S, so that S(1), S(2), S(3), ..., S(1000) might have the values 4, 14, 9, ..., 7. We will tally the 16 possible values in an array N, and then print out the tally as follows:

```
670 DIM N(16)
680 FOR J = 1 TO 1000
690    X = S(J)
700    N(X) = N(X) + 1
710 NEXT J
720 PRINT "INTEGER",  "TALLY"
730 FOR J = 1 TO 16
740    PRINT J, N(J)
750 NEXT J
```

Such sorting and counting methods are valuable for the Monte Carlo simulations considered in this book. These simulations often must be run a great many times to give meaningful results, and tabulating routines are quite useful.

A1.21 Routines for Matrix Manipulation

Some dialects of BASIC have built-in functions for performing multiplication, addition, inversion, etc. of a matrix (two-by-two array). The following routines may be used to replace or substitute for these functions. The various matrices are assumed to have been previously set up with a DIM statement. In all of these, R is a previously defined variable representing number of rows, and C is the number of columns of the matrix.

A. Reading data into a matrix **A**:

```
760 DIM A(R,C)
770 FOR I = 1 TO R          : REM ROWS
780    FOR J = 1 TO C       : REM COLUMNS
790       READ A(I,J)
800    NEXT J
810 NEXT I
820 DATA ...
```

B. Printing a matrix **A**:

```
830 FOR I = 1 TO R
840    FOR J = 1 TO C
850       PRINT A(I,J);" ";
860    NEXT J
870    PRINT " "
880 NEXT I
```

C. Scalar multiplication of matrix **A** by constant K, with results to matrix **B**:

```
890 FOR I = 1 TO R
900    FOR J = 1 TO C
910       B(I,J) = A(I,J) * K
920    NEXT J
930 NEXT I
```

D. Post-multiplication of matrix **A** by column vector **X** (length C) to give vector **Y**:

```
940 FOR I = 1 TO R
950    Y(I) = 0
960    FOR J = 1 TO C
970       Y(I) = Y(I) + A(I,J) * X(J)
980    NEXT J
990 NEXT I
```

A1.22 Important Immediate Level Commands

Several commands are essential or useful in working with BASIC programs, although they may not be part of the programs. They will help in editing and writing the programs. These commands can be quite machine- and dialect-specific, so that users may have to discover the equivalent commands for their specific installation. Frequently, BASICs have built-in editing capabilities to make it easy to modify a line of code instead of retyping it completely.

Some useful commands are given below for the two most common BASICs on microcomputers: Applesoft on Apple II machines, and BASICA or GW-BASIC on IBM and compatible machines. The majority of commands with Microsoft Macintosh BASIC are like those for BASICA; they are also accessible with pull-down menus.

Operation	Applesoft	BASICA
1) Clears screen display	HOME	CLS
2) Begins execution of current program	RUN	RUN
3) Halts execution of a program	ctrl-C	ctrl-break
4) Lists a program from beginning to end	LIST	LIST
5) Lists a program from lines 100 to 200	LIST 100-200	LIST 100-200
6) Lists a program from line 300 to the end	LIST 300-	LIST 300-
7) Lists a program from beginning to line 400	LIST -400	LIST -400
8) Deletes line 310 from the program	310	310
9) Continues a program run after STOP or keyboard halt	CONT	CONT
10) Saves a program called NAME to storage (e.g. disk)	SAVE NAME	SAVE"NAME"
11) Recalls program NAME from the storage medium	LOAD NAME	LOAD"NAME"
12) Removes program NAME from storage (lost forever...)	DELETE NAME	KILL"NAME"

APPENDIX 2

GRAPH: A SIMPLE PROGRAM FOR PLOTTING SIMULATION DATA

Most exercises in this book produce simulation data that describe relationships between dependent and independent variables. These x-y data sets may be printed out as columns of numbers, but a graph is more quickly grasped than a table. The data from simulation programs may be plotted with a number of techniques. One obvious method is to plot the simulation data by hand on graph paper; however, unless a person has abundant free time, there seems little point to this procedure. A method of producing output in graphic form on a simple line printer is illustrated in Figure 1.2; resolution is coarse.

A third possibility is to put the simulation data into one of the computer's storage devices (e.g. a disk file). A short BASIC routine for data-file storage can be attached easily to BASIC simulation programs. Dozens of more or less elaborate data-plotting program packages are available for most large and small computers. These are capable of taking data files from the computer to produce attractive graphs of high resolution on plotters or other graphic devices.

All of the methods above are slow when compared to production of instantaneous graphical output as the simulation program runs. There are two benefits to working with immediate graphic output of a simulation. The obvious one is that many mistakes in implementing the simulation program can be detected and corrected immediately. The second and more important benefit is that "What if..." questions about altering the simulation can be asked and answered quickly.

This appendix describes a BASIC program that produces an x-y graph, designed for use with the chapter exercises of this textbook. Versions for the three most numerous microcomputers in North American colleges and universities are listed in Figures A2.1-A2.4. The programs may be adapted easily for other machines, assuming that BASIC interpreters are available for working with graphics on a video output device.

A2.1 General Procedure for the GRAPH Program

The GRAPH program is intended to be used as an appendage of BASIC programs that produce x-y data. (About 90 percent of the exercises of this text will require such programs.) These BASIC programs should be written as if x-y data were to be printed sequentially as x-y pairs. Then, with the addition of four lines of elementary BASIC code, the routines of the GRAPH program may be called to produce a graph of the data.

The program may be combined with simulation programs in a variety of ways. In the simplest procedure, the GRAPH program is loaded into the computer with BASIC running. Then Lines 60 through 200 containing the sample demonstration are deleted, and the lines of the new BASIC program are written in place of the sample demonstration.

A2.2 General Comments on the GRAPH Program

Some general features of the GRAPH program will be described here, and notes on the machine-specific implementations of the program will accompany the separate listings below. The program has been kept as simple and short as possible consistent with clarity. Internal documentation of the program is minimal. While not good programming practice generally, this shortens the program considerably. (As different text exercises are saved to the storage devices, the shorter program saves a lot of storage space.) The routines are clear enough that many persons with no prior programming experience have modified the GRAPH routines to produce more elaborate output after working through three or four exercises from Chapter 1.

The program performs much as a person would proceed in producing a graph from a set of x-y data. (1) the axes are drawn, with tic-marks added for scaling; (2) the unit values are written on the axes; (3) the axes are labelled with the names assigned to the x-y variables; (4) the x-y points are plotted, and perhaps connected with lines. Multiple lines on the same graph may be drawn with differently shaped points.

A2.3 Drawing and Labelling GRAPH Axes

The GRAPH program will draw and label a pair of x-y coordinates with one or two lines of BASIC code. Users need to define the following variables, either with assignment (LET) statements, or with READ-DATA statements:

XM: the maximum value for units on the x-axis.

XN: the minimum value of the x-axis. This value is set at the origin.

YM: the maximum value for units on the y-axis.

YN: the minimum value for the y-axis. This value is set at the origin.

X$:   the label for the x-axis.

Y$:   the label for the y-axis.

The statement GOSUB 3000 will cause the graph axes to be drawn based on the assigned values.

A2.4 Fixing and Plotting Points on the GRAPH Axes

After the axes have been drawn with GOSUB 3000, pairs of x-y points may be plotted by assigning values to the variables X and Y, and then using the statement GOSUB 4000. For example, the following segment of a program will send a series of 10 points to the subroutine to be plotted:

```
160 FOR Q = 1 TO 10
170    R2 = P * EXP(Q)
180    X = Q : Y = R2 :   GOSUB 4000
200 NEXT Q
```

Without further instructions the points will be plotted as rather small single points. The points may be plotted as open circles if the statement XC = 1 is placed just before each GOSUB 4000 statement. In the program segment above, such a statement might be inserted as Line 175. Alternatively, the points may be plotted as a small closed box if the statement XB = 1 is used in place of XC = 1.

The points may be plotted as a single connected line if the pair of statements XA = 1 and XD = 0 are inserted into the program before the series of x-y points is plotted. In the program segment above, the series of 10 points will be connected if the two statements were to be written as Lines 140 and 150, for example. In effect, XA = 1 issues the command "connect the following points with a line" and XD = 0 indicates "begin the line with the next point".

A2.5 Limits and Labelling

Any value acceptable to the computer may be used for the values and labels on the x and y axes. However, all such values may not be printed in their entirety. Only the left-most 4 or 5 characters will be printed for the values of XM, XN, YM, YN. If the assignment YM = 500000 is made, the graphic output may show only 5000 for the maximum y-value. However, x-y points will be plotted according to the assigned value of 500000. There are several ploys for overcoming this label limit. As one example, set XM = 50 and divide the x-value by 10000 just before plotting it with the GOSUB 4000 statement; inserting "x10000" into the label for the x-axis will make the graph comprehensible to readers.

Besides labelling the axes with the maximum and minimum values, the program also labels values at (max - min)/2 and (max - min)/4. Care in selecting maximum and minimum values will produce suitably rounded intermediate values for the axes.

The number of characters in the X$ and Y$ labels for the graph axes are also limited; the number depends upon the machine and the axis. All the versions of the GRAPH program have provision for making general headings for the graphs, and for labelling individual lines when more than two are plotted on the same axes. The procedures for these labels vary with machines; specific procedures are discussed below.

A2.6 Restricted Variables

The GRAPH program uses several variables, all beginning with X or Y. A simple and practical procedure in writing programs is to avoid using variables that begin with X or Y, except when addressing the GRAPH routines.

A2.7 Using GRAPH with the IBM Personal Computer

Using GRAPH with an IBM Personal Computer or a compatible machine requires a BASIC interpreter capable of screen graphics; most versions of BASICA (Advanced BASIC) and GW-BASIC are suitable and widely available. The machine must also have a video card capable of graphic output. Figures A2.1 and A2.2 show listings for two widely available graphics cards. The different dimensions of the graphics screens are the only occasion for differences in the listings. Other graphics conventions may be incorporated in the program easily. The programs work with the usual PC-DOS or MS-DOS operating systems.

Figure A2.1. Listing of GRAPH program for IBM-PC or compatible machine with Color Graphics Adapter (CGA) or similar graphics card (640 × 200 points). This program may be used with Advanced BASIC (BASICA) or GW-BASIC or similar BASIC interpreters.

```
10 REM EXERCISE #0.0
20 REM SINE WAVE DEMO
30 REM GRAPH: IBM-CGA
40 REM 30 FEBRUARY 1990
50 :
60 REM DEFINE GRAPH AXES
70 READ XM,XN,YM,YN,X$,Y$
80 DATA 2,0,1,-1,VALUE,SINE
90 GOSUB 3000:REM DRAW GRAPH
100 :
110 PI=3.14159
120 XA=1:XD=0:REM LINE PLOT
130 FOR I=0 TO 160
140     V=I/80
150     S=SIN (2*PI*V)
160     X=V:Y=S
170     GOSUB 4000:REM PLOT
180 NEXT I
190 LOCATE 3,30:PRINT"SINE WAVE"
200 LOCATE 23,1:END
3000 :
3010 CLS:SCREEN 2:KEY OFF:WIDTH 80
3020 LINE(84,10)-(84,170)
3030 LINE(84,170)-(564,170)
3040 FOR XI= 10 TO 150 STEP 20
3050 LINE(84,XI)-(89,XI):NEXT
3060 FOR XI=144 TO 564 STEP 60
3070 LINE(XI,170)-(XI,167):NEXT
3095 XL$=LEFT$(X$,12)
3100 LOCATE 23,57-LEN(XL$)/2:PRINT XL$
3105 XL$=LEFT$(STR$(XN),5)
3110 LOCATE 23,11-LEN(XL$)/2:PRINT XL$
3115 XL$=LEFT$(STR$((XM+3*XN)/4),5)
3120 LOCATE 23,26-LEN(XL$)/2:PRINT XL$
3125 XL$=LEFT$(STR$((XM+XN)/2),5)
3130 LOCATE 23,41-LEN(XL$)/2:PRINT XL$
3135 XL$=LEFT$(STR$(XM),5)
3140 LOCATE 23,71-LEN(XL$)/2:PRINT XL$
3145 XL$=LEFT$(Y$,10)
3150 LOCATE 7,11-LEN(XL$):PRINT XL$
3155 XL$=LEFT$(STR$(YN),5)
3160 LOCATE 22,11-LEN(XL$):PRINT XL$
3165 XL$=LEFT$(STR$((YM+3*YN)/4),5)
3170 LOCATE 17,11-LEN(XL$):PRINT XL$
3175 XL$=LEFT$(STR$((YM+YN)/2),5)
3180 LOCATE 12,11-LEN(XL$):PRINT XL$
3185 XL$=LEFT$(STR$(YM),5)
3190 LOCATE 2,11-LEN(XL$):PRINT XL$
3200 RETURN
4000 :
4010 X0=84+480*(X-XN)/(XM-XN)
4020 Y0=170-160*(Y-YN)/(YM-YN)
4030 IF X0<2 OR X0>635 THEN XD=0:GOTO 4090
4040 IF Y0<2 OR Y0>197 THEN XD=0:GOTO 4090
```

```
4050 IF XB=1 THEN LINE(XO-2,YO-1)-(XO+2,YO+1),1,BF:GOTO 4090
4060 IF XC=1 THEN CIRCLE(XO,YO),3,1:GOTO 4090
4070 IF XD=1 THEN LINE-(XO,YO):GOTO 4090
4080 PSET(XO,YO):IF XA=1 THEN XD=1
4090 XB=0:XC=0:RETURN
```

Figure A2.1. GRAPH listing for IBM-PC. (concluded)

Figure A2.2. Listing of GRAPH program for IBM-PC or compatible machine with a Hercules graphics card or equilavent (720 × 349 points). The program may be used with various versions of BASIC suitable for use with this graphics adapter.

```
10 REM EXERCISE #0.0
20 REM SINE WAVE DEMO
30 REM GRAPH: IBM-HERC
40 REM 30 FEBRUARY 1990
50 :
60 REM DEFINE GRAPH AXES
70 READ XM,XN,YM,YN,X$,Y$
80 DATA 2,0,1,-1,VALUE,SINE
90 GOSUB 3000:REM DRAW GRAPH
100 :
110 PI=3.14159
120 XA=1:XD=0:REM LINE PLOT
130 FOR I=0 TO 160
140     V=I/80
150     S=SIN (2*PI*V)
160     X=V:Y=S
170     GOSUB 4000:REM PLOT
180 NEXT I
190 LOCATE 3,30:PRINT"SINE WAVE"
200 LOCATE 23,1:END
3000 :
3010 CLS:SCREEN 2:KEY OFF:WIDTH 80
3020 LINE(95,20)-(95,300)
3030 LINE(95,300)-(639,300)
3040 FOR XI= 20 TO 265 STEP 35
3050 LINE(95,XI)-(101,XI):NEXT
3060 FOR XI=163 TO 639 STEP 68
3070 LINE(XI,300)-(XI,295):NEXT
3095 XL$=LEFT$(X$,12)
3100 LOCATE 23,57-LEN(XL$)/2:PRINT XL$
3105 XL$=LEFT$(STR$(XN),5)
3115 XL$=LEFT$(STR$((XM+3*XN)/4),5)
3120 LOCATE 23,26-LEN(XL$)/2:PRINT XL$
3125 XL$=LEFT$(STR$((XM+XN)/2),5)
3130 LOCATE 23,41-LEN(XL$)/2:PRINT XL$
3135 XL$=LEFT$(STR$(XM),5)
3140 LOCATE 23,71-LEN(XL$)/2:PRINT XL$
```

```
3145 XL$=LEFT$(Y$,10)
3150 LOCATE 7,11-LEN(XL$):PRINT XL$
3155 XL$=LEFT$(STR$(YN),5)
3160 LOCATE 22,11-LEN(XL$):PRINT XL$
3165 XL$=LEFT$(STR$((YM+3*YN)/4),5)
3170 LOCATE 17,11-LEN(XL$):PRINT XL$
3175 XL$=LEFT$(STR$((YM+YN)/2),5)
3180 LOCATE 12,11-LEN(XL$):PRINT XL$
3185 XL$=LEFT$(STR$(YM),5)
3190 LOCATE 2,11-LEN(XL$):PRINT XL$
3200 RETURN
4000 :
4010 X0=95+544*(X-XN)/(XM-XN)
4020 Y0=300-280*(Y-YN)/(YM-YN)
4030 IF X0<2 OR X0>716 THEN XD=0:GOTO 4090
4040 IF Y0<2 OR Y0>347 THEN XD=0:GOTO 4090
4050 IF XB=1 THEN LINE(X0-2,Y0-1)-(X0+2,Y0+1),1,BF:GOTO 4090
4060 IF XC=1 THEN CIRCLE(X0,Y0),3,1:GOTO 4090
4070 IF XD=1 THEN LINE-(X0,Y0):GOTO 4090
4080 PSET(X0,Y0):IF XA=1 THEN XD=1
4090 XB=0:XC=0:RETURN
```

Figure A2.2. GRAPH listing for IBM-PC with Hercules graphics card. (concluded)

The graph axes are drawn and labelled with statements from Lines 3000-3200. Lines 4000-4130 are used in plotting the points. Line 180 of the sample program shows how labels may be placed on the graph using LOCATE statements. The GRAPH program uses an 80-column × 25-row character screen; text labels can be fixed with fairly good precision using the LOCATE statement for labeling lines or titling the graph.

The LOCATE 23,1:END statement of Line 190 should be used generally to terminate programs. It positions the cursor out of the way in the lower left corner of the screen, and halts the program without a message.

Listings of the program on a printer may be obtained using the LLIST command, assuming the printer is set up as the BASIC line printer. A "screen-dump" of the graph may be obtained with many printers using the shift-PrtSc key combination. This ability will depend upon the printer's capability, and will require the execution of the GRAPHICS.COM file before BASIC is loaded into the machine.

In Line 3010, the SCREEN 2 statement calls up the high-resolution graphics screen for drawing the graph. On some machines, editing and listing of the program will proceed slowly on this screen; typing in SCREEN 0 at the command level (immediate mode) will recall the normal text screen.

A2.8 Using GRAPH with the Apple II Computer

The Apple II version of GRAPH (Figure A2.3) uses the Applesoft BASIC interpreter that is built into the Apple II-series of computers and compatible machines. It also uses the high-resolution graphics screen built into Applesoft. GRAPH may be used with both the DOS and ProDOS operating systems for these computers; Line 3013 will need to be changed appropriately for ProDOS.

The graph axes are drawn and labelled with statements from Lines 3000-3200. Lines 4000-4090 are used in plotting the points. Titles and line labels may be put on the graphic output with GOSUB 3500 statements as in Line 190 of the sample program. The position of the label is fixed by specifying the coordinates for the beginning of the message, using the variables X0 and Y0, and the message as a string variable XL$. The coordinates of the graphics screen run from 0 to 279 horizontally, and from 0 to 159 vertically. The origin (0,0) is located at the top left corner of the screen.

Figure A2.3. Listing of GRAPH program for the Apple II-series of computers using Applesoft BASIC and DOS 3.3.

```
10 REM EXERCISE #0.0
20 REM SINE WAVE DEMO
30 REM GRAPH: APPLE II
40 REM 30 FEBRUARY 1990
50 :
60 REM DEFINE GRAPH AXES
70 READ XM,XN,YM,YN,X$,Y$
80 DATA 2,0,+1,-1,VALUE,SINE
90 GOSUB 3000:  REM DRAW GRAPH
100 :
110 PI = 3.14159
120 XA = 1:XD = 0:  REM LINE PLOT
130 FOR I = 0 TO 160
140     V = I / 80
150     S = SIN (2 * PI * V)
160     X = V:Y = S
170     GOSUB 4000
180 NEXT I
190 X0 = 85:Y0 = 10:XL$ = "SINE WAVE":  GOSUB 3500
200 END
3000 :
3010 HGR : IF PEEK (16384) = 127 THEN 3020
3013 PRINT CHR$ (13) + CHR$ (4)"BLOAD SHAPES,A$4000"
3015 POKE 232,0:  POKE 233,64:  HCOLOR= 3:  SCALE= 1
3020 HPLOT 30,0 TO 30,149
3030 HPLOT 30,149 TO 279,149
3040 FOR XI = 9 TO 116 STEP 35
3050 HPLOT 31,XI TO 34,XI: NEXT
3060 FOR XI = 60 TO 270 STEP 30
3070 HPLOT XI,146 TO XI,149:  NEXT
3095 XL$ = LEFT$ (X$,10)
3100 X0 = 175:Y0 = 152:  GOSUB 3500
3105 XL$ = LEFT$ ( STR$ (XN),4)
3110 X0 = 31 - 3.5 * LEN (XL$):Y0 = 152:  GOSUB 3500
3115 XL$ = LEFT$ ( STR$ ((XM + XN * 3) / 4),4)
3120 X0 = 91 - 3.5 * LEN (XL$):Y0 = 152:  GOSUB 3500
3125 XL$ = LEFT$ ( STR$ ((XM + XN) / 2),4)
3130 X0 = 151 - 3.5 * LEN (XL$):Y0 = 152:  GOSUB 3500
3135 XL$ = LEFT$ ( STR$ (XM),4)
3140 X0 = 268 - 3.5 * LEN (XL$):Y0 = 152:  GOSUB 3500
3145 XL$ = LEFT$ (Y$,8)
3150 X0 = 10:Y0 = 68:XR = 1:  GOSUB 3500
3155 XL$ = LEFT$ ( STR$ (YM),4)
3160 X0 = 28 - LEN (XL$) * 7:Y0 = 6:  GOSUB 3500
3165 XL$ = LEFT$ ( STR$ ((YM + YN) / 2),4)
3170 X0 = 28 - LEN (XL$) * 7:Y0 = 76:  GOSUB 3500
3175 XL$ = LEFT$ ( STR$ ((YM + YN * 3) / 4),4)
3180 X0 = 28 - LEN (XL$) * 7:Y0 = 111:  GOSUB 3500
3185 XL$ = LEFT$ ( STR$ (YN),4)
3190 X0 = 28 - LEN (XL$) * 7:Y0 = 143:  GOSUB 3500
3200 RETURN
3500 :
3510 FOR XI = 1 TO LEN (XL$)
3520     X3 = ASC ( MID$ (XL$,XI,1))
3530     XU = XI * 7 - 7
3540     IF XR = 0 THEN ROT= 0:  DRAW X3 AT X0 + XU,Y0
3550     IF XR = 1 THEN ROT= 48:  DRAW X3 AT X0,Y0 - XU
3560 NEXT XI
```

```
3590 XR = 0:   RETURN
4000 :
4010 X0 = 30 + 240 * (X - XN) / (XM - XN)
4020 Y0 = 149 - 140 * (Y - YN) / (YM - YN)
4030 IF X0 < 2 OR X0 > 276 THEN XD = 0:   GOTO 4090
4040 IF Y0 < 4 OR Y0 > 149 THEN XD = 0:   GOTO 4090
4050 IF XB = 1 THEN DRAW 2 AT X0,Y0:   GOTO 4090
4060 IF XC = 1 THEN DRAW 3 AT X0,Y0:   GOTO 4090
4070 IF XD = 1 THEN HPLOT TO X0,Y0:   GOTO 4090
4080 HPLOT X0,Y0:   IF XA = 1 THEN XD = 1
4090 XB = 0:XC = 0:   RETURN
```

Figure A2.3. GRAPH listing for Apple II. (concluded)

Methods of obtaining a printout of a program listing will depend upon the specific printer installed with the computer. Conventionally, a printer interface is installed in Slot 1 of the Apple II. Output is directed to the printer with a PR#1 command in the immediate mode. To obtain a listing of a program, the PR#1 command would be followed by any commands needed to initialize the printer, and then by LIST. The sequence of commands for obtaining a copy of the graphic output of the high-resolution screen is completely dependent on the attached printer.

The alphanumeric characters used with GRAPH for the Apple II high-resolution screen are obtained with the DRAW command (Lines 3560 and 3570) from a shape table that is loaded into memory in Lines 3013 and 3015. The equivalent shape numbers correspond to the ASCII character code numbers for the Apple II. These numbers are given in Table A2.1. These equivalents are useful for short labels. For example, a BASIC statement DRAW 42 AT 100,90 will produce an asterisk at horizontal position 100 and vertical position 90 on the high-resolution screen. The DRAW command can be used with the Apple II+ to produce lower-case letters on the high-resolution screen even though they are not available on the keyboard.

32 = space	64 = @	96 = `
33 = !	65 = A	97 = a
34 = "	66 = B	98 = b
35 = #	67 = C	99 = c
36 = $	68 = D	100 = d
37 = %	69 = E	101 = e
38 = &	70 = F	102 = f
39 = ´	71 = G	103 = g
40 = (	72 = H	104 = h
41 =)	73 = I	105 = i
42 = *	74 = J	106 = j
43 = +	75 = K	107 = k
44 = ,	76 = L	108 = l
45 = -	77 = M	109 = m
46 = .	78 = N	110 = n
47 = /	79 = O	111 = o
48 = 0	80 = P	112 = p
49 = 1	81 = Q	113 = q
50 = 2	82 = R	114 = r
51 = 3	83 = S	115 = s
52 = 4	84 = T	116 = t
53 = 5	85 = U	117 = u
54 = 6	86 = V	118 = v
55 = 7	87 = W	119 = w
56 = 8	88 = X	120 = x
57 = 9	89 = Y	121 = y
58 = :	90 = Z	122 = z
59 = ;	91 = [	123 = {
60 = <	92 = \	124 =	
61 = =	93 =]	125 = }
62 = >	94 = ^	126 = ~
63 = ?	95 = _	

Table A2.1. List of alphanumeric characters and corresponding numbers in the SHAPES file, for Apple II computers. These numbers are used with the Applesoft DRAW statement in the Apple II GRAPH program.

A2.9 Using GRAPH with the Apple Macintosh Computer

A version of GRAPH is shown in Figure A2.4 for the Apple Macintosh computer. It has been used successfully with the Microsoft BASIC interpreter, Versions 1.0 and 2.0, on 128K, 512K and Plus models of the Macintosh. It may be readily adapted to other BASICs and other machines. (The GRAPH program uses little of the Macintosh's notable graphic point-and-click mouse interface; this is not a problem because the interface was designed for program users, not program writers.) The graph axes are drawn and labelled with Lines 3000-3200. Lines 4000-4090 are used in plotting the points.

Before running the program, the output window for the BASIC interpreter should be made to fill the Macintosh screen by dragging the size box of the output window. GRAPH assumes that the output window will have x-y dimensions of about 500 points by 250 points.

Titles and line labels may be placed on the graphic output with GOSUB 3500 statements as in Line 190 of the sample program. The position of the label is fixed by specifying the coordinates for the beginning of the message, using the variables X0 and Y0, and the message as a string variable XL$. The coordinates of the output window have the x-y origin (0,0) in the upper left corner.

Methods of obtaining a printout of a program listing will depend upon the specific printer and its method of connection with the computer. The BASIC immediate command LLIST works with printers attached to the serial interface printer port. The BASIC commands for obtaining a copy of the output window is dependent on the version of BASIC, the model of the Macintosh, the printer, and its mode of attachment. In some cases the usual Macintosh methods of performing a "screen-dump" to the printer or disk will work. Likewise, the mechanics of directing tabular output from a program to a printer will vary with the installation of computer and printer.

Figure A2.4. Listing of GRAPH program for Apple Macintosh computer using Microsoft BASIC versions 1.0 and 2.0.

```
10 REM EXERCISE #0.0
20 REM SINE WAVE DEMO
30 REM GRAPH: MACINTOSH/MS BASIC
40 REM 30 FEBRUARY 1990
50 :
60 REM DEFINE GRAPH AXES
70 READ XM,XN,YM,YN,X$,Y$
80 DATA 2,0,+1,-1,VALUE,SINE
90 GOSUB 3000:  REM DRAW GRAPH
100 :
110 PI = 3.14159
120 XA=1 :  XD=0:  REM LINE PLOT
130 FOR I = 0 TO 160
140   V = I / 80
150   S = SIN (2 * PI * V)
160   X = V : Y = S
170     GOSUB 4000
180 NEXT I
190 X0=180 :  Y0=15 :  XL$="SINE WAVE":  GOSUB 3500
200 END
3000 :
3010 CLS
3020 LINE(80,10)-(80,210)
3030 LINE(80,210)-(480,210)
3040 FOR XI= 10 TO 185 STEP 25
3050     LINE(80,XI)-(85,XI) : NEXT
3060 FOR XI= 130 TO 480 STEP 50
3070     LINE(XI,210)-(XI,205) :  NEXT
3095 XL$= LEFT$(X$,10)
3100 X0= 350 :  Y0= 225 :  GOSUB 3500
3105 XL$= LEFT$(STR$(XN),5)
3110 X0= 81 - 3.5* LEN(XL$) :  Y0= 225 :  GOSUB 3500
3115 XL$= LEFT$(STR$((XM+XN*3)/4),5)
3120 X0= 181 - 3.5* LEN(XL$) :  Y0= 225 :  GOSUB 3500
3125 XL$= LEFT$(STR$((XM+XN)/2),5)
3130 X0= 281 - 3.5* LEN(XL$) :  Y0= 225 :  GOSUB 3500
3135 XL$= LEFT$(STR$(XM),5)
3140 X0= 471 - 3.5*LEN(XL$) :   Y0 = 225 :  GOSUB 3500
3145 XL$= LEFT$(Y$,8)
3150 X0= 70 - 8*LEN(XL$) :  Y0 = 65 :  GOSUB 3500
3155 XL$= LEFT$(STR$(YM),5)
3160 X0= 70 - 8*LEN(XL$) :   Y0 = 15 :  GOSUB 3500
3165 XL$= LEFT$(STR$((YM+YN)/2),5)
3170 X0= 70 - 8*LEN(XL$) :   Y0 = 115 :  GOSUB 3500
3175 XL$= LEFT$(STR$((YM+YN*3)/4),5)
3180 X0= 70 - 8*LEN(XL$) :   Y0 = 165 :  GOSUB 3500
3185 XL$= LEFT$(STR$(YN),5)
3190 X0= 70 - 8*LEN(XL$) :  Y0 = 215 :  GOSUB 3500
3200 RETURN
3500 :
3510 CALL MOVETO (X0,Y0)
3520 PRINT XL$;
3530 RETURN
4000 :
4010 X0 = 80 + 400 * (X-XN)/(XM-XN)
4020 Y0 = 210 - 200 * (Y-YN)/(YM-YN)
4030 IF X0 < 2 OR X0 > 490 THEN XD = 0:  GOTO 4090
4040 IF Y0 < 3 OR Y0 > 215 THEN XD = 0:  GOTO 4090
```

```
4050 IF XB = 1 THEN LINE(X0-1,Y0-1)-(X0+1,Y0+1),,BF : GOTO 4090
4060 IF XC = 1 THEN CIRCLE(X0,Y0),2 :  GOTO 4090
4070 IF XD = 1 THEN LINE -(X0,Y0) :  GOTO 4090
4080 PSET (X0,Y0) :  IF XA = 1 THEN XD = 1
4090 XB = 0 :  XC = 0 :  RETURN
```

Figure A2.4. GRAPH program for Apple Macintosh computer. (concluded)

APPENDIX 3

DESCRIPTION OF CURFIT PROGRAM

CURFIT is a BASIC program written specifically for this textbook. It is intended for use with x-y data to make it easy to fit any of 10 equations that often occur in biological research. The program applies the technique of linear transformation of data to fit a straight line with least-squares regression. The methods and the equations used in the program are discussed in Chapter 3. This appendix will describe the mechanics of using CURFIT: how to work through the program, and how to read data into the program. Some specifics of its implementation on two common microcomputers are discussed separately below, including the procedures for producing printed output from the program.

The requirements for hardware and software to run the CURFIT program are the same as for the GRAPH program.

A3.1 General Approach

The CURFIT program is intended to provide an interactive method for transformation of data and fitting of equations. The program is not fail-safe. It is written so that most casual mistyping does not halt execution. There are some built-in error checking procedures, but these are not comprehensive. It is possible to enter data incorrectly into CURFIT, or to set constants so that the program fails in its calculations, and produces BASIC error messages. For example, a negative number can be entered and the program will halt when taking the log of the datum. CURFIT is intended to be used by individuals who have worked their way through some exercises and BASIC programs of Chapters 1 and 2.

A3.2 Data Input

The program is written so that data are entered from a data file from a storage device (e.g. a file on disk). These data files can be created in a variety of ways for any given computer. Such methods include word processing programs, file line-editors, spreadsheet programs, etc. (If these methods are used, the data should be saved in standard "ASCII Format".)

Figures A3.1 and A3.2 show listings of a simple BASIC program for creating data files from a set of x-y data for common microcomputers. If this program is selected as the method for making data files, then the program is loaded and the x-y data are entered as DATA statements. (The sample data given in the programs should be removed, of course.) Running the program will create a data file with the name provided in the program.

```
10 REM * * * * * * * * * * * * * * * * * * * * * * * * * * * * * * * * * * * * *
20 REM * SAMPLE BASIC PROGRAM FOR MAKING A DATA FILE FOR                    *
30 REM * CURFIT, POLYFIT, AND CURNLFIT PROGRAMS (MS/PC-DOS MACHINES         *
40 REM *                                                                    *
50 REM * REPLACE NAME IN QUOTES WITH YOUR DATA FILE NAME (LINE 170)         *
60 REM * REPLACE DATA STATEMENTS WITH YOUR DATA (LINE 1000)                 *
70 REM *      N: NUMBER OF X-Y PAIRS (MAXIMUM 50)                           *
80 REM *         TYPE IN X & Y AS PAIRS, WITH COMMAS                        *
90 REM * * * * * * * * * * * * * * * * * * * * * * * * * * * * * * * * * * * * *
100 :
110 DIM X(50),Y(50)
120 READ N
130 FOR I = 1 TO N
140     READ X(I), Y(I)
150 NEXT I
160 :
170 OPEN "TRIALFIT" FOR OUTPUT AS #1
180 WRITE #1,N
190 FOR I = 1 TO N
200 WRITE #1, X(I), Y(I)
210 NEXT I
220 CLOSE #1
230 END
999 :
1000 DATA 8
1010 DATA 1,2.02
1020 DATA 2,8.05
1030 DATA 3,18.03
1040 DATA 4,32.02
1050 DATA 5,50.04
1060 DATA 6,72.07
1070 DATA 7,98.06
1080 DATA 8,128.01
```

Figure A3.1. Sample BASIC program for creating data files on disk for IBM personal computers using MS-DOS and BASICA or GW-BASIC interpreters. The data files are usable with the CURFIT, POLYFIT, and CURNLFIT programs.

```
10 REM * * * * * * * * * * * * * * * * * * * * * * * * * * * * * * * * * * * * * * *
20 REM * SAMPLE BASIC PROGRAM FOR MAKING A DATA FILE FOR              *
30 REM * CURFIT, POLYFIT, AND CURNLFIT PROGRAMS (APPLE II/DOS3.3      *
40 REM *                                                              *
50 REM * REPLACE NAME IN QUOTES WITH YOUR DATA FILE NAME (LINE 170)   *
60 REM * REPLACE DATA STATEMENTS WITH YOUR DATA (LINE 1000)           *
70 REM *     N: NUMBER OF X-Y PAIRS (MAXIMUM 50)                      *
80 REM *     TYPE IN X & Y AS PAIRS, WITH COMMAS                      *
90 REM * * * * * * * * * * * * * * * * * * * * * * * * * * * * * * * * * * * * * * *
100 :
110   DIM X(50),Y(50)
120   READ N
130   FOR I = 1 TO N
140       READ X(I),Y(I)
150   NEXT I
160 :
170 D$ = CHR$ (13) + CHR$ (4) :  N$ = "TRIALFIT"
180   PRINT D$"OPEN"N$
190   PRINT D$"DELETE"N$
200   PRINT D$"OPEN"N$
210   PRINT D$"WRITE"N$
220   PRINT N
230   FOR I = 1 TO N
240       PRINT X(I)","Y(I)
250   NEXT I
260   PRINT D$"CLOSE"N$
270   END
999 :
1000   DATA 8
1010   DATA 1, 2.02
1020   DATA 2, 8.05
1030   DATA 3, 18.03
1040   DATA 4, 32.02
1050   DATA 5, 50.04
1060   DATA 6, 72.07
1070   DATA 7, 98.06

1080   DATA 8,128.01
```

Figure A3.2. Sample Applesoft BASIC program for creating data files on disk for Apple II computers using Apple DOS 3.3. The data files are usable with the CURFIT, POLYFIT, and CURNLFIT programs.

This file-making program allows the data to be modified before being used by the CURFIT program. For example, data could be adjusted by adding a constant value to the x- or y-values to remove negative values. A couple of lines of BASIC code can be added to the program to perform such operations easily.

The CURFIT program may be modified readily to take data from other sources, including DATA statements in the CURFIT program, or with INPUT statements during execution of the program. The lines for input of data are found at about Line 5200 in all versions of CURFIT.

A3.3 Using CURFIT

When CURFIT is LOADed and RUN, the initial title display should show the correct name of the file holding the data for analysis and curve fitting. If not, the program should be halted and the correct name of the data file inserted in Line 100, following the format of the sample name in the program. With the correct data file, the user should proceed following the diagram in Figure A3.3. The program begins with a scatter plot, a list of the equations available for fitting with linear transformation, etc.

The program allows the user to move backwards through the program. This is a useful feature particularly when fitting an equation like the logistic with a fixed constant in the equation. Backing up allows the analysis to be repeated rapidly, to find the value of the constant that produces the best fit, judged by the F-value or the variation of the residuals.

*1. Title and Data File Check

   [P: go to *2]

   [Q: exit]

*2. Scatter Plot

   [P: go to *3]

   [Z: go to *1]

   [(S: go to *9)]

   [Q: exit]

*3. List of Equations #1-10

? Select Eqn #

   [#1- 6: go to *5]

   [#7-10: go to *4]

   [Z: go to *2]

   [(S: go to *9)]

   [Q: exit]

*4. Select a Constant #

   [#: go to *5]

   [Z: go to *3]

   [(S: go to *9)]

   [Q: exit]

*5. Curve-Fitting Statistics

? See predicted-y list (Y/N)

   [Y: go to *6]

   [N: go to ?6]

   [Z: go to *3 or *4]

   [S: go to *9]

   [Q: exit]

*6. List of x, y, pred-y

? See plot of residuals (Y/N)

   [Y: go to *7]

   [N: go to ?7]

   [Z: go to *5]

   [S: go to *9]

   [Q: exit]

*7. Plot of Residuals   *8. Plot of Data and Curve

? See Plot of Data & Curve (Y/N)?   Go to Summary &/or another Eqn

[Y: go to *8]   [S: go to *9]

[N: go to ?8]   [Z: go to ?7]

[Z: go to *6]   [Q: exit]

[S: go to *9]

[Q: exit]

*9. List of Summary Statistics

[E: go to *3]

[Z: go to ?7]

[Q: exit]

Figure A3.3. List of flow and options for using CURFIT program. Video presentation is underlined; user reply is bracketed. (* indicates screens keyed to listing in Figures A3.4 - A3.5.)

A3.4 CURFIT on the IBM Personal Computer

The machine requirements for running CURFIT (Figure A3.4) are the same as for the GRAPH program (Appendix 2). With BASIC running, the user should LOAD the CURFIT program, and then insert the name of the data file in Line 100. (It may be necessary to include as a part of the name a label for a specific disk drive or other input device.) Then the program is started as usual with RUN.

Printed output of the graphs and of the summary statistics may be obtained with the shift-PrtSc convention. However, if more than about 15 pairs of x-y data points are put into the analysis, the listing of the observed and predicted data points will not be complete ("Screen 6"). The data sets for the exercises in this book are not so long that this will be a problem. For longer data sets, the PRINT statements could be changed to LPRINT statements in Lines 1600, 1610, and 1650 to produce a listing on an attached printer. Alternatively, the program may be modified easily to permit the user to select a "PRINT" option for directing the list of data to a printer; see the Apple II listing for an example.

Other methods for directing output to a printer are possible, depending upon the specific machinery and programs available with any given machine.

Figure A3.4. Listing of CURFIT program for IBM-PC and compatible machines, using BASICA or similar interpreters. The graphic subroutines of Lines 3000-4090 from the appropriate GRAPH program of Appendix 2 should be inserted into this program.

```
10 REM
20 REM     CURFIT.....WRITTEN BY ROBERT E. KEEN & JAMES D. SPAIN
30 REM     COPYRIGHT (C) 1981,1990 BY JOHN WILEY & SONS, INC. NYC
40 REM
50 REM
60 REM     REPLACE "TRIALFIT" IN LINE
70 REM     100 WITH YOUR DATA FILE NAME
80 REM
90 :
100 DATA "TRIALFIT"
110 :
120 DIM X(50), Y(50), YP(50)
130 :
140 REM****SCREEN 1 ************************************
150 GOSUB 5000 :   REM TITLE & DATA ENTRY******
160 :
170 REM****SCREEN 2 ************************************
180 REM FIND MAX X, MIN Y & MAX Y
190 YL=9.900001E+37
200 FOR J=1 TO N
210     IF XH < X(J) THEN XH = X(J)
220     IF YH < Y(J) THEN YH = Y(J)
230     IF YL > Y(J) THEN YL = Y(J)
240 NEXT J
250 X$="X-VALUE":Y$="Y-VALUE":XN=0:YN=0:YM=YH:XM=XH
260 GOSUB 3000 :   REM DRAW & LABEL AXES******
270 FOR J=1 TO N
280     X=X(J) : Y=Y(J) : XC=1 :   GOSUB 4000 :   REM PLOT WITH CIRCLES
290 NEXT J
300 LOCATE 24,1:PRINT"ENTER 'P' TO PROCEED WITH MODEL SELECTION ";
310 PRINT"(Q=QUIT, Z=BACKUP"; :   IF QS<>0 THEN PRINT", S=SUMMARY";
320 INPUT") ";QM$
330 IF QM$="Q" THEN SCREEN 0 :   CLS : END
340 IF QM$="Z" THEN 150
350 IF QM$="S" THEN 2080
360 IF QM$<>"P" THEN 300
370 :
380 REM****SCREEN 3 ************************************
390 CLS : SCREEN 0
400 PRINT"THE FOLLOWING EQUATIONS MAY BE USED WITH THIS PROGRAM:"
410 FOR J=1 TO 10
420     LOCATE (J*2+1),1:IF J<10 THEN PRINT" ";
430     PRINT J"...";N$(J);TAB(30);E$(J)
440 NEXT J
450 LOCATE 24,1:PRINT"(OR Q=QUIT, Z=BACKUP";:IF QS<>0 THEN PRINT",
        S=SUMMARY";
460 PRINT")";
470 LOCATE 23,1:INPUT"ENTER NUMBER OF EQUATION TO BE USED";QM$
480 IF QM$="Z" THEN 250
490 IF QM$="Q" THEN CLS : END
500 IF QM$="S" THEN 2080
510 Q=INT(VAL(QM$)):   IF Q<1 OR Q>10 THEN 470
520 IF Q < 7 GOTO 750
530 :
540 REM****SCREEN 4 ************************************
```

```
550 CLS : PRINT: PRINT: PRINT"EQUATION NO."Q" "N$(Q);":   ";E$(Q):PRINT
560 IF Q = 9 GOTO 620
570 PRINT"EQUATION"Q"REQUIRES AN ESTIMATE FOR THE MAXIMUM VALUE, ";
580 IF Q=7 THEN PRINT"A":PRINT
590 IF Q<>7 THEN PRINT"K":PRINT
600 PRINT" -ESTIMATE MUST BE LARGER THAN THE LARGEST VALUE OF Y (="YH")"
    :PRINT
610 GOTO 660
620 :
630 PRINT"EQUATION 9 REQUIRES A VALUE FOR THE INTERCEPT, K":  PRINT
640 PRINT" -THE VALUE MAY BE ZERO, OR ANY POSITIVE VALUE BETWEEN ZERO
    AND"
650 PRINT" THE MINIMUM VALUE OF Y (="YL")":  PRINT
660 PRINT"ENTER AN ESTIMATED VALUE (OR Q=QUIT Z=BACKUP";
670 IF QS<>0 THEN PRINT " S=SUMMARY";
680 INPUT") >>>";QM$
690 IF QM$="S" THEN 2080
700 IF QM$="Z" THEN 390
710 IF QM$="Q" THEN CLS : END
720 K=VAL(QM$)  :  AE=K
730 IF Q=9 AND (K>=YL AND K<0) THEN 640
740 IF (Q<>9) AND (K<=YH) THEN 600
750 :
760 X1=0: X2=0: XY=0: Y1=0: Y2=0: REM INITIALIZE SUMS
770 :
780 REM****SCREEN 5 ***********************************
790 CLS : SCREEN 0
800 PRINT"FITTING EQUATION #"Q" "N$(Q)
810 FOR J=1 TO N:PRINT".";  : REM DATA TRANSFORMATION****
820   X=X(J):Y=Y(J)
830   ON Q GOSUB 850,860,870,880,890,900,910,920,930,940
840   GOTO 960
850   : RETURN
860   Y=X/Y : RETURN
870   Y=1/Y : RETURN
880   Y=LOG(Y) : RETURN
890   Y=LOG(Y) : X=1/X : RETURN
900   Y=LOG(Y/X) : RETURN
910   Y=LOG(AE-Y) : RETURN
920   Y=LOG(K/Y-1) :  RETURN
930   Y=LOG(Y-K) : X=LOG(X) : RETURN
940   Y=LOG(K/Y-1) :  X=LOG(X) : RETURN
950   REM SUMS & SUM OF SQUARES******
960   X1 = X1 + X
970   X2 = X2 + X*X
980   XY = XY + X*Y
990   Y1 = Y1 + Y
1000      Y2 = Y2 + Y*Y
1010 NEXT J:PRINT
1020 REM L.S. INTERCEPT(I) AND SLOPE(S)
1030 I = (Y1*X2 - X1*XY) / (N*X2 - X1*X1)
1040 S = (N*XY - X1*Y1) / (N*X2-X1*X1)
1050 R2 = ((XY - X1*Y1/N)^2) / (X2 - (X1*X1)/N) / (Y2 - (Y1*Y1)/N)
1060 PRINT"STATISTICS FOR TRANSFORMED DATA:"
1070 PRINT" Y = "; FN RO(I); " + "; FN RO(S); " * X"
1080 PRINT" COEF. OF DETERMINATION (R-SQ) = "; FN RO(R2)
1090 IF Q = 1 THEN A=I : B=S
1100 IF Q > 1 THEN A=1/S : B=I*A
1110 IF Q > 3 THEN A=EXP(I) : B=S
1120 PRINT:PRINT"THE "N$(Q)" EQUATION IS "E$(Q) : PRINT
```

Figure A3.4. CURFIT program for IBM-PC and compatible machines. (continued)

```
1130 IF Q=7 THEN PRINT" YOUR ESTIMATED VALUE OF A = "; AE
1140 IF Q > 7 THEN PRINT" YOUR ESTIMATED VALUE OF K = ";K :PRINT
1150 PRINT" THE CALCULATED VALUE OF A = " FN RO(A) :PRINT
1160 PRINT" THE CALCULATED VALUE OF B = " FN RO(B) :PRINT
1170 :
1180 Y3=0 :  Y4=0 :  RS=0 :  Y5=0 :  Y6=0
1190 FOR J=1 TO N
1200    X=X(J)
1210    ON Q GOSUB 1230,1240,1250,1260,1270,1280,1290,1300,1310,1320
1220    GOTO 1330
1230    Y=A+B*X : RETURN
1240    Y=A*X/(B+X) : RETURN
1250    Y=A/(B+X) : RETURN
1260    Y=A*EXP(B*X) : RETURN
1270    Y=A*EXP(B/X) : RETURN
1280    Y=A*X*EXP(B*X) : RETURN
1290    Y=A*(1-EXP(B*X)) : RETURN
1300    Y=K/(1+A*EXP(B*X)) : RETURN
1310    Y=K+A*X^B : RETURN
1320    Y=K/(1+A*X^B) : RETURN
1330    REM RESIDUALS
1340    R = Y(J) - Y : YP(J) = Y
1350    Y3 = Y3 + Y : Y4 = Y4 + Y*Y
1360    Y5 = Y5 + Y(J) : Y6 = Y6 + Y(J)*Y(J)
1370    RS = RS + R*R
1380 NEXT J
1390 PRINT"SUM OF SQUARES OF RESIDUALS = ";FN RO(RS)
1400 PRINT : P=2 :   REM # OF EST. PARAMETERS
1410 VR = RS/(N-P)
1420 SE(Q) = SQR(VR) : SE(Q) = FN RO(SE(Q))
1430 F(Q) = (Y4 - (2*Y5/N)*Y3 + (Y5*Y5)/N) / (VR)
1440 F(Q)=FN RO(F(Q))
1450 QS=1 :   REM SUMMARY FLAG
1460 PRINT"VARIANCE OF RESIDUALS FOR";N-P;"DF = ";
1470 PRINT FN RO(VR) : PRINT
1480 PRINT"THE STD. ERROR OF RESIDUALS =";SE(Q)
1490 PRINT:PRINT"FOR AN F-TEST WITH 1 & "(N-P)" DF, F = "F(Q):PRINT
1500 LOCATE 24,1:PRINT"(OR Z=BACKUP S=SUMMARY Q=QUIT)";:LOCATE 23,1
1510 INPUT"DO YOU WANT TO SEE A LIST OF DATA & THE ESTIMATED POINTS
        (Y/N)" ;QM$
1520 IF QM$="Z" AND Q<7 THEN 390
1530 IF QM$="Z" AND Q>6 THEN 550
1540 IF QM$="Q" THEN CLS : END
1550 IF QM$="S" THEN 2080
1560 IF QM$<>"Y" THEN 1670
1570 :
1580 REM****SCREEN 6 ************************************
1590 CLS:SCREEN 0:  PRINT"COMPARISON OF ORIGINAL DATA WITH ESTIMATED
        POINTS"
1600 PRINT"EQUATION #"Q;N$(Q)
1610 PRINT:PRINT" X Y(DATA) Y(PREDICTED) DIFFERENCE"
1620 FOR J=1 TO N
1630    X=X(J)
1640    R = Y(J) - YP(J)
1650    PRINT X, Y(J), FN RO(YP(J)), FN RO(R)
1660 NEXT J
1670 PRINT:PRINT:PRINT
1680 LOCATE 24,1:PRINT"(OR Z=BACKUP S=SUMMARY Q=QUIT)";:LOCATE 23,1
1690 INPUT"DO YOU WISH TO SEE A PLOT OF THE RESIDUALS (Y/N)";QM$
1700 IF QM$="Z" THEN 750
```

Figure A3.4. CURFIT program for IBM-PC and compatible machines. (continued)

```
1710 IF QM$="Q" THEN CLS : END
1720 IF QM$="S" THEN 2080
1730 IF QM$<>"Y" THEN PRINT:PRINT:PRINT : GOTO 1830
1740 :
1750 REM****SCREEN 7 ************************************
1760 Y$="RESIDUAL-%" :   YM=+100 :   YN=-100 :   GOSUB 3000
1770 LOCATE 1,1:PRINT"EQN #"Q
1780 XA=1:  XD=0 :   REM LINE DRAWING
1790 X=0:Y=0 :   GOSUB 4000:  X=XM : GOSUB 4000
1800 FOR J=1 TO N
1810 X=X(J) :  Y=(Y(J)-YP(J))*100/YH :  XB=1 :   GOSUB 4000
1820 NEXT J
1830 LOCATE 24,1:PRINT"DO YOU WANT THE MODEL PLOTTED WITH THE DATA
       (Y/N; ";
1840 INPUT"Z=BACKUP S=SUMMARY Q=QUIT)";QM$
1850 IF QM$="Z" THEN 1590
1860 IF QM$="Q" THEN SCREEN 0 :   CLS : END
1870 IF QM$="S" THEN 2080
1880 IF QM$<>"Y" THEN PRINT:PRINT:PRINT : GOTO 2020
1890 :
1900 REM****SCREEN 8 ************************************
1910 Y$="Y-VALUE" :   YN=0 :   YM=YH : GOSUB 3000
1920 LOCATE 1,1:PRINT"EQN #"Q
1930 XA=1 :   XD=0
1940 FOR J=1 TO 100 STEP 2
1950     X=J*XM/100
1960     ON Q GOSUB 1230,1240,1250,1260,1270,1280,1290,1300,1310,1320
1970     GOSUB 4000
1980 NEXT J
1990 FOR J=1 TO N
2000     X=X(J) :  Y=Y(J) :  XC=1 :   GOSUB 4000
2010 NEXT J
2020 LOCATE 24,1:PRINT"ENTER S FOR SUMMARY &/OR ANOTHER EQN (Z=BACKUP, ";
2030 INPUT"Q=QUIT)";QM$
2040 IF QM$="Q" THEN SCREEN 0 :   CLS : END
2050 IF QM$="Z" THEN PRINT:PRINT:PRINT : GOTO 1830
2060 :
2070 REM****SCREEN 9 ************************************
2080 CLS : SCREEN 0 :   PRINT"SUMMARY OF CURVE-FITTING STATISTICS
2090 PRINT:PRINT"EQUATION STD. ERROR F-VALUE"
2100 PRINT:FOR J=1 TO 10:IF J<10 THEN PRINT" ";
2110     PRINT J"..."N$(J);TAB(28);
2120     IF F(J)=0 THEN PRINT" - - -"," - - -":GOTO 2140
2130     PRINT SE(J),F(J)
2140 NEXT J:PRINT:PRINT
2150 INPUT"ENTER E TO SELECT ANOTHER EQUATION (Z=BACKUP Q=QUIT)";QM$
2160 IF QM$="E" THEN 390
2170 IF QM$="Q" THEN CLS : END
2180 IF QM$="Z" THEN 1830
2190 GOTO 2150
2200 :
2210 REM****END OF PROGRAM *******************************
3000 :
  .
  .
  .
```

Insert lines 3000-4090 from the **GRAPH** program here

```
  .
  .
5000 :
5010 REM***** START SUBROUTINE ***************************
```

Figure A3.4. CURFIT program for IBM-PC and compatible machines. (continued)

```
5020 SCREEN 0 :   CLS : WIDTH 80 :   KEY OFF : COLOR 7
5030 LOCATE 4,37:PRINT"CURFIT"
5040 LOCATE 7,13:PRINT" Program for Fitting 10 Common Model Equations to
     Data"
5050 LOCATE 9,13:PRINT"Information about the program may be found in the
     book"
5060 LOCATE 10,16:COLOR 9
5070 PRINT"Computer Simulation in Biology:  A BASIC Approach":COLOR 7
5080 LOCATE 11,24:PRINT"by Robert E. Keen & James D. Spain"
5090 LOCATE 12,14:PRINT"Published by John Wiley & Sons, Inc.  Copyright
     1990"
5100 READ F$ :   RESTORE
5110 LOCATE 16,11:PRINT"This program is currently set to read data from
     a ";
5120 PRINT"file named":  LOCATE 18,30:PRINT">"F$:LOCATE 20,12
5130 PRINT"If this is not the name of the diskfile holding the data,"
     :LOCATE,9
5140 PRINT"then quit the program and change the name of the file in
     Line 100"
5150 PRINT:INPUT"Enter P to read & plot data from this file, Q to quit"
     ;QM$
5160 IF QM$="Q" THEN CLS:END
5170 IF QM$="p" OR QM$="q" THEN PRINT"CAPITAL LETTERS PLEASE"
5180 IF QM$<>"P" THEN 5150
5190 OPEN"I",1,F$
5200 INPUT #1,N
5210 FOR J=1 TO N:INPUT #1,X(J),Y(J):NEXT J
5220 CLOSE 1
5230 DEF FN RO(X) = INT(10000*X + .5)/10000
5240 N$( 1)="STRAIGHT LINE"        :  E$( 1)="Y = A + B* X"
5250 N$( 2)="HYPERBOLIC"           :  E$( 2)="Y = (A*X) / (B+X)"
5260 N$( 3)="MODIFIED INVERSE"     :  E$( 3)="Y = A / (B + X)"
5270 N$( 4)="EXPONENTIAL"          :  E$( 4)="Y = A * EXP(B*X)"
5280 N$( 5)="EXP. RECIPROCAL"      :  E$( 5)="Y = A * EXP(B/X)"
5290 N$( 6)="MAXIMA FUNCTION"      :  E$( 6)="Y = A * X * EXP(B*X)"
5300 N$( 7)="EXP. SATURATION"      :  E$( 7)="Y = A * [1 - EXP(B*X)]"
5310 N$( 8)="LOGISTIC"             :  E$( 8)="Y = K / [1 + A * EXP(B*X)]"
5320 N$( 9)="LOGARITHMIC"          :  E$( 9)="Y = (A * X^B) + K"
5330 N$(10)="'SIGMOID'"            :  E$(10)="Y = K / (1 + A * X^B)"
5340 RETURN
```

Figure A3.4. CURFIT program for IBM-PC and compatible machines. (concluded)

A3.5 CURFIT on Apple II Computers

The machine requirements for running CURFIT (Figure A3.5) are the same as for the GRAPH program (Appendix 2). Rather than LOADing the program and then altering Line 100 for the data-file name, first the program should be RUN. When "Screen 1" appears with the incorrect name, the program should be halted with the "quit" option, and the file name changed in Line 100. (This change in procedure is needed because of the double-loading relocation discussed below.)

NOTE: If other BASIC programs are LOADed and RUN after CURFIT has been used, the relocation can cause problems. It is probably best to reboot the computer after running CURFIT.

Printed output of the summary statistics ("Screen 5") and of the list of data and predicted y-values ("Screen 6") may be obtained by selecting the "Hardcopy" option for these screens, if a printer is attached. Should they need modification, the BASIC commands for turning the printer on and off are found beginning at Lines 5400.

The possibility of obtaining printed output of graphs will depend upon the printer. Lines 5460-5490 hold the code for the Apple Silentype printer. Code for other printers may be substituted if appropriate. Other methods for directing output to a printer are possible, depending upon the specific machinery and programs available for any given machine. For example, the execution of CURFIT may be halted with the "quit" option when the desired graph is on the screen; then the screen may be either printed using a "screen-dump" program appropriate for the printer, or the screen image may be saved to a disk using the BSAVE command and printed later.

The CURFIT program is so long that it cannot fit into the normal memory reserved for BASIC programs without interfering with the high-resolution graphics screen. This requires the program to be located above the screen. This relocation is accomplished in Lines 5021-5023, which causes the program to be loaded twice.

CURFIT uses the same shape-table that GRAPH uses, but the table is placed low in memory. If you want to modify the BASIC code of CUR-FIT, you should note that Line 3013 of CURFIT substitutes "A$0800" for "A$4000" used in the same line of GRAPH. Similarly, Line 3015 of CUR-FIT uses a "POKE 233,08" rather than the "POKE 233,64" of GRAPH.

Figure A3.5. Listing of CURFIT program for Apple II machines, using Applesoft BASIC. The graphic subroutines from Lines 3020-4090 from the GRAPH program (Appendix 2) must be inserted into this program.

```
10    REM
20    REM CURFIT..WRITTEN BY ROBERT E. KEEN & JAMES D. SPAIN
30    REM COPYRIGHT (C) 1981,1990 BY JOHN WILEY & SONS, INC. NYC
40    REM
50    REM
60    REM      REPLACE "TRIALFIT" IN LINE
70    REM      100 WITH YOUR DATA FILE NAME
80    REM
90  :
100   DATA "TRIALFIT"
110 :
120   DIM X(50),Y(50),YP(50)
130 :
140   REM *****SCREEN 1 ********************************
150   GOSUB 5000:  REM TITLE & DATA ENTRY *****
160 :
170   REM *****SCREEN 2 ********************************
180   REM FIND MAX X, MIN Y & MAX Y
190 YL = 9.9E + 36
200   FOR J = 1 TO N
210   IF XH < X(J) THEN XH = X(J)
220   IF YH < Y(J) THEN YH = Y(J)
230   IF YL > Y(J) THEN YL = Y(J)
240   NEXT J
250 X$ = "X-VALUE":Y$ = "Y-VALUE":XN = 0:YN = 0:YM = YH:XM = XH
260   GOSUB 3000:  REM DRAW & LABEL AXES
270   FOR J = 1 TO N
280 X = X(J):Y = Y(J):XC = 1:  GOSUB 4000:  REM PLOT WITH CIRCLES
290   NEXT J
300   VTAB 24:  PRINT "ENTER 'P' =PROCEED W/MODEL SELECTION"
310   PRINT "(Q=QUIT Z=BACKUP";:  IF QS < > 0 THEN PRINT " S=SUMMARY";
320   INPUT ") ";QM$
330   IF QM$ = "Q" THEN TEXT : HOME : END
340   IF QM$ = "Z" THEN 150
350   IF QM$ = "S" THEN 2080
360   IF QM$ < > "P" THEN 300
370 :
380   REM *****SCREEN 3 ********************************
390 TEXT : HOME
400   PRINT "THESE EQNS MAY BE FIT TO THE DATA:": PRINT
410   FOR J = 1 TO 10
420   IF J < 10 THEN PRINT " ";
430   PRINT J".."N$(J); TAB( 20)E$(J)
440   NEXT J
450   VTAB 23:  PRINT "(OR Q=QUIT Z=BACKUP";:  IF QS < > 0 THEN PRINT "
      S=SUMMARY";
460   PRINT ")";
470   VTAB 22:  HTAB 1:  INPUT "ENTER NUMBER OF EQN TO BE USED ";QM$
480   IF QM$ = "Z" THEN HOME : GOTO 250
490   IF QM$ = "Q" THEN HOME : END
500   IF QM$ = "S" THEN 2080
510 Q = INT ( VAL (QM$)):  IF Q < 1 OR Q > 10 THEN 470
520   IF Q < 7 GOTO 750
530 :
540   REM *****SCREEN 4 ********************************
```

```
550    HOME : PRINT "EQN #"Q" "N$(Q)" "E$(Q): PRINT
560    IF Q = 9 GOTO 620
570    PRINT "EQN "Q" REQUIRES AN ESTIMATE FOR THE":  PRINT "MAXIMUM
       VALUE, ";
580    IF Q = 7 THEN PRINT "A":  PRINT
590    IF Q < > 7 THEN PRINT "K":  PRINT
600    PRINT " -ESTIMATE MUST BE GREATER THAN THE":  PRINT " LARGEST
       VALUE OF Y (="YH")":  PRINT
610    GOTO 660
620 :
630    PRINT "EQN 9 REQUIRES AN ESTIMATE FOR THE":  PRINT "INTERCEPT, K":
       PRINT
640    PRINT " -THE VALUE MAY BE ZERO, OR ANY":  PRINT "POSITIVE VALUE
       BETWEEN ZERO AND THE"
650    PRINT " MINIMUM VALUE OF Y (="YL")":  PRINT
660    VTAB 23:  PRINT "(OR Q=QUIT Z=BACKUP";:  IF QS < > 0 THEN PRINT
       " S=SUMMARY)";
670    PRINT ")";
680    VTAB 20:  HTAB 1:  INPUT "ENTER AN ESTIMATED VALUE >>> ";QM$
690    IF QM$ = "S" THEN 2080
700    IF QM$ = "Z" THEN 390
710    IF QM$ = "Q" THEN HOME : END
720 K = VAL (QM$):AE = K
730    IF Q = 9 AND (K > = YL AND K < 0) THEN 640
740    IF (Q < > 9) AND (K < = YH) THEN 600
750 :
760 X1 = 0:X2 = 0:XY = 0:Y1 = 0:Y2 = 0:  REM INITIALIZE SUMS
770 :
780    REM *****SCREEN 5 *********************************
790    TEXT : HOME
800    PRINT "FITTING EQN #"Q" "N$(Q)
810    FOR J = 1 TO N: PRINT ".";:  REM DATA TRANSFORMATION***
820 X = X(J):Y = Y(J)
830    ON Q   GOSUB 850,860,870,880,890,900,910,920,930,940
840    GOTO 960
850 :  RETURN
860 Y = X / Y: RETURN
870 Y = 1 / Y: RETURN
880 Y =  LOG (Y): RETURN
890 Y =  LOG (Y):X = 1 / X: RETURN
900 Y =  LOG (Y / X): RETURN
910 Y =  LOG (AE - Y): RETURN
920 Y =  LOG (K / Y - 1):  RETURN
930 Y =  LOG (Y - K):X = LOG (X): RETURN
940 Y =  LOG (K / Y - 1):X = LOG (X): RETURN
950    REM SUMS & SUMS OF SQUARES***
960 X1 = X1 + X
970 X2 = X2 + X * X
980 XY = XY + X * Y
990 Y1 = Y1 + Y
1000 Y2 = Y2 + Y * Y
1010   NEXT J: PRINT
1020   REM L.S. INTERCEPT (I) & SLOPE(S)
1030 I = (Y1 * X2 - X1 * XY) / (N * X2 - X1 * X1)
1040 S = (N * XY - X1 * Y1) / (N * X2 - X1 * X1)
1050 R2 = ((XY - X1 * Y1 / N) ^2) / (X2 - (X1 * X1) / N) /
       (Y2 - (Y1 * Y1) / N)
1060   PRINT "STATISTICS FOR TRANSFORMED DATA:"
1070   PRINT " Y =" FN RO(I)" + " FN RO(S)" * X"
1080   PRINT " COEF. OF DETERMINATION (R-SQ)=" FN RO(R2)
```

Figure A3.5. CURFIT program for Apple II machines. (continued)

```
1090   IF Q = 1 THEN A = I:B = S
1100   IF Q > 1 THEN A = 1 / S:B = I * A
1110   IF Q > 3 THEN A = EXP (I):B = S
1120   PRINT : PRINT "THE EQN IS "E$(Q): PRINT
1130   IF Q = 7 THEN PRINT " YOUR ESTIMATE OF A = ";AE
1140   IF Q > 7 THEN PRINT " YOUR ESTIMATE OF K = ";K: PRINT
1150   PRINT " THE CALCULATED VALUE OF A = " FN RO(A): PRINT
1160   PRINT " THE CALCULATED VALUE OF B = " FN RO(B): PRINT
1170   :
1180   Y3 = 0:Y4 = 0:RS = 0:Y5 = 0:Y6 = 0
1190   FOR J = 1 TO N
1200   X = X(J)
1210   ON Q GOSUB 1230,1240,1250,1260,1270,1280,1290,1300,1310,1320
1220   GOTO 1330
1230   Y = A + B * X: RETURN
1240   Y = A * X / (B + X): RETURN
1250   Y = A / (B + X): RETURN
1260   Y = A * EXP (B * X): RETURN
1270   Y = A * EXP (B / X): RETURN
1280   Y = A * X * EXP (B * X): RETURN
1290   Y = A * (1 - EXP (B * X)): RETURN
1300   Y = K / (1 + A * EXP (B * X)): RETURN
1310   Y = K + A * X ^B: RETURN
1320   Y = K / (1 + A * X ^B): RETURN
1330   REM RESIDUALS
1340   R = Y(J) - Y:YP(J) = Y
1350   Y3 = Y3 + Y:Y4 = Y4 + Y * Y
1360   Y5 = Y5 + Y(J):Y6 = Y6 + Y(J) * Y(J)
1370   RS = RS + R * R
1380   NEXT J
1390   PRINT "SUM OF SQ OF RESIDUALS = "; FN RO(RS)
1400   P = 2:   REM # OF EST. PARAMETERS
1410   VR = RS / (N - P)
1420   SE(Q) = SQR (VR):SE(Q) = FN RO(SE(Q))
1430   F(Q) = (Y4 - (2 * Y5 / N) * Y3 + (Y5 * Y5) / N) / (VR)
1440   F(Q) = FN RO(F(Q))
1450   QS = 1:   REM SUMMARY FLAG
1460   PRINT "VAR OF RESIDUALS("(N - P)" DF) = ";
1470   PRINT FN RO(VR)
1480   PRINT "STD ERROR OF RESIDUALS = ";SE(Q)
1490   PRINT : PRINT "FOR AN F-TEST W/ 1&"(N - P)" DF, F= "F(Q): PRINT
1495   IF HP = 1 THEN 5500
1500   VTAB 23:   PRINT "(Q=QUIT Z=BACKUP S=SUMMARY H=HARDCOPY)";:
       VTAB 22:   HTAB 1
1510   INPUT "LIST DATA & PRED. Y-VALUES? (Y/N)";QM$
1520   IF QM$ = "Z" AND Q < 7 THEN 390
1530   IF QM$ = "Z" AND Q > 6 THEN 550
1540   IF QM$ = "Q" THEN HOME : END
1550   IF QM$ = "S" THEN 2080
1555   IF QM$ = "H" THEN HL = 1:   GOTO 5400
1560   IF QM$ < > "Y" THEN 1670
1570   :
1580   REM ****SCREEN 6 ********************************
1590   TEXT : HOME : PRINT "COMPARING DATA WITH PREDICTED Y-VALUES"
1595   HF = 1
1600   PRINT "EQN #"Q" "N$(Q)
1610   PRINT : PRINT " X";:   HTAB 10:   PRINT " Y";:   HTAB 20:   PRINT
       "PRED Y";:   HTAB 30:   PRINT "D   IFF"
1620   FOR J = 1 TO N
1630   X = X(J)
```

Figure A3.5. CURFIT program for Apple II machines. (continued)

```
1640 R = Y(J) - YP(J)
1650    PRINT X;:  HTAB 10:  PRINT Y(J);:  HTAB 20:  PRINT FN RO(YP(J));:
        HTAB 30:  PRINT FN RO(R)
1660    NEXT J
1670    PRINT : PRINT : PRINT : PRINT
1675    IF HP = 1 THEN 5500
1680    VTAB 23:  PRINT "(Q=QUIT Z=BACKUP S=SUMMARY";
1685    IF HF = 1 THEN PRINT " H=HARDCOPY";:HF = 0
1686    PRINT ")";:  VTAB 22:  HTAB 1
1690    INPUT "SEE A PLOT OF RESIDUALS? (Y/N) ";QM$
1700    IF QM$ = "Z" THEN 750
1710    IF QM$ = "Q" THEN TEXT : HOME : END
1720    IF QM$ = "S" THEN 2080
1725    IF QM$ = "H" THEN HL = 2:  GOTO 5400
1730    IF QM$ < > "Y" THEN PRINT : PRINT : PRINT : PRINT : GOTO 1830
1740 :
1750    REM ****SCREEN 7 *********************************
1760 Y$ = "RESID. %":YM = + 100:YN = - 100:  GOSUB 3000
1765 HF = 1
1770 XO = 150:YO = 1:XL$ = "EQN #" + STR$ (Q): GOSUB 3500
1780 XA = 1:XD = 0:  REM LINE DRAWING
1790 X = 0:Y = 0:  GOSUB 4000:X = XM: GOSUB 4000
1800    FOR J = 1 TO N '
1810 X = X(J):Y = (Y(J) - YP(J)) * 100 / YH:XB = 1:  GOSUB 4000
1820    NEXT J
1830    VTAB 23:  PRINT "(Q=QUIT Z=BACKUP S=SUMMARY";
1835    IF HF = 1 THEN PRINT " H=HARDCOPY";:HF = 0
1836    PRINT ")";:  VTAB 22:  HTAB 1
1840    INPUT "SEE MODEL PLOTTED W/DATA? (Y/N) ";QM$
1850    IF QM$ = "Z" THEN 1590
1860    IF QM$ = "Q" THEN TEXT : HOME : END
1870    IF QM$ = "S" THEN 2080
1875    IF QM$ = "H" THEN HL = 4:  GOTO 5400
1880    IF QM$ < > "Y" THEN PRINT : PRINT : PRINT : PRINT : GOTO 2020
1890 :
1900    REM ****SCREEN 8 *********************************************
1910 Y$ = "Y-VALUE":YN = 0:YM = YH: GOSUB 3000
1915 HF = 1
1920 XO = 150:YO = 1:XL$ = "EQN #" + STR$ (Q): GOSUB 3500
1930 XA = 1:XD = 0
1940    FOR J = 1 TO 100 STEP 2
1950 X = J * XM / 100
1960    ON Q GOSUB 1230,1240,1250,1260,1270,1280,1290,1300,1310,1320
1970    GOSUB 4000
1980    NEXT J
1990    FOR J = 1 TO N
2000 X = X(J):Y = Y(J):XC = 1:  GOSUB 4000
2010    NEXT J
2020    VTAB 23:  PRINT "(Q=QUIT Z=BACKUP S=SUMMARY";
2025    IF HF = 1 THEN PRINT " H=HARDCOPY";:HF = 0
2026    PRINT ")";:  VTAB 22:  HTAB 1
2030    INPUT "ENTER S FOR SUMMARY &/OR ANOTHER EQN ";QM$
2040    IF QM$ = "Q" THEN TEXT : HOME : END
2045    IF QM$ = "H" THEN HL = 5:  GOTO 5400
2050    IF QM$ = "Z" THEN PRINT : PRINT : PRINT : PRINT : GOTO 1830
2060 :
2070    REM ****SCREEN 9 *********************************
2080    TEXT : HOME : PRINT "SUMMARY OF CURVE-FITTING STATISTICS"
2090    PRINT : PRINT " EQUATION STD ERROR F-VALUE"
2100    PRINT : FOR J = 1 TO 10: IF J < 10 THEN PRINT " ";
```

Figure A3.5. CURFIT program for Apple II machines. (continued)

```
2110    PRINT J" "N$(J);:  HTAB 19
2120    IF F(J) = 0 THEN PRINT "- - - - - -":  GOTO 2140
2130    PRINT SE(J);:  HTAB 30:  PRINT F(J)
2140    NEXT J: PRINT : PRINT
2145    IF HP = 1 THEN 5500
2150    VTAB 20:  PRINT "(Z=BACKUP Q=QUIT H=HARDCOPY)";:  VTAB 19:
        HTAB 1:  INPUT "ENTER E = ANOTHER EQN ";QM$
2160    IF QM$ = "E" THEN 390
2170    IF QM$ = "Q" THEN TEXT : HOME : END
2180    IF QM$ = "Z" THEN 1830
2185    IF QM$ = "H" THEN HL = 3:  GOTO 5400
2190    GOTO 2150
2200 :
2210    REM ****END OF PROGRAM ****************************
3000 :
3010    HGR : IF PEEK (2048) = 127 THEN 3020
3013    PRINT CHR$ (13) + CHR$ (4)"BLOAD SHAPES,A$0800"
3015    POKE 232,0:  POKE 233,8:  HCOLOR= 3:  SCALE= 1
.
.
. Insert lines 3020-4090 from the GRAPH program here
.
.
5000 :
5010    REM **** START SUBROUTINE ****************************
5020    TEXT : HOME
5021    IF PEEK (104) = 64 THEN 5030
5022 P$ = CHR$ (13) + CHR$ (4) + "RUN CURFIT"
5023    POKE 103,0:  POKE 104,64:  POKE 16383,0:  PRINT P$
5030    VTAB 1:  HTAB 17:  PRINT "CURFIT"
5040    VTAB 3:  HTAB 8:  PRINT "A PROGRAM FOR FITTING 10":  HTAB 8:
        PRINT "COMMON EQUATIONS TO DATA"
5050    VTAB 6:  PRINT "PROGRAM IN FORMATION IS FOUND IN THE BOOK"
5060    VTAB 7:  HTAB 5:  INVERSE
5070    PRINT "COMPUTER SIMULATION IN BIOLOGY:":  HTAB 12:  PRINT "A
        BASIC APPROACH":  NORMAL
5080    PRINT " BY ROBERT E. KEEN & JAMES D. SPAIN"
5090    PRINT " (C) 1990 BY JOHN WILEY & SONS, INC"
5100    READ F$:  RESTORE
5110    VTAB 12:  PRINT "THIS PROGRAM IS SET TO READ DATA FROM"
5120    PRINT "A FILE NAMED":  PRINT : HTAB 10:  PRINT ">>>" ;F$:  PRINT
5130    PRINT "IF THIS IS NOT THE NAME OF THE DISKFILE":  PRINT "HOLDING
        THE DATA THEN QUIT THE PROGRAM"
5140    PRINT "& CHANGE THE FILENAME IN LINE 100":  PRINT
5150    INPUT "ENTER 'P' TO PROCEED, 'Q' TO QUIT ";QM$
5160    IF QM$ = "Q" THEN HOME : TEXT : END
5170    IF QM$ = CHR$ (112) OR QM$ = CHR$ (113) THEN PRINT "CAPITAL
        LETTERS PLEASE"
5180    IF QM$ < > "P" THEN 5150
5190 D$ = CHR$ (13) + CHR$ (4):  PRINT D$"VERIFY"F$:
        PRINT D$"OPEN"F$:  PRINT D$"READ"F$
5200    INPUT N
5210    FOR J = 1 TO N: INPUT X(J),Y(J): NEXT J
5220    PRINT D$"CLOSE"F$
5230    DEF FN RO(X) = INT (10000 * X + 0.5) / 10000
5240 N$(1) = "STRAIGHT LINE":E$(1) = "Y = A + B*X"
5250 N$(2) = "HYPERBOLIC":E$(2) = "Y = (A*X) / (B+X)"
5260 N$(3) = "MOD. INVERSE":E$(3) = "Y = A / (B + X)"
5270 N$(4) = "EXPONENTIAL":E$(4) = "Y = A * EXP(B*X)"
5280 N$(5) = "EXP. RECIP.":E$(5) = "Y = A * EXP(B/X)"
```

Figure A3.5. CURFIT program for Apple II machines. (continued)

```
5290 N$(6) = "MAXIMA FUNC.":E$(6) = "Y = A * X * EXP(B*X)"
5300 N$(7) = "EXP. SAT'N":E$(7) = "Y = A*[1 - EXP(B*X)]"
5310 N$(8) = "LOGISTIC":E$(8) = "Y = K/[1+A*EXP(B*X)]"
5320 N$(9) = "LOGARITHMIC":E$(9) = "Y = (A * X^B) + K"
5330 N$(10) = "SIGMOID":E$(10) = "Y = K / (1 + A * X^B)"
5340    RETURN
5400 :
5410    REM **** HARDCOPY SUBROUTINE ************************
5420    IF HL > 3 THEN 5460
5430    PRINT CHR$ (13) + CHR$ (4) + "PR#1"
5440 HP = 1
5450    ON HL GOTO 750,1590,2060
5460    REM SCREEN-DUMP FOR APPLE SILENTYPE PRINTER
5470    POKE - 12528,7:  POKE - 12529,255:  POKE - 12524,0
5480    PRINT CHR$ (13) + CHR$ (4) + "PR#1"
5490    PRINT CHR$ (17)
5500    PRINT CHR$ (13) + CHR$ (4) + "PR#0"
5510 HP = 0
5520    ON HL GOTO 750,1590,2060,1740,1890
5530 END
```

Figure A3.5. CURFIT program for Apple II machines. (concluded)

A3.6   CURFIT on the Apple Macintosh Computer

CURFIT is available for use on Macintosh computers with Microsoft BASIC interpreters. Appendix 6 describes how to obtain copies of this program.

APPENDIX 4

DESCRIPTION OF POLYFIT PROGRAM

Like the CURFIT program in Appendix 3, POLYFIT is a BASIC program written specifically for this textbook. It is intended to be used in fitting polynomial equations through order 5 to sets of x-y data.

The program is based on the method for calculating the coefficients of a cubic equation given in Croxton and Cowden (1955). The technique involves the simultaneous solution of $r+1$ equations which define Σy, Σxy, $\Sigma x^2 y$, $\Sigma x^3 y$, ..., $\Sigma x^{r+1} y$, where r is the order of the polynomial being fitted. POLYFIT solves the equations using a standard matrix inversion and multiplication procedure. The value of n and the sums of x, x^2, x^3, x^4, ..., x^{2r} are stored as elements of a square matrix, A. The sums of y, xy, $x^2 y$, $x^3 y$, ..., $x^{r+1} y$ are stored as elements of a vector, S. The calculations proceed with matrix inversion and a vector multiplication in which the coefficients of the polynomial become the elements of the product vector, C. Other procedures used in polynomial analysis are discussed in Chapter 3.

The mechanics of using POLYFIT are almost identical with those of CURFIT. The general approach and machine requirements are the same for both programs. Like CURFIT, the program is set up to read data from a disk data file. These files may be created using the program given in Figure A3.1 and A3.2. POLYFIT will read the same data files as the CURFIT program. POLYFIT is slightly more flexible, permitting the input and analysis of negative and zero values. These values will be used successfully in calculations, although they may not be plotted successfully on the graphical output.

The diagram showing flow of operation of the program is given in Figure A4.1. Listings of POLYFIT for the IBM-PC and Apple II personal computers are given in Figures A4.2 and A4.3. A version of POLYFIT for Macintosh computers is available for Microsoft BASIC interpreters; see Appendix 6.

*1. Title and Data File Check
   [P: go to *2]
   [Q: exit]

*2. Scatter Plot
   [P: go to *3]
   [Z: go to *1]
   [(S: go to *7)]
   [Q: exit]

*3. List of Polynomials #1-6
   [#1- 6: go to *4]
   [Z: go to *2]
   [(S: go to *7)]
   [Q: exit]

*4. Curve-Fitting Statistics
   ? See predicted-y list (Y/N)
   [Y: go to *5]
   [N: go to ?5]
   [Z: go to *3]
   [S: go to *7]
   [Q: exit]

*5. List of x, y, pred-y
   ? See plot of residuals (Y/N)
   [Y: go to *6]
   [N: go to ?6]
   [Z: go to *4]
   [S: go to *7]
   [Q: exit]

*6. Plot of Data and Curve
   ? Go to Summary &/or another Eqn
   [S: go to *7]
   [Z: go to ?5]
   [Q: exit]

*7. List of Summary Statistics
   [E: go to *3]
   [Z: go to ?5]
   [Q: exit]

Figure A4.1. List of flow and options for using POLYFIT program. Video presentation is underlined; user reply is bracketed. (* indicates screens keyed to listing in Figures A4.2 and A4.3).

Figure A4.2. Listing of POLYFIT program for IBM-PC and compatible machines, using BASICA or similar interpreters. The graphic subroutines of Lines 3000-4090 from the appropriate GRAPH program of Appendix 2 should be inserted into this program.

```
10 REM
20 REM      POLYFIT....WRITTEN BY JAMES D. SPAIN & ROBERT E. KEEN
30 REM      COPYRIGHT (C) 1981,1990 BY JOHN WILEY & SONS, INC. NYC
40 REM
50 REM
60 REM      REPLACE "TRIALFIT" IN LINE
70 REM      100 WITH YOUR DATA FILE NAME
80 REM
90 :
100 DATA "TRIALFIT"
110 :
120 DIM X#(50), Y#(50), YP#(50)
130 DIM A#(6,12), S#(11), Z#(12), C#(10)
140 :
150 REM*****SCREEN 1 ***************************************
160 GOSUB 5000:REM TITLE & DATA ENTRY******
170 :
180 REM*****SCREEN 2 ***************************************
190 REM FIND MAX X, MIN Y & MAX Y
200 YL=9.900001E+37
210 FOR I=1 TO N
220     IF XH < X#(I) THEN XH=X#(I)
230     IF YH < Y#(I) THEN YH=Y#(I)
240     IF YL > Y#(I) THEN YL=Y#(I)
250 NEXT I
260 FOR I=1 TO 10 :  Z#(I)=0 :  S#(I)=0 :  NEXT I
270 FOR I = 1 TO N
280     FOR J = 1 TO 10
290         Z#(J) = Z#(J) + X#(I)^J
300     NEXT J
310     FOR J = 1 TO 6
320         S#(J) = S#(J) + Y#(I)*X#(I)^(J-1)
330     NEXT J
340     S#(9) = S#(9) + Y#(I)^2
350 NEXT I
360 X$="X-VALUE": Y$="Y-VALUE":  XN=0:  YN=0:  YM=YH: XM=XH
370 GOSUB 3000 :  REM DRAW & LABEL AXES******
380 FOR I=1 TO N
390 X = X#(I) : Y = Y#(I) : XC = 1 :  GOSUB 4000 :  REM PLOT WITH CIRCLES
400 NEXT I
410 LOCATE 24,1:PRINT"ENTER 'P' TO PROCEED WITH EQUATION SELECTION
        (Q=QUIT, ";
420 PRINT"Z=BACKUP"; :  IF Q<>0 THEN PRINT", S=SUMMARY";
430 INPUT") ";QM$
440 IF QM$ = "Q" THEN SCREEN 0 :  CLS : END
450 IF QM$ = "Z" THEN 160
460 IF QM$ = "S" AND Q<>0 THEN 1860
470 IF QM$<> "P" THEN 410
480 :
490 REM*****SCREEN 3 ***************************************
500 CLS : SCREEN 0
510 PRINT"THE FOLLOWING EQUATIONS MAY BE USED WITH THIS PROGRAM:"
520 FOR I=1 TO 5
530     LOCATE (I*2+1),1
```

```
540     PRINT I "..."; N$(I); TAB(25); E$(I)
550 NEXT I
560 LOCATE 14,1:PRINT"(OR Q=QUIT, Z=BACKUP"; :IF Q<>0 THEN PRINT",S=
    SUMMARY";
570 PRINT")";
580 LOCATE 13,1:INPUT"ENTER NUMBER OF EQUATION TO BE USED"; QM$
590 IF QM$ = "Z" THEN 190
600 IF QM$ = "Q" THEN CLS : END
610 IF QM$ = "S" AND Q <> 0 THEN 1860
620 Q = INT(VAL(QM$)):  IF Q<1 OR Q>5 THEN 580
630 IF N > Q+1 THEN 680
640 LOCATE 17,1:PRINT"OOPS! YOU MUST HAVE AT LEAST"Q+2" DATA POINTS"
650 PRINT"TO FIT A POLYNOMIAL OF DEGREE"Q:GOTO 580
660 :
670 REM*****SCREEN 4 ****************************************
680 CLS : SCREEN 0
690 PRINT"FITTING POLYNOMIAL OF DEGREE"Q :   "N$(Q)
700 PRINT"--"; :   REM SET MATRIX TO 0
710 FOR I = 1 TO 6 :  FOR J = 1 TO 12 :  A#(I,J) = 0 :   NEXT J : NEXT I
720 PRINT"--"; :   REM LOAD MATRIX WITH SUMS SQS
730 A#(1,1) = N :  A#(1,2) = Z#(1) :   A#(2,1) = Z#(1) :   A#(2,2) = Z#(2)
740 IF Q = 1 THEN 870
750 A#(3,1)=Z#(2) :   A#(1,3)=Z#(2) :   A#(2,3)=Z#(3)
760 A#(3,2)=Z#(3) :   A#(3,3)=Z#(4)
770 IF Q = 2 THEN 870
780 A#(4,1)=Z#(3) :   A#(1,4)=Z#(3) :   A#(2,4)=Z#(4) :   A#(4,2)=Z#(4)
790 A#(3,4)=Z#(5) :   A#(4,3)=Z#(5) :   A#(4,4)=Z#(6)
800 IF Q = 3 THEN 870
810 A#(5,1)=Z#(4) :   A#(1,5)=Z#(4) :   A#(2,5)=Z#(5) :   A#(5,2)=Z#(5)
820 A#(3,5)=Z#(6):A#(5,3)=Z#(6):A#(4,5)=Z#(7) :   A#(5,4)=Z#(7):A#(5,5)=
    Z#(8)
830 IF Q = 4 THEN 870
840 A#(6,1)=Z#(5) :   A#(1,6)=Z#(5) :   A#(2,6)=Z#(6) :   A#(6,2)=Z#(6)
850 A#(3,6)=Z#(7) :   A#(6,3)=Z#(7) :   A#(4,6)=Z#(8) :   A#(6,4)=Z#(8)
860 A#(5,6)=Z#(9) :   A#(6,5)=Z#(9) :   A#(6,6)=Z#(10)
870 PRINT "--"; :   REM IDENTITY MATRIX FOR INVERSION
880 M = Q + 1
890 FOR I = 1 TO M : A#(I,I+M) = 1 :   NEXT I
900 REM INVERSION OF MATRIX A****************
910 FOR I = 1 TO M : PRINT "--";
920     FOR J = 1 TO M : F#(J) = A#(J,I) : NEXT J
930     FOR J = I TO 2*M : A#(I,J) = A#(I,J) / F#(I) : NEXT J
940     IF I = M THEN 1010
950     K = I + 1
960     FOR J = K TO M
970         FOR L = 1 TO 2*M
980             A#(J,L) = A#(J,L) - A#(I,L)*F#(J)
990         NEXT L
1000     NEXT J
1010 NEXT I
1020 FOR I = M TO 2 STEP -1 :  PRINT "--";
1030     FOR J = M TO I STEP -1 :  F#(J) = A#(I-1,J) : NEXT J
1040     FOR J = M TO I STEP -1
1050         FOR L = I TO 2*M
1060             A#(I-1,L) = A#(I-1,L) - A#(J,L)*F#(J)
1070         NEXT L
1080     NEXT J
1090 NEXT I
1100 PRINT"--";
1110 FOR I = 1 TO M
```

Figure A4.2. POLYFIT listing for IBM-PC. (continued)

```
1120 FOR J = M+1 TO 2*M : A#(I,J-M) = A#(I,J) : NEXT J
1130 NEXT I
1140 PRINT"--"; :  REM MULT OF MAT C = A * S
1150 FOR I = 1 TO 6 :  C#(I) = 0 :  NEXT I
1160 FOR I = 1 TO M
1170    FOR J = 1 TO M : C#(I) = C#(I) + A#(I,J)*S#(J) : NEXT J
1180 NEXT I
1190 R2# = 0
1200 FOR I = 1 TO 6 :  R2# = R2# + C#(I)*S#(I) : EC(Q,I)=C#(I) : NEXT I
1210 R2# = (R2# - (S#(1)^2)/N) / (S#(9) - (S#(1)^2)/N) : R2(Q) = R2#
1220 PRINT : PRINT "COEFFICIENTS FOR THE EQN. ";E$(Q): PRINT
1230 PRINT A$(1); EC(Q,1)
1240 PRINT A$(2); EC(Q,2)
1250 IF Q > 1 THEN PRINT A$(3); EC(Q,3)
1260 IF Q > 2 THEN PRINT A$(4); EC(Q,4)
1270 IF Q > 3 THEN PRINT A$(5); EC(Q,5)
1280 IF Q > 4 THEN PRINT A$(6); EC(Q,6)
1290 PRINT : PRINT"COEFFICIENT OF DETERMINATION (R-SQ.) = ";FN RO(R2#)
1300 RS# = 0
1310 FOR I = 1 TO N : REM FIND PREDICTED Y
1320    YP#(I) = C#(1) + C#(2)*X#(I) + C#(3)*X#(I)^2 + C#(4)*X#(I)^3
1330    YP#(I) = YP#(I) + C#(5)*X#(I)^4 + C#(6)*X#(I)^5
1340    R# = Y#(I) - YP#(I)
1350    RS# = RS# + R#*R#
1360 NEXT I
1370 PRINT:PRINT"SUM OF SQUARES OF RESIDUALS = ";FN RO(RS#)
1380 PRINT
390 VR#=RS#/(N-M) : REM M = NO. OF ESTIMATED PARAMETERS
1400 SE#(Q) = SQR(VR#)
1410 FV#(Q) = R2# * (S#(9) - (S#(1)^2)/N) / VR#
1420 PRINT"VARIANCE OF RESIDUALS FOR";N-M;"DF = "; FN RO(VR#):PRINT
1430 PRINT"THE STD. ERROR OF RESIDUALS = "; FN RO(SE#(Q))
1440 PRINT:PRINT"FOR AN F-TEST WITH 1 & "(N-M)" DF, F = ";FN RO(FV#(Q))
1450 LOCATE 24,1:PRINT"(OR Z=BACKUP S=SUMMARY Q=QUIT)";:  LOCATE 23,1
1460 INPUT"DO YOU WANT TO SEE A LIST OF DATA & THE ESTIMATED POINTS (Y/N)";
     QM$
1470 IF QM$="Z" THEN GOTO 500
1480 IF QM$="Q" THEN CLS:END
1490 IF QM$="S" THEN 1860
1500 IF QM$<>"Y" THEN 1600
1510 :
1520 REM*****SCREEN 5 **************************************
1530 CLS : PRINT"COMPARISON OF ORIGINAL DATA WITH ESTIMATED POINTS"
1540 PRINT"EQUATION #"Q;N$(Q)
1550 PRINT:PRINT" X Y(DATA) Y(PREDICTED) DIFFERENCE"
1560 FOR I=1 TO N
1570 R# = Y#(I) - YP#(I)
1580 PRINT FN RO(X#(I)), FN RO(Y#(I)), FN RO(YP#(I)), FN RO(R#)
1590 NEXT I
1600 PRINT:PRINT:PRINT
1610 LOCATE 24,1:PRINT"DO YOU WANT THE MODEL PLOTTED WITH THE DATA (Y/N; ";
1620 INPUT"Z=BACKUP S=SUMMARY Q=QUIT)";QM$
1630 IF QM$="Z" THEN 680
1640 IF QM$="Q" THEN SCREEN 0 :  CLS : END
1650 IF QM$="S" THEN 1860
1660 IF QM$<>"Y" THEN PRINT:PRINT:PRINT:GOTO 1800
1670 :
1680 REM*****SCREEN 6 **************************************
1690 Y$="Y-VALUE" :  X$="X-VALUE" :   GOSUB 3000
1700 LOCATE 1,1 :  PRINT"ORDER"Q
```

Figure A4.2. POLYFIT listing for IBM-PC. (continued)

```
1710 XA=1:XD=0
1720 FOR I=1 TO 100 STEP 2
1730    X=I*XM/100
1740    Y=C#(1) + C#(2)*X + C#(3)*X^2 + C#(4)*X^3 + C#(5)*X^4 + C#(6)*X^5
1750    GOSUB 4000
1760 NEXT I
1770 FOR J=1 TO N
1780    X=X#(J) : Y=Y#(J) : XC=1 :  GOSUB 4000
1790 NEXT J
1800 LOCATE 24,1:PRINT"ENTER S FOR SUMMARY &/OR ANOTHER EQN (Z=BACKUP, ";
1810 INPUT"Q=QUIT)";QM$
1820 IF QM$="Q" THEN SCREEN 0:CLS:END
1830 IF QM$="Z" THEN PRINT:PRINT:PRINT:GOTO 1610
1840 :
1850 REM*****SCREEN 7 ***********************************
1860 CLS : SCREEN 0
1870 PRINT "SUMMARY OF POLYNOMIAL FIT OF DATA":PRINT
1880 T(1)=7:  T(2)=21:  T(3)=35:  T(4)=49:  T(5)=63:  X2$="-----"
1890 FOR I = 1 TO 5 :  PRINT TAB(T(I)+1); N$(I); :  NEXT I : PRINT:
     PRINT
1900 FOR I = 1 TO 6 :  PRINT A$(I); :  M = 1
1910    IF I > 2 THEN M = I-1
1920    FOR J = M TO 5
1930       IF FV#(J)=0 THEN PRINT TAB(T(J)); X2$; :  GOTO 1950
1940       PRINT TAB(T(J)); FN RO(EC(J,I));
1950    NEXT J:PRINT
1960 NEXT I
1970 PRINT:PRINT:PRINT:PRINT "R-SQ= ";
1980 FOR I = 1 TO 5 :  PRINT TAB(T(I))
1990    IF FV#(I)=0 THEN PRINT X2$; :  GOTO 2010
2000    PRINT FN RO(R2(I));
2010 NEXT I:PRINT
2020 PRINT:PRINT "SE = ";
2030 FOR I = 1 TO 5 :  PRINT TAB(T(I))
2040 IF FV#(I)=0 THEN PRINT X2$; :  GOTO 2060
2050 PRINT FN RO(SE#(I));
2060 NEXT I:PRINT
2070 PRINT:PRINT "F = ";
2080 FOR I = 1 TO 5 :  PRINT TAB(T(I))
2090    IF FV#(I)=0 THEN PRINT X2$; :  GOTO 2110
2100    PRINT FN RO(FV#(I));
2110 NEXT I
2120 LOCATE 21,1
2130 INPUT"ENTER E TO SELECT ANOTHER EQUATION (Z=BACKUP Q=QUIT)";
     QM$
2140 IF QM$="E" THEN 500
2150 IF QM$="Q" THEN CLS:END
2160 IF QM$="Z" THEN 1610
2170 GOTO 2120
2180 :
2190 REM*****END OF PROGRAM *****************************
3000 :
 .
 .
 .
```

Insert lines 3000-4090 from the GRAPH program here

```
 .
 .
5000 :
5010 REM***** STARTUP SUBROUTINE **************************
5020 SCREEN 0 :  CLS : WIDTH 80 :  KEY OFF : COLOR 7
```

Figure A4.2. POLYFIT listing for IBM-PC. (continued)

```
5030 LOCATE 4,36 :  PRINT"POLYFIT"
5040 LOCATE 7,15:PRINT"A Program for Fitting Polynomial Equations to Data"
5050 LOCATE 9,13:PRINT"Information about the program may be found in the
     book"
5060 LOCATE 10,14:COLOR 9
5070 PRINT"Computer Simulation in Biology:  A BASIC Introduction":COLOR 7
5080 LOCATE 11,24:PRINT"by Robert E. Keen & James D. Spain"
5090 LOCATE 12,14:PRINT"Published by John Wiley & Sons, Inc.  Copyright 1990"
5100 READ F$:RESTORE
5110 LOCATE 16,11:PRINT"This program is currently set to read data from a
     ";
5120 PRINT"file named":  LOCATE 18,30:PRINT">"F$:LOCATE 20,12
5130 PRINT"If this is not the name of the diskfile holding the data,":
     LOCATE,9
5140 PRINT"then quit the program and change the name of the file in Line
     100"
5150 PRINT:INPUT"Enter P to read & plot data from this file, Q to quit";QM$
5160 IF QM$="Q" THEN CLS:END
5170 IF QM$="p" OR QM$="q" THEN PRINT"CAPITAL LETTERS PLEASE"
5180 IF QM$<>"P" THEN 5150
5190 OPEN"I",1,F$
5200 INPUT #1,N
5210 FOR J=1 TO N:INPUT #1,X#(J),Y#(J):NEXT J
5220 CLOSE 1
5230 DEF FN RO(J) = INT(J * 10000 + .5)/10000
5240 N$(1)="LINEAR" :  N$(2)="QUADRATIC" :  N$(3)="CUBIC"
5250 N$(4)="QUARTIC" :  N$(5)="QUINTIC"
5260 A$(1)="AO = " :  A$(2)="A1 = " :  A$(3) = "A2 = "
5270 A$(4)="A3 = " :  A$(5)="A4 = " :  A$(6) = "A5 = "
5280 E$(1)="Y = AO + A1*X" :  E$(2)=E$(1)+" + A2*X^2"
5290 E$(3)=E$(2)+" + A3*X^3" :  E$(4)=E$(3)+" + A4*X^4"
5300 E$(5)=E$(4)+" + A5*X^5"
5310 RETURN
```

Figure A4.2. POLYFIT listing for IBM-PC. (concluded)

Figure A4.3. Listing of POLYFIT program for Apple II machines using Applesoft BASIC. The graphic subroutines from Lines 3020-4090 from the GRAPH program (Appendix 2) must be inserted into this program.

```
10   REM
20   REM POLYFIT..WRITTEN BY R. E. KEEN & J. D. SPAIN
30   REM COPYRIGHT (C) 1981,1990  BY JOHN WILEY & SONS, INC. NYC
40   REM
50   REM
60   REM    REPLACE "TRIALFIT" IN LINE
70   REM    100 WITH YOUR DATA FILE NAME
80   REM
90  :
100  DATA "TRIALFIT"
110 :
120  DIM X(50  ),Y(50),YP(50)
130  DIM A(6,12),S(11),Z(12),C(10)
140 :
150  REM *****SCREEN 1 ********************************
160  GOSUB 5000:  REM TITLE & DATA ENTRY****
170 :
180  REM *****SCREEN 2 ********************************
190  REM FIND MAX X, MIN Y & MAX Y
200 YL = 9.9E + 36
210  FOR I = 1 TO N
220  IF XH < X(I) THEN XH = X(I)
230  IF YH < Y(I) THEN YH = Y(I)
240  IF YL > Y(I) THEN YL = Y(I)
250  NEXT I
260  FOR I = 1 TO 10:Z(I) = 0:S(I) = 0:  NEXT I
270  FOR I = 1 TO N: PRINT " -";
280  FOR J = 1 TO 10
290 Z(J) = Z(J) + X(I) ^J
300  NEXT J
310  FOR J = 1 TO 6
320 S(J) = S(J) + Y(I) * X(I) ^(J - 1)
330  NEXT J
340 S(9) = S(9) + Y(I) ^2
350  NEXT I
360 X$ = "X-VALUE":Y$ = "Y-VALUE":XN = 0:YN = 0:YM = YH:XM = XH
370  GOSUB 3000:  REM DRAW & LABEL AXES
380  FOR I = 1 TO N
390 X = X(I):Y = Y(I):XC = 1:  GOSUB 4000:  REM PLOT WITH CIRCLES
400  NEXT I
410  VTAB 24:  PRINT "ENTER 'P' = PROCEED W/MODEL SELECTION"
420  PRINT "(Q=QUIT Z=BACKUP";:  IF Q < > 0 THEN PRINT " S=SUMMARY";
430  INPUT ") ";QM$
440  IF QM$ = "Q" THEN TEXT : HOME : END
450  IF QM$ = "Z" THEN 160
460  IF QM$ = "S" AND Q < > 0 THEN 1860
470  IF QM$ < > "P" THEN 410
480 :
490  REM *****SCREEN 3 ********************************
500  TEXT : HOME
510  PRINT "THESE EQNS MAY BE FIT TO THE DATA:": PRINT
520  FOR I = 1 TO 5
530 :
540  PRINT I") "N$(I): PRINT " "E$(I);:  PRINT : PRINT
550  NEXT I
```

```
560   VTAB 19:   PRINT "(OR Q=QUIT Z=BACKUP";:  IF Q < > O THEN PRINT " S=
      SUMMARY";
570   PRINT ")";
580   VTAB 18:   HTAB 1:   INPUT "ENTER NUMBER OF EQUATION TO BE USED ";QM$
590   IF QM$ = "Z" THEN HOME :  GOTO 190
600   IF QM$ = "Q" THEN HOME :  END
610   IF QM$ = "S" AND Q < > O THEN 1860
620   Q = INT ( VAL (QM$)):  IF Q < 1 OR Q > 5 THEN 580
630   IF N > Q + 1 THEN 680
640   VTAB 22:   PRINT "OOPS! YOU MUST HAVE AT LEAST "Q + 2" DATA"
650   PRINT "POINTS TO FIT A POLYNOMIAL OF DEGREE "Q";:GOTO 580
660  :
661  VTAB 23:   PRINT "(OR Q=QUIT Z=BACKUP";:  IF QS < > O THEN PRINT " S=
      SUMMARY)";
670   REM *****SCREEN 4 ***********************************
680   TEXT : HOME
690   PRINT "FIT OF "N$(Q)" POLYNOMIAL OF DEGREE "Q
700   PRINT "--";:  REM SET DOT MATRIX TO 0
710   FOR I = 1 TO 6:   FOR J = 1 TO 12:A(I,J) = 0:  NEXT J: NEXT I
720   PRINT "--";:  REM LOAD MATRIX WITH SUMS SQS
730   A(1,1) = N:A(1,2) = Z(1):A(2,1) = Z(1):A(2,2) = Z(2)
740   IF Q = 1 THEN 870
750   A(3,1) = Z(2):A(1,3) = Z(2):A(2,3) = Z(3)
760   A(3,2) = Z(3):A(3,3) = Z(4)
770   IF Q = 2 THEN 870
780   A(4,1) = Z(3):A(1,4) = Z(3):A(2,4) = Z(4):A(4,2) = Z(4)
790   A(3,4) = Z(5):A(4,3) = Z(5):A(4,4) = Z(6)
800   IF Q = 3 THEN 870
810   A(5,1) = Z(4):A(1,5) = Z(4):A(2,5) = Z(5):A(5,2) = Z(5)
820   A(3,5) = Z(6):A(5,3) = Z(6):A(4,5) = Z(7):A(5,4) = Z(7):A(5,5) = Z(8)
830   IF Q = 4 THEN 870
840   A(6,1) = Z(5):A(1,6) = Z(5):A(2,6) = Z(6):A(6,2) = Z(6)
850   A(3,6) = Z(7):A(6,3) = Z(7):A(4,6) = Z(8):A(6,4) = Z(8)
860   A(5,6) = Z(9):A(6,5) = Z(9):A(6,6) = Z(10)
870   PRINT "--";:  REM IDENTITY MATRIX FOR INVERSION
880   M = Q + 1
890   FOR I = 1 TO M:A(I,I + M) = 1:   NEXT I
900   REM INVERSION OF MATRIX A**********
910   FOR I = 1 TO M: PRINT "--";
920   FOR J = 1 TO M:F(J) = A(J,I): NEXT J
930   FOR J = 1 TO 2 * M:A(I,J) = A(I,J) / F(I): NEXT J
940   IF I = M THEN 1010
950   K = I + 1
960   FOR J = K TO M
970   FOR L = 1 TO 2 * M
980   A(J,L) = A(J,L) - A(I,L) * F(J)
990   NEXT L
1000  NEXT J
1010  NEXT I
1020   FOR I = M TO 2 STEP - 1:   PRINT "-";
1030   FOR J = M TO I STEP - 1:F(J) = A(I - 1,J): NEXT J
1040   FOR J = M TO I STEP - 1
1050   FOR L = I TO 2 * M
1060  A(I - 1,L) = A(I - 1,L) - A(J,L) * F(J)
1070   NEXT L
1080   NEXT J
1090   NEXT I
1100   PRINT "--";
1110   FOR I = 1 TO M
1120   FOR J = M + 1 TO 2 * M:A(I,J - M) = A(I,J): NEXT J
```

Figure A4.3. POLYFIT listing for Apple II. (continued)

```
1130   NEXT I
1140   PRINT "--";:   REM MULT OF MAT C=A*S
1150   FOR I = 1 TO 6:C(I) = 0:   NEXT I
1160   FOR I = 1 TO M
1170   FOR J = 1 TO M:C(I) = C(I) + A(I,J) * S(J): NEXT J
1180   NEXT I
1190   R2 = 0
1200   FOR I = 1 TO 6:R2 = R2 + C(I) * S(I):EC(Q,I) = C(I): NEXT I
1210   R2 = (R2 - (S(1) ^2) / N) / (S(9) - (S(1) ^2) / N):R2(Q) = R2
1220   PRINT : PRINT "COEFFICIENTS FOR THE EQN."
1230   PRINT A$(1)" ";EC(Q,1)
1240   PRINT A$(2)" ";EC(Q,2)
1250   IF Q > 1 THEN PRINT A$(3)" ";EC(Q,3)
1260   IF Q > 2 THEN PRINT A$(4)" ";EC(Q,4)
1270   IF Q > 3 THEN PRINT A$(5)" ";EC(Q,5)
1280   IF Q > 4 THEN PRINT A$(6)" ";EC(Q,6)
1290   PRINT : PRINT "COEFF. OF DETERMINATION (R-SQ)= "; FN RO(R2)
1300   RS = 0
1310   FOR I = 1 TO N: REM FIND PREDICTED Y
1320   YP(I) = C(1) + C(2) * X(I) + C(3) * X(I) ^2 + C(4) * X(I) ^3
1330   YP(I) = YP(I) + C(5) * X(I) ^4 + C(6) * X(I) ^5
1340   R = Y(I) - YP(I)
1350   RS = RS + R * R
1360   NEXT I
1370   PRINT : PRINT "SUM OF SQ OF RESIDUALS = "; FN RO(RS)
1380   PRINT
1390   VR = RS / (N - M): REM M=# OF EST. PARAMETERS
1400   SE(Q) = SQR (VR)
1410   FV(Q) = R2 * (S(9) - (S(1) ^2) / N) / VR
1420   PRINT "VAR OF RESIDUALS("(N - M)" DF) = "; FN RO(VR)
1430   PRINT "STD ERROR OF RESIDUALS = "; FN RO(SE(Q))
1440   PRINT : PRINT "FOR AN F-TEST W/ 1&"(N - M)" DF, F= "FN RO(FV(Q))
1445   IF HP = 1 THEN 5500
1450   VTAB 23:   PRINT "(Q=QUIT Z=BACKUP S=SUMMARY H=HARDCOPY)";:   VTAB 22:
       HTAB 1
1460   INPUT "LIST DATA & PRED. Y-VALUES? (Y/N)";QM$
1470   IF QM$ = "Z" THEN 500
1480   IF QM$ = "Q" THEN HOME : END
1490   IF QM$ = "S" THEN 1860
1495   IF QM$ = "H" THEN HL = 1:   GOTO 5400
1500   IF QM$ < > "Y" THEN 1600
1510   :
1520   REM ****SCREEN 5 *************************************
1530   TEXT : HOME : PRINT "COMPARING DATA WITH PREDICTED Y-VALUES"
1535   HF = 1
1540   PRINT "EQN #"Q" "N$(Q)
1550   PRINT : PRINT " X";:   HTAB 10:   PRINT " Y";:   HTAB 20:   PRINT "PRED
       Y";:   HTAB 30:   PRINT "DIFF"
1560   FOR I = 1 TO N
1570 R = Y(I) - YP(I)
1580   PRINT X(I);:   HTAB 10:   PRINT Y(I);:   HTAB 20:   PRINT FN RO(YP(I));:
       HTAB 30:   PRINT FN RO(R)
1590   NEXT I
1600   PRINT : PRINT : PRINT : PRINT
1605   IF HP = 1 THEN 5500
1610   VTAB 23:   PRINT "(Q=QUIT Z=BACKUP S=SUMMARY";
1615   IF HF = 1 THEN PRINT " H=HARDCOPY";:HF = 0
1616   PRINT ")";:   VTAB 22:   HTAB 1
1620   INPUT "SEE MODEL PLOTTED W/DATA? (Y/N) ";QM$
1630   IF QM$ = "Z" THEN 680
```

Figure A4.3. POLYFIT listing for Apple II. (continued)

```
1640    IF QM$ = "Q" THEN TEXT : HOME : END
1650    IF QM$ = "S" THEN 1860
1655    IF QM$ = "H" THEN HL = 2:  GOTO 5400
1660    IF QM$ < > "Y" THEN PRINT : PRINT : PRINT : PRINT : GOTO 1800
1670    :
1680    REM ****SCREEN 6 **********************************
1690    Y$ = "Y-VALUE":YN = 0:YM = YH: GOSUB 3000
1695    HF = 1
1700    XO = 150:YO = 1:XL$ = "ORDER " + STR$ (Q): GOSUB 3500
1710    XA = 1:XD = 0
1720    FOR I = 1 TO 100 STEP 2
1730    X = I * XM / 100
1740    Y = C(1) + C(2) * X + C(3) * X ^2 + C(4) * X ^3 + C(5) * X ^4 + C(6)
        * X ^5
1750    GOSUB 4000
1760    NEXT I
1770    FOR J = 1 TO N
1780    X = X(J):Y = Y(J):XC = 1:  GOSUB 4000
1790    NEXT J
1800    VTAB 23:  PRINT "(Q=QUIT Z=BACKUP";
1805    IF HF = 1 THEN PRINT " H=HARDCOPY";:HF = 0
1806    PRINT ")";:  VTAB 22:  HTAB 1
1810    INPUT "ENTER S FOR SUMMARY &/OR ANOTHER EQN ";QM$
1820    IF QM$ = "Q" THEN TEXT : HOME : END
1825    IF QM$ = "H" THEN HL = 4:  GOTO 5400
1830    IF QM$ = "Z" THEN PRINT : PRINT : PRINT : PRINT : GOTO 1610
1840    :
1850    REM ****SCREEN 7 **************************************
1860    TEXT : HOME : PRINT "SUMMARY OF POLYNOMIAL FIT OF DATA"
1865    X2$ = "- - - - - - - - - -"
1870    PRINT : HTAB 11:  PRINT "R-SQ STD ERR F-VALUE"
1880    FOR I = 1 TO 5
1890    PRINT : PRINT N$(I);:  HTAB 11
1900    IF FV(I) = 0 THEN PRINT X2$:  GOTO 1940
1910    PRINT FN RO(R2(I));:  HTAB 20
1920    PRINT FN RO(SE(I));:  HTAB 30
1930    PRINT FN RO(FV(I))
1940    NEXT I
1945 IF HP = 1 THEN 5500
1950    REM LINES 1960-2120 OMITTED**************
2130    VTAB 20:  PRINT "(Z=BACKUP Q=QUIT H=HARDCOPY)";:  VTAB 19:  HTAB 1:
        INPUT "ENTER E = ANOTHER EQN ";QM$
2140    IF QM$ = "E" THEN 500
2150    IF QM$ = "Q" THEN TEXT : HOME : END
2160    IF QM$ = "Z" THEN 1610
2165    IF QM$ = "H" THEN HL = 3:  GOTO 5400
2170    GOTO 2130
2180    :
2190    REM ****END OF PROGRAM ****************************
3000    :
3010    HGR : IF PEEK (2048) = 127 THEN 3020
3013    PRINT CHR$ (13) + CHR$ (4)"BLOAD SHAPES,A$0800"
3015    POKE 232,0:  POKE 233,8:  HCOLOR= 3:  SCALE= 1
  .
  .
  .     Insert lines 3020-4090 from the GRAPH program here
  .
  .
5000    :
5010    REM **** START SUBROUTINE ***************************
```

Figure A4.3. POLYFIT listing for Apple II. (continued)

```
5020    TEXT : HOME
5021    IF PEEK (104) = 64 THEN 5030
5022    P$ = CHR$ (13) + CHR$ (4) + "RUN POLYFIT"
5023    POKE 103,0:  POKE 104,64:  POKE 16383,0:  PRINT P$
5030    VTAB 1:  HTAB 17:  PRINT "POLYFIT"
5040    VTAB 3:  HTAB 10:  PRINT "A PROGRAM FOR FITTING":  HTAB 6:  PRINT
        "POLYNOMIAL EQUATIONS TO DATA"
5050    VTAB 6:  PRINT "PROGRAM INFORMATION IS FOUND IN THE BOOK"
5060    VTAB 7:  HTAB 5:  INVERSE
5070    PRINT "COMPUTER SIMULATION IN BIOLOGY:":  HTAB 12:  PRINT "A BASIC
        APPROACH":  NORMAL
5080    PRINT " BY ROBERT E. KEEN & JAMES D. SPAIN"
5090    PRINT " (C) 1990 BY JOHN WILEY & SONS, INC"
5100    READ F$:  RESTORE
5110    VTAB 12:  PRINT "THIS PROGRAM IS SET TO READ DATA FROM"
5120    PRINT "A FILE NAMED":  PRINT :  HTAB 10:  PRINT ">>>";F$:  PRINT
5130    PRINT "IF THIS IS NOT THE NAME OF THE DISKFILE":  PRINT "HOLDING THE
        DATA THEN QUIT THE PROGRAM"
5140    PRINT "& CHANGE THE FILENAME IN LINE 100":  PRINT
5150    INPUT "ENTER 'P' TO PROCEED, 'Q' TO QUIT ";QM$
5160    IF QM$ = "Q" THEN HOME : TEXT : END
5170    IF QM$ = CHR$ (112) OR QM$ = CHR$ (113) THEN PRINT "CAPITAL LETTERS
        PLEASE"
5180    IF QM$ < > "P" THEN 5150
5190    D$ = CHR$ (13) + CHR$ (4):  VTAB 20:  PRINT
        D$"VERIFY"F$:  PRINT D$"OPEN"F$:  PRINT D$"READ"F$
5200    INPUT N
5210    FOR J = 1 TO N: INPUT X(J),Y(J): NEXT J
5220    PRINT D$"CLOSE"F$
5230    DEF FN RO(X) = INT (10000 * X + 0.5) / 10000
5240    N$(1) = "LINEAR":N$(2) = "QUADRATIC":N$(3) = "CUBIC"
5250    N$(4) = "QUARTIC":N$(5) = "QUINTIC"
5260    A$(1) = "AO = ":A$(2) = "A1 = ":A$(3) = "A2 = "
5270    A$(4) = "A3 = ":A$(5) = "A4 = ":A$(6) = "A5 ="
5280    E$(1) = "Y = AO + A1*X":E$(2) = E$(1) + " + A2*X^2"
5290    E$(3) = E$(2) + " + A3*X^3":E$(4) = "Y=AO+A1*X+A2*X^2+A3*X^3+A4*X^4"
5300    E$(5) = E$(4) + "+A5*X^5"
5310    RETURN
5400    :
5410    REM **** HARDCOPY SUBROUTINE ************************
5420    IF HL > 3 THEN 5460
5430    PRINT CHR$ (13) + CHR$ (4) + "PR#1"
5440    HP = 1
5450    ON HL GOTO 680,1530,1840
5460    REM SCREEN-DUMP FOR SILENTYPE
5470    POKE - 12528,7:  POKE - 12529,255:  POKE - 12524,0
5480    PRINT CHR$ (13) + CHR$ (4) + "PR#1"
5490    PRINT CHR$ (17)
5500    PRINT CHR$ (13) + CHR$ (4) + "PR#0"
5510    HP = 0
5520    ON HL GOTO 680,1530,1840,1670
5530    END
```

Figure A4.3. POLYFIT listing for Apple II. (concluded)

APPENDIX 5

DESCRIPTION OF
CURNLFIT PROGRAM

Like the CURFIT and POLYFIT programs in the two preceding appendices, CURNLFIT is a BASIC program written specifically for this textbook. It is intended to be used for fitting a nonlinear curve to a set of x-y data. As such, it is a complement to the CURFIT program, working toward the same objective.

The program is based on the nonlinear regression program by Duggleby (1981), which uses the Gauss-Newton method for calculating the coefficients of a function. The method works well and quickly with reasonable initial estimates of the coefficients. The program allows a choice of three different weighting schemes for variation of y-values with increasing values of x (see Chapter 3). It also permits selection of a method for weighting the effect of values lying well away from the best-fit line. As discussed in Chapter 3, the program uses an iterative technique to find values of the coefficients that minimize the residual sum of squares. The criterion used to assess convergence is the change in standard deviation of the coefficients between iterations. Iteration halts and values for the coefficients are assumed to have converged when the summed absolute values of the relative changes of the coefficients do not differ more than 10^{-5} between iterations. In the case of nonconvergence, a generous limit of ten iterations is allowed before the program terminates with a warning message.

CURNLFIT is used in much the same way as POLYFIT and CURFIT. The general approach and machine requirements are the same as those for CURFIT given in Appendix 3. The program is set up to read data from a disk data file, and can read those used for both CURFIT and POLYFIT.

The diagram showing the flow of operation of the program is given in Figure A5.1. Operation resembles that of CURFIT. Estimates of coefficients must be given to the program for starting values for iteration to the best estimate. Weighting schemes for variation in y must also be selected.

Listings of CURNLFIT for the IBM-PC and Apple II computers are given in Figures A5.2 and A5.3. A version of CURNLFIT for Macintosh computers is available for Microsoft BASIC interpreters; see Appendix 6.

The program is set up to estimate coefficients for the 9 curvilinear functions used in the CURFIT program. In addition, any equation with a dependent variable y expressed as a function of an independent variable x may be inserted into the program. The equation may include two parameters, A and B, as well as a constant K. The user's equation may be written into the program as shown in the program listings. The general format for writing the equation is indicated by inspection of the other nonlinear equations in the program.

The routines used by Duggleby (1981) are efficient, particularly when the initial estimates are close to the final value. However, the program may fail to converge on a solution because of division by zero, or because the magnitude of some values may exceed the capacity of the computer. This can occur for several reasons, including incorrect initial estimates of coefficients, and selection of an unreasonable equation. These "fatal errors" may halt the execution of the program with some BASIC interpreters.

*1. Title and Data File Check
  [P: go to *2]
  [Q: exit]

*2. Scatter Plot
  [P: go to *3]
  [Z: go to *1]
  [Q: exit]

*3. List of Equations #1-10
  [#1-10: go to *4]
  [Z: go to *2]
  [Q: exit]

*4. Pick Wtng, Initial A, B, (K)
  [Values: go thru *4+]
  [P: go to *5]
  [Z: go to *3]
  [Q: exit]

*5. Iterations display
  [P: go to *6]
  [Z: go to *3]
  [Q: exit]

*6. Curve- Fitting Statistics
  ? See predicted-y list (Y/N)
  [Y: go to *7]
  [N: go to ?7]
  [Z: go to *4]
  [Q: exit]

*7. List of x, y, pred-y
? See plot of residuals (Y/N)
  [Y: go to *8]
  [N: go to ?8]
  [Z: go to *6]
  [Q: exit]

*8. Plot of Residuals
  ? See Plot of Data & Curve (Y/N)
  [Y: go to *9]
  [N: go to ?9]
  [Z: go to *7]
  [Q: exit]

*9. Plot of Data and Curve
? Proceed
  [P: go to *10]
  [Z: go to ?8]
  [Q: exit]

*10 Termination of Program
  [Z: go to ?8]
  [Q: exit]

Figure A5.1. List of flow and options for using CURNLFIT program. Video presentation is underlined; user reply is bracketed. (* indicates screens keyed to listing in Figures A5.2 and A5.3).

Figure A5.2. Listing of CURNLFIT program for IBM-PC and compatible machines, using BASICA or similar interpreters. The graphic subroutines of Lines 3000-4090 from the appropriate GRAPH program of Appendix 2 should be inserted into this program.

```
10 REM
20 REM      CURNLFIT...WRITTEN BY ROBERT E. KEEN & JAMES D. SPAIN
30 REM      COPYRIGHT (C) 1990 BY JOHN WILEY & SONS, INC. NYC
40 REM
50 REM
60 REM      REPLACE "TRIALFIT" IN LINE
70 REM      100 WITH YOUR DATA FILE NAME
80 REM
90 :
100 DATA "TRIALFIT"
110 :
120 REM
130 REM       EQUATION IN LINE 1710 MAY BE REPLACED WITH USER'S EQUATION
140 REM       - SEE TEXTBOOK FOR MORE INFORMATION
150 REM
160 :
170 DIM X(50), Y(50), YP(50), W(50)
180 :
190 REM****SCREEN 1 ************************************************
200 GOSUB 5000 :   REM TITLE & DATA ENTRY******
210 :
220 REM****SCREEN 2 ************************************************
230 REM FIND MAX X, MIN Y & MAX Y
240 YL=9.900001E+37
250 FOR J=1 TO N
260     IF XH < X(J) THEN XH = X(J)
270     IF YH < Y(J) THEN YH = Y(J)
280     IF YL > Y(J) THEN YL = Y(J)
290 NEXT J
300 X$="X-VALUE":Y$="Y-VALUE":XN=0:YN=0:YM=YH:XM=XH
310 GOSUB 3000 :   REM DRAW & LABEL AXES******
320 FOR J=1 TO N
330     X=X(J) : Y=Y(J) : XC=1 :   GOSUB 4000 :   REM PLOT WITH CIRCLES
340 NEXT J
350 LOCATE 24,1:PRINT"ENTER 'P' TO PROCEED WITH MODEL SELECTION ";
360 INPUT"(Q=QUIT, Z=BACKUP)";QM$
370 IF QM$="Q" THEN SCREEN 0 :   CLS : END
380 IF QM$="Z" THEN 200
390 IF QM$<>"P" THEN 350
400 :
410 REM****SCREEN 3 ************************************************
420 CLS : SCREEN 0
430 PRINT"THE FOLLOWING EQUATIONS MAY BE USED WITH THIS PROGRAM:"
440 FOR J=1 TO 10
450     LOCATE (J*2+1),1:IF J<10 THEN PRINT" ";
460 PRINT J"...";N$(J);TAB(30);E$(J)
470 NEXT J
480 LOCATE 24,1:PRINT"(OR Q=QUIT, Z=BACKUP)";
490 LOCATE 23,1:INPUT"ENTER NUMBER OF EQUATION TO BE USED";QM$
500 IF QM$="Z" THEN 300
510 IF QM$="Q" THEN CLS : END
520 Q=INT(VAL(QM$)):   IF Q<1 OR Q>10 THEN 490
530 :
540 REM****SCREEN 4 ************************************************
```

```
550 CLS : PRINT"EQUATION NO."Q" "N$(Q);":   ";E$(Q):PRINT
560 PRINT"WEIGHTING SCHEME TO BE APPLIED:"
570 PRINT" -CONSTANT STD. DEVIATION......(1) +++++++++++++++++++++++"
580 PRINT" -STD. DEV. PROPORTIONAL TO Y..(2) + Z=BACKUP Q=QUIT +"
590 PRINT" -BETWEEN THESE TWO............(3) +++++++++++++++++++++++"
600 INPUT"ENTER NUMBER OF YOUR CHOICE (OR Z/Q)";QM$
610 IF QM$="Q" THEN CLS : END
620 IF QM$="Z" THEN 420
630 WG=INT(VAL(QM$)):  IF WG<1 OR WG>3 THEN 600
640 PRINT :PRINT"USE BISQUARE WEIGHTING FOR OUTLYING VALUES? [Y/N] ";
650 INPUT"(OR Q/Z)";QM$
660 IF QM$="Q" OR QM$="Z" THEN 610
670 IF QM$<>"Y" AND QM$<>"N" THEN 640
680 WB=1 :  IF QM$="N" THEN WB=0
690 PRINT:PRINT:PRINT"ENTER AN INITIAL ESTIMATE FOR THE VALUE OF A ";
700 INPUT "(OR Q/Z)";QM$
710 IF QM$="Q" OR QM$="Z" THEN 610
720 A=VAL(QM$)
730 PRINT:PRINT"ENTER AN INITIAL ESTIMATE FOR THE VALUE OF B ";
740 INPUT "(OR Q/Z)";QM$
750 IF QM$="Q" OR QM$="Z" THEN 610
760 B=VAL(QM$)
770 IF Q<> 1 THEN 800
780 PRINT:PRINT"IF YOUR EQUATION REQUIRES A CONSTANT K, THEN ENTER ITS
        VALUE"
790 PRINT"OTHERWISE, ENTER ZERO "; :  GOTO 820
800 IF Q<8 AND Q>1 THEN 850
810 PRINT:PRINT"EQUATION"Q"REQUIRES A VALUE FOR K - ENTER IT HERE ";
820 INPUT"(OR Q/Z)";QM$
830 IF QM$="Q" OR QM$="Z" THEN 610
840 K=VAL(QM$)
850 PRINT:INPUT"ENTER P TO PROCEED WITH THE ANALYSIS (OR Q/Z)";QM$
860 IF QM$="Q" OR QM$="Z" THEN 610
870 IF QM$<>"P" THEN 850
880 :
890 WF = 0
900 :
910 REM****SCREEN 5 ***********************************************
920 CLS : SCREEN 0
930 PRINT"FITTING COEFFICIENTS FOR EQN #"Q" "N$(Q)" :   "E$(Q)
940 IF Q=1 OR Q>7 THEN PRINT:PRINT" K-VALUE SET AT"K:PRINT
950 FOR I=1 TO N
960     W(I)=1
970     IF WG=1 THEN 1010
980     W(I)= 1/Y(I)
990     IF WG=3 THEN 1010
1000    W(I)= W(I)*W(I)
1010 NEXT I
1020 :
1030 I=0 :  SO=1 :  PRINT:PRINT"--------- BEGINNING ITERATIONS --------":
        PRINT
1040 REM *** ITERATIONS START HERE
1050 IF I=0 THEN 1360
1060     S1=0:  S2=0:  S3=0:  S4=0:  S5=0:  S6=0:  T1=0:  REM INITIALIZE
            SUMS
1070     FOR J = 1 TO N
1080       X = X(J)
1090       ON Q GOSUB 1710,1740,1750,1760,1770,1780,1790,1800,1810,1820
1100       YD = Y(J) - Y
1110       T5 = SQR(W(J)) * YD
```

Figure A5.2. CURNLFIT listing for IBM-PC. (continued)

```
1120        T1 = T1 + ABS(T5)
1130        A = 1.02*A
1140        ON Q GOSUB 1710,1740,1750,1760,1770,1780,1790,1800,1810,1820
1150        YU = Y : A = A * .98/1.02
1160        ON Q GOSUB 1710,1740,1750,1760,1770,1780,1790,1800,1810,1820
1170        A = A/.98 :  PA = (YU - Y)/(.04*A)
1180        B = 1.02*B
1190        ON Q GOSUB 1710,1740,1750,1760,1770,1780,1790,1800,1810,1820
1200        YU = Y : B = B * .98/1.02
1210        ON Q GOSUB 1710,1740,1750,1760,1770,1780,1790,1800,1810,1820
1220        B = B/.98 :  PB = (YU - Y)/(.04*B)
1230        T3=1
1240        IF WB = 0 THEN 1280 :   REM SKIP BISQUARE WTNG
1250        IF I = 1 THEN 1280
1260        T3 = 0 :   T4 = (T5/T2)^2
1270        IF T4 < 1 THEN T3 = (1 - T4)^2
1280        REM SUMS
1290        S1 = S1 + (T3 * W(J) * PA * PA) : S2 = S2 + (T3 * W(J) * PA *
            PB)
1300        S3 = S3 + (T3 * W(J) * PB * PB) : S4 = S4 + (T3 * W(J) * PA *
            YD)
1310        S5 = S5 + (T3 * W(J) * PB * YD) : S6 = S6 + (T3 * W(J) * YD *
            YD)
1320     NEXT J
1330     T2 = 6 * T1/N : S7 = S1*S3 - S2*S2
1340     SA = (S3*S4 - S2*S5) /S7 :   SB = (S1*S5 - S2*S4) /S7
1350     SO = ABS(SA/A) + ABS(SB/B) : A = A + SA : B = B + SB
1360     PRINT"ITERATION"I;TAB(15);"A= "A;TAB(35);"B= "B;TAB(55);
            "CRITERION=";
1370     IF I=0 THEN PRINT" -----"; :  GOTO 1390
1380     PRINT SO;
1390      I = I + 1 :  PRINT
1400 IF SO < 9.999999E-06 THEN PRINT:PRINT"SUCCESSFUL CONVERGENCE":
     GOTO 1430
1410 IF I < 11 THEN 1040 :   REM DO ANOTHER ITERATION
1420 PRINT : WF$="WARNING: CONVERGENCE NOT OBTAINED" :  WF=1 :  PRINT WF$
1430 REM END OF ITERATIONS
1440 PRINT:PRINT:PRINT:PRINT:INPUT"ENTER P TO PROCEED (Z=BACKUP, Q=QUIT)";
     QM$
1450 IF QM$="Z" OR QM$="Q" THEN 610
1460 IF QM$<>"P" THEN 1440
1470 :
1480 REM****SCREEN 6 ****************************************************
1490 CLS:SCREEN 0
1500 PRINT"RESULTS OF NONLINEAR CURVE FITTING: EQUATION NO."Q
1510 PRINT:PRINT"THE "N$(Q)" EQUATION IS "E$(Q) : PRINT
1520 PRINT: PRINT"WEIGHTING SCHEME USED: ";
1530 IF WG=1 THEN PRINT "CONSTANT STD. DEVIATION"
1540 IF WG=2 THEN PRINT "STD. DEV. PROPORTIONAL TO Y"
1550 IF WG=3 THEN PRINT "BETWEEN CONSTANT AND PROPORTIONAL STD. DEV."
1560 IF WB=1 THEN PRINT " -BISQUARE WEIGHTING FOR OUTLIERS WAS APPLIED"
1570 PRINT : IF Q=1 OR Q>7 THEN PRINT" THE K-VALUE WAS SET ="K:PRINT
1580 PRINT" THE CALCULATED VALUE OF A = " FN RO(A);
1590 PRINT TAB(47)"+ "; FN RO(SA);" (S.D.)":PRINT
1600 PRINT" THE CALCULATED VALUE OF B = " FN RO(B);
1610 PRINT TAB(47)"+ "; FN RO(SB);" (S.D.)":PRINT
1620 IF WF=1 THEN PRINT WF$:PRINT
1630 :
1640 Y3=0 :  Y4=0 :  RS=0 :  Y5=0 :  Y6=0
1650 FOR J=1 TO N
```

Figure A5.2. CURNLFIT listing for IBM-PC. (continued)

```
1660    X=X(J)
1670    ON Q GOSUB 1710,1740,1750,1760,1770,1780,1790,1800,1810,1820
1680    GOTO 1830
1690 :
1700    REM THE EQN ON THE NEXT LINE MAY BE REPLACED WITH USER'S EQN
1710    Y = A + B*X
1720 :
1730    RETURN
1740    Y=A*X/(B+X) : RETURN
1750    Y=A/(B+X) : RETURN
1760    Y=A*EXP(B*X) : RETURN
1770    Y=A*EXP(B/X) : RETURN
1780    Y=A*X*EXP(B*X) : RETURN
1790    Y=A*(1-EXP(B*X)) : RETURN
1800    Y=K/(1+A*EXP(B*X)) : RETURN
1810    Y=K+A*X^B : RETURN
1820    Y=K/(1+A*X^B) : RETURN
1830    REM RESIDUALS
1840    R = Y(J) - Y : YP(J) = Y
1850    Y3 = Y3 + Y : Y4 = Y4 + Y*Y
1860    Y5 = Y5 + Y(J) : Y6 = Y6 + Y(J)*Y(J)
1870    RS = RS + R*R
1880 NEXT J
1890 PRINT"SUM OF SQUARES OF RESIDUALS = ";FN RO(RS)
1900 PRINT : P=2 :  REM # OF EST. PARAMETERS
1910 VR = RS/(N-P)
1920 SE = SQR(VR) : SE = FN RO(SE)
1930 PRINT"VARIANCE OF RESIDUALS FOR";N-P;"DF = ";
1940 PRINT FN RO(VR) : PRINT
1950 PRINT"THE STD. ERROR OF RESIDUALS =";SE
1960 LOCATE 24,1:PRINT"(OR Z=BACKUP Q=QUIT)";:LOCATE 23,1
1970 INPUT"DO YOU WANT TO SEE A LIST OF DATA & THE ESTIMATED POINTS (Y/N)";
        QM$
1980 IF QM$="Z" THEN 550
1990 IF QM$="Q" THEN CLS : END
2000 IF QM$<>"Y" THEN 2110
2010 :
2020 REM****SCREEN 7 ****************************************************
2030 CLS : SCREEN 0 :PRINT"COMPARISON OF ORIGINAL DATA WITH ESTIMATED
        POINTS"
2040 PRINT"EQUATION #"Q;N$(Q)
2050 PRINT:PRINT" X Y(DATA) Y(PREDICTED) DIFFERENCE"
2060 FOR J=1 TO N
2070    X=X(J)
2080    R = Y(J) - YP(J)
2090    PRINT X, Y(J), FN RO(YP(J)), FN RO(R)
2100 NEXT J
2110 PRINT:PRINT:PRINT
2120 LOCATE 24,1:PRINT"(OR Z=BACKUP Q=QUIT)";:LOCATE 23,1
2130 INPUT"DO YOU WISH TO SEE A PLOT OF THE RESIDUALS (Y/N)";QM$
2140 IF QM$="Z" THEN 1490
2150 IF QM$="Q" THEN CLS : SCREEN 0 :  END
2160 IF QM$<>"Y" THEN PRINT:PRINT:PRINT : GOTO 2260
2170 :
2180 REM****SCREEN 8 ****************************************************
2190 Y$="RESIDUAL-%" :  YM=+100 :  YN=-100 :  GOSUB 3000
2200 LOCATE 1,1:PRINT"EQN #"Q
2210 XA=1:  XD=0 :  REM LINE DRAWING
2220 X=0:Y=0 :  GOSUB 4000:  X=XM : GOSUB 4000
2230 FOR J=1 TO N
```

Figure A5.2. CURNLFIT listing for IBM-PC. (continued)

```
2240 X=X(J) : Y=(Y(J)-YP(J))*100/YH : XB=1 :   GOSUB 4000
2250 NEXT J
2260 LOCATE 24,1:PRINT"DO YOU WANT THE MODEL PLOTTED WITH THE DATA (Y/N; ";
2270 INPUT"Z=BACKUP Q=QUIT)";QM$
2280 IF QM$="Z" THEN 2030
2290 IF QM$="Q" THEN SCREEN 0 :   CLS : END
2300 IF QM$<>"Y" THEN PRINT:PRINT:PRINT : GOTO 2440
2310 :
2320 REM****SCREEN 9 *****************************************************
2330 Y$="Y-VALUE" :   YN=0 :   YM=YH : GOSUB 3000
2340 LOCATE 1,1:PRINT"EQN #"Q
2350 XA=1 :   XD=0
2360 FOR J=1 TO 100 STEP 2
2370     X=J*XM/100
2380     ON Q GOSUB 1710,1740,1750,1760,1770,1780,1790,1800,1810,1820
2390     GOSUB 4000
2400 NEXT J
2410 FOR J=1 TO N
2420     X=X(J) : Y=Y(J) : XC=1 :   GOSUB 4000
2430 NEXT J
2440 LOCATE 24,1:PRINT"ENTER P TO PROCEED (Z=BACKUP, ";
2450 INPUT"Q=QUIT)";QM$
2460 IF QM$="Q" THEN SCREEN 0 :   CLS : END
2470 IF QM$="Z" THEN PRINT:PRINT:PRINT : GOTO 2260
2480 :
2490 REM****SCREEN 10 ****************************************************
2500 CLS:SCREEN 0
2510 PRINT:PRINT:PRINT:PRINT"END OF NONLINEAR CURVE-FITTING PROGRAM"
2520 PRINT:PRINT:PRINT:INPUT"ENTER Q=QUIT, Z=BACKUP";QM$
2530 IF QM$="Q" THEN CLS : END
2540 IF QM$="Z" THEN 2260
2550 GOTO 2520
2560 :
2570 REM****END OF PROGRAM **********************************************
3000 :
  .
  .
  .  Insert lines 3000-4090 from the GRAPH program here
  .
  .
5000 :
5010 REM***** START SUBROUTINE *******************************************
5020 SCREEN 0 :   CLS : WIDTH 80 :   KEY OFF : COLOR 7
5030 LOCATE 3,36:PRINT"CURNLFIT"
5040 LOCATE 6,13:PRINT" A Program for Fitting Equations to Curvelinear "
5050 LOCATE 7,13:PRINT" Data using Methods of Nonlinear Regression "
5060 LOCATE 9,13:PRINT"Information about the program may be found in the
     book"
5070 LOCATE 10,16:COLOR 9
5080 PRINT"Computer Simulation in Biology:  A BASIC Approach":COLOR 7
5090 LOCATE 11,24:PRINT"by Robert E. Keen & James D. Spain"
5100 LOCATE 12,14:PRINT"Published by John Wiley & Sons, Inc.  Copyright
     1990"
5110 READ F$ :   RESTORE
5120 LOCATE 16,11:PRINT"This program is currently set to read data from a";
5130 PRINT"file named":  LOCATE 18,30:PRINT">"F$:LOCATE 20,12
5140 PRINT"If this is not the name of the disk file holding the data,":
     LOCATE,9
5150 PRINT"then quit the program and change the name of the file in Line
     100"
```

Figure A5.2. CURNLFIT listing for IBM-PC. (continued)

```
5160 PRINT:INPUT"Enter P to read & plot data from this file, Q to quit";
     QM$
5170 IF QM$="Q" THEN CLS:END
5180 IF QM$="p" OR QM$="q" THEN PRINT"CAPITAL LETTERS PLEASE":GOTO 5160
5190 IF QM$<>"P" THEN 5160
5200 OPEN"I",1,F$
5210 INPUT #1,N
5220 FOR J=1 TO N:INPUT #1,X(J),Y(J):NEXT J
5230 CLOSE 1
5240 DEF FN RO(X) = INT(10000*X + .5)/10000
5250 N$( 1)="USER'S EQUATION"  :  E$( 1)="Y = f(X)"
5260 N$( 2)="HYPERBOLIC"  :  E$( 2)="Y = (A*X) / (B+X)"
5270 N$( 3)="MODIFIED INVERSE"  :  E$( 3)="Y = A / (B + X)"
5280 N$( 4)="EXPONENTIAL"  :  E$( 4)="Y = A * EXP(B*X)"
5290 N$( 5)="EXP. RECIPROCAL"  :  E$( 5)="Y = A * EXP(B/X)"
5300 N$( 6)="MAXIMA FUNCTION"  :  E$( 6)="Y = A * X * EXP(B*X)"
5310 N$( 7)="EXP. SATURATION"  :  E$( 7)="Y = A * [1 - EXP(B*X)]"
5320 N$( 8)="LOGISTIC"  :  E$( 8)="Y = K / [1 + A * EXP(B*X)]"
5330 N$( 9)="LOGARITHMIC"  :  E$( 9)="Y = (A * X^B) + K"
5340 N$(10)="'SIGMOID'"  :  E$(10)="Y = K / (1 + A * X^B)"
5350 RETURN
```

Figure A5.2. CURNLFIT listing for IBM-PC. (concluded)

Figure A5.3. Listing of CURNLFIT program for Apple II machines using Applesoft BASIC. The graphic subroutines from Lines 3020-4090 from the GRAPH program (Appendix 2) must be inserted into this program.

```
10   REM
20   REM CURNLFIT......WRITTEN BY ROBERT KEEN AND JAMES SPAIN
30   REM COPYRIGHT (C) 1981,1990 BY JOHN WILEY & SONS, INC. NYC
40   REM
50   REM
60   REM      REPLACE "TRIALFIT" IN LINE
70   REM      100 WITH YOUR DATA FILE NAME
80   REM
90   :
100  DATA "TRIALFIT"
110  :
120  REM
130  REM      EQN IN LINE 1710 MAY BE REPLACED WITH USER'S EQN
140  REM      -SEE TEXTBOOK FOR MORE INFORMATION
150  REM
160  :
170  DIM X(50),Y(50),YP(50),W(50)
180  :
190  REM *****SCREEN 1 ******************************************************
200  GOSUB 5000:   REM TITLE & DATA ENTRY *****
210  :
220  REM *****SCREEN 2 ******************************************************
230  REM FIND MAX X, MIN Y & MAX Y
240  YL = 9.9E + 36
250  FOR J = 1 TO N
260  IF XH < X(J) THEN XH = X(J)
270  IF YH < Y(J) THEN YH = Y(J)
280  IF YL > Y(J) THEN YL = Y(J)
290  NEXT J
300  X$ = "X-VALUE":Y$ = "Y-VALUE":XN = 0:YN = 0:YM = YH:XM = XH
310  GOSUB 3000:   REM DRAW & LABEL AXES
320  FOR J = 1 TO N
330  X = X(J):Y = Y(J):XC = 1:   GOSUB 4000:   REM PLOT WITH CIRCLES
340  NEXT J
350  VTAB 24:   PRINT "ENTER 'P' =PROCEED W/MODEL SELECTION"
360  INPUT "(Q=QUIT Z=BACKUP) ";QM$
370  IF QM$ = "Q" THEN TEXT : HOME : END
380  IF QM$ = "Z" THEN 200
390  IF QM$ < > "P" THEN 350
400  :
410  REM *****SCREEN 3 ******************************************************
420  TEXT : HOME
430  PRINT "THESE EQNS MAY BE FIT TO THE DATA:":   PRINT
440  FOR J = 1 TO 10
450  IF J < 10 THEN PRINT " ";
460  PRINT J".."N$(J); TAB( 20)E$(J)
470  NEXT J
480  VTAB 23:   PRINT "(OR Q=QUIT Z=BACKUP);
490  VTAB 22:   HTAB 1:   INPUT "ENTER NUMBER OF EQN TO BE USED ";QM$
500  IF QM$ = "Z" THEN HOME : GOTO 300
510  IF QM$ = "Q" THEN HOME : END
520  Q = INT ( VAL (QM$)):   IF Q < 1 OR Q > 10 THEN 490
530  :
540  REM *****SCREEN 4 ******************************************************
550  HOME : PRINT "EQN #"Q" "N$(Q)" "E$(Q):   PRINT
```

```
560   PRINT "WEIGHTING SCHEME TO BE APPLIED:"
570   PRINT " -CONSTANT STD DEVIATION.....(1)"
580   PRINT " -STD DEV PROPORTIONAL TO Y..(2)"
590   PRINT " -BETWEEN THESE TWO..........(3)"
600   INPUT "ENTER # OF CHOICE (OR Q=QUIT Z=BACKUP) ";QM$
610   IF QM$ = "Q" THEN HOME : END
620   IF QM$ = "Z" THEN 420
630 WG = INT ( VAL (QM$)):  IF WG < 1 OR WG > 3 THEN 600
640   PRINT "USE BISQUARE WEIGHTING FOR OUTLYING":  PRINT " VALUES? [Y/N] ";
650   INPUT "(OR Q/Z) ";QM$
660   IF QM$ = "Q" OR QM$ = "Z" THEN 610
670   IF QM$ < > "Y" AND QM$ < > "N" THEN 640
680 WB = 1:  IF QM$ = "N" THEN WB = 0
690   PRINT : PRINT : PRINT "ESTIMATED VALUE OF A ";
700   INPUT "(OR Q/Z) =";QM$
710   IF QM$ = "Q" OR QM$ = "Z" THEN 610
720 A = VAL (QM$)
730   PRINT : PRINT "ESTIMATED VALUE OF B ";
740   INPUT "(OR Q/Z) =";QM$
750   IF QM$ = "Q" OR QM$ = "Z" THEN 610
760 B = VAL (QM$)
770   IF Q < > 1 THEN 800
780   PRINT : PRINT "IF YOUR EQN USES A CONSTANT K, ENTER"
790   PRINT "A VALUE NOW; OTHERWISE ENTER 0 ";:  GOTO 820
800   IF Q < 8 AND Q > 1 THEN 850
810   PRINT : PRINT "EQN #"Q" REQUIRES A VALUE FOR K":  PRINT " -ENTER IT
         HERE ";
820   INPUT "(OR Q/Z) ";QM$
830   IF QM$ = "Q" OR QM$ = "Z" THEN 610
840 K = VAL (QM$)
850   PRINT : INPUT "ENTER P TO PROCEED (OR Q/Z): ";QM$
860   IF QM$ = "Q" OR QM$ = "Z" THEN 610
870   IF QM$ < > "P" THEN 850
880 :
890 WF = 0
900 :
910   REM *****SCREEN 5 *****************************************************
920   TEXT : HOME
930   PRINT "FITTING EQN #"Q" "N$(Q)
940   IF Q = 1 OR Q > 7 THEN PRINT : PRINT "K-VALUE SET AT ";K: PRINT
950   FOR I = 1 TO N
960 W(I) = 1
970   IF WG = 1 THEN 1010
980 W(I) = 1 / Y(I)
990   IF WG = 3 THEN 1010
1000  W(I) = W(I) * W(I)
1010   NEXT I
1020 :
1030  I = 0:SO = 1:  PRINT : PRINT "------- BEGINNING ITERATIONS---------":
         PRINT
1040   REM ****ITERATIONS START HERE
1050   IF I = 0 THEN 1360
1060  S1 = 0:S2 = 0:S3 = 0:S4 = 0:S5 = 0:S6 = 0:T1 = 0:  REM INITIALIZE SUMS
1070   FOR J = 1 TO N
1080  X = X(J)
1090   ON Q GOSUB 1710,1740,1750,1760,1770,1780,1790,1800,1810,1820
1100  YD = Y(J) - Y
1110  T5 = SQR (W(J)) * YD
1120  T1 = T1 + ABS (T5)
1130  A = 1.02 * A
```

Figure A5.3. CURNLFIT listing for Apple II. (continued)

```
1140  ON Q GOSUB 1710,1740,1750,1760,1770,1780,1790,1800,1810,1820
1150  YU = Y:A = A * 0.98 / 1.02
1160  ON Q GOSUB 1710,1740,1750,1760,1770,1780,1790,1800,1810,1820
1170  A = A / 0.98:PA = (YU - Y) / (0.04 * A)
1180  B = 1.02 * B
1190  ON Q GOSUB 1710,1740,1750,1760,1770,1780,1790,1800,1810,1820
1200  YU = Y:B = B * 0.98 / 1.02
1210  ON Q GOSUB 1710,1740,1750,1760,1770,1780,1790,1800,1810,1820
1220  B = B / 0.98:PB = (YU - Y) / (.04 * B)
1230  T3 = 1
1240  IF WB = 0 THEN 1280:  REM SKIP BISQ WTNG
1250  IF I = 1 THEN 1280
1260  T3 = 0:T4 = (T5 / T2) ^2
1270  IF T4 < 1 THEN T3 = (1 - T4) ^2
1280  REM SUMS
1290  S1 = S1 + (T3 * W(J) * PA * PA):S2 = S2 + (T3 * W(J) * PA * PB)
1300  S3 = S3 + (T3 * W(J) * PB * PB):S4 = S4 + (T3 * W(J) * PA * YD)
1310  S5 = S5 + (T3 * W(J) * PB * YD):S6 = S6 + (T3 * W(J) * YD * YD)
1320  NEXT J
1330  T2 = 6 * T1 / N:S7 = S1 * S3 - S2 * S2
1340  SA = (S3 * S4 - S2 * S5) / S7:SB = (S1 * S5 - S2 * S4) / S7
1350  S0 = ABS (SA / A) + ABS (SB / B):A = A + SA:B = B + SB
1360  PRINT "ITERATION "I;:  HTAB 14:  PRINT "CRITERION= ";
1370  IF I = 0 THEN PRINT " - - -":  GOTO 1385
1380  PRINT S0
1385  : PRINT " A= "A" B= "B
1390  I = I + 1:  PRINT
1400  IF S0 < .00001 THEN PRINT : PRINT "SUCCESSFUL CONVERGENCE":
         GOTO 1430
1410  IF I < 11 THEN 1040:  REM DO ANOTHER ITERATION
1420  PRINT :WF$ = "WARNING:CONVERGENCE NOT OBTAINED":WF = 1:  PRINT WF$
1430  REM END OF ITERATIONS
1440  PRINT : PRINT : INPUT "ENTER P TO PROCEED (Z=BACKUP Q=QUIT)";QM$
1450  IF QM$ = "Z" OR QM$ = "Q" THEN 610
1460  IF QM$ < > "P" THEN 1440
1470  :
1480  REM *****SCREEN 6 *********************************************
1490  TEXT : HOME :
1500  PRINT "NONLINEAR CURVE FITTING RESULTS: EQN "Q
1510  PRINT N$(Q)":  "E$(Q): PRINT
1520  PRINT "WTG SCHEME USED: ";
1530  IF WG = 1 THEN PRINT "CONSTANT STD DEV"
1540  IF WG = 2 THEN PRINT "STD DEV PROP. TO Y"
1550  IF WG = 3 THEN PRINT "BETWEEN CONST & PROP"
1560  IF WB = 1 THEN PRINT "BISQUARE WTG FOR OUTLIERS USED"
1570  PRINT : IF Q = 1 OR Q > 7 THEN PRINT " THE K-VALUE WAS SET = ";K:
         PRINT
1580  PRINT " CALC VALUE OF A= " FN RO(A);
1590  PRINT " +/- " FN RO(SA)" SD":  PRINT
1600  PRINT " CALC VALUE OF B= " FN RO(B);
1610  PRINT " +/- " FN RO(SB)" SD":  PRINT
1620  IF WF = 1 THEN PRINT WF$
1630  :
1640  Y3 = 0:Y4 = 0:RS = 0:Y5 = 0:Y6 = 0
1650  FOR J = 1 TO N
1660  X = X(J)
1670  ON Q GOSUB 1710,1740,1750,1760,1770,1780,1790,1800,1810,1820
1680  GOTO 1830
1690  :
```

Figure A5.3. CURNLFIT listing for Apple II. (continued)

```
1700   REM THE EQN ON THE NEXT LINE MAY BE REPLACED W/ USER'S EQN
1710   Y = A + B * X
1720   :
1730   RETURN
1740   Y = A * X / (B + X): RETURN
1750   Y = A / (B + X): RETURN
1760   Y = A * EXP (B * X): RETURN
1770   Y = A * EXP (B / X): RETURN
1780   Y = A * X * EXP (B * X): RETURN
1790   Y = A * (1 - EXP (B * X)): RETURN
1800   Y = K / (1 + A * EXP (B * X)): RETURN
1810   Y = K + A * X ^B: RETURN
1820   Y = K / (1 + A * X ^B): RETURN
1830   REM RESIDUALS
1840   R = Y(J) - Y:YP(J) = Y
1850   Y3 = Y3 + Y:Y4 = Y4 + Y * Y
1860   Y5 = Y5 + Y(J):Y6 = Y6 + Y(J) * Y(J)
1870   RS = RS + R * R
1880   NEXT J
1890   PRINT "SUM OF SQ OF RESIDUALS = "; FN RO(RS)
1900   P = 2:   REM # OF EST. PARAMETERS
1910   VR = RS / (N - P)
1920   SE = SQR (VR):SE = FN RO(SE)
1930   PRINT "VAR OF RESIDUALS("(N - P)" DF) = ";
1940   PRINT FN RO(VR)
1950   PRINT "STD ERROR OF RESIDUALS = ";SE
1955   IF HP = 1 THEN 5500
1960   VTAB 23:   PRINT "(Q=QUIT Z=BACKUP H=HARDCOPY)";:   VTAB 22:   HTAB 1
1970   INPUT "LIST DATA & PRED. Y-VALUES? (Y/N)";QM$
1980   IF QM$ = "Z" THEN 550
1990   IF QM$ = "Q" THEN HOME : END
1995   IF QM$ = "H" THEN HL = 1:   GOTO 5400
2000   IF QM$ < > "Y" THEN 2110
2010   :
2020   REM ****SCREEN 7 *****************************************************
2030   TEXT : HOME : PRINT "COMPARING DATA WITH PREDICTED Y-VALUES"
2035   HF = 1
2040   PRINT "EQN #"Q" "N$(Q)
2050   PRINT : PRINT "X";:   HTAB 10: PRINT "Y";:   HTAB 20: PRINT "PRED Y";:
       HTAB 30:   PRINT "DIFF"
2060   FOR J = 1 TO N
2070   X = X(J)
2080   R = Y(J) - YP(J)
2090   PRINT X;:   HTAB 10:   PRINT Y(J);:   HTAB 20:   PRINT FN RO(YP(J));:
       HTAB 30:   PRINT FN RO(R)
2100   NEXT J
2110   PRINT : PRINT : PRINT : PRINT
2115   IF HP = 1 THEN 5500
2120   VTAB 23:   PRINT "(Q=QUIT Z=BACKUP";
2125   IF HF = 1 THEN PRINT " H=HARDCOPY";:HF = 0
2126   PRINT ")";:   VTAB 22:   HTAB 1
2130   INPUT "SEE A PLOT OF RESIDUALS? (Y/N) ";QM$
2140   IF QM$ = "Z" THEN 1490
2150   IF QM$ = "Q" THEN TEXT : HOME : END
2155   IF QM$ = "H" THEN HL = 2:   GOTO 5400
2160   IF QM$ < > "Y" THEN PRINT : PRINT : PRINT : PRINT : GOTO 2260
2170   :
2180   REM ****SCREEN 8 *****************************************************
2190   Y$ = "RESID. %":YM = + 100:YN = - 100:   GOSUB 3000
2195   HF = 1
```

Figure A5.3. CURNLFIT listing for Apple II. (continued)

```
2200 XO = 150:YO = 1:XL$ = "EQN #" + STR$ (Q): GOSUB 3500
2210 XA = 1:XD = 0:   REM LINE DRAW
2220 X = 0:Y = 0:   GOSUB 4000:X = XM: GOSUB 4000
2230   FOR J = 1 TO N
2240   X = X(J):Y = (Y(J) - YP(J)) * 100 / YH:XB = 1:   GOSUB 4000
2250   NEXT J
2260   VTAB 23:   PRINT "(Q=QUIT Z=BACKUP";
2265   IF HF = 1 THEN PRINT " H=HARDCOPY";:HF = 0
2266   PRINT ")";:   VTAB 22:   HTAB 1
2270   INPUT "SEE MODEL PLOTTED W/DATA? (Y/N) ";QM$
2280   IF QM$ = "Z" THEN 2030
2290   IF QM$ = "Q" THEN TEXT : HOME : END
2295   IF QM$ = "H" THEN HL = 3:   GOTO 5400
2300   IF QM$ < > "Y" THEN PRINT : PRINT : PRINT : PRINT : GOTO 2440
2310 :
2320   REM ****SCREEN 9 ************************************************
2330 Y$ = "Y-VALUE":YN = 0:YM = YH: GOSUB 3000
2335 HF = 1
2340 XO = 150:YO = 1:XL$ = "EQN #" + STR$ (Q): GOSUB 3500
2350 XA = 1:XD = 0
2360   FOR J = 1 TO 100 STEP 2
2370 X = J * XM / 100
2380   ON Q GOSUB 1710,1740,1750,1760,1770,1780,1790,1800,1810,1820
2390   GOSUB 4000
2400   NEXT J
2410   FOR J = 1 TO N
2420   X = X(J):Y = Y(J):XC = 1:   GOSUB 4000
2430   NEXT J
2440   VTAB 23:   PRINT "(Q=QUIT Z=BACKUP";
2445   IF HF = 1 THEN PRINT " H=HARDCOPY";:HF = 0
2446   PRINT ")";:   VTAB 22:   HTAB 1
2447   PRINT "                ":   VTAB 22:   HTAB 1
2450   INPUT "ENTER P TO PROCEED ";QM$
2460   IF QM$ = "Q" THEN TEXT : HOME : END
2465   IF QM$ = "H" THEN HL = 4:   GOTO 5400
2470   IF QM$ = "Z" THEN PRINT : PRINT : PRINT : PRINT : GOTO 2260
2480 :
2490   REM ****SCREEN 10 ***********************************************
2500   TEXT : HOME
2510   PRINT : PRINT : PRINT : PRINT "END OF NONLINEAR CURVE-FITTING
       PROGRAM"
2520   PRINT : PRINT : PRINT : INPUT "ENTER Q=QUIT, Z=BACKUP";QM$
2530   IF QM$ = "Q" THEN TEXT : HOME : END
2540   IF QM$ = "Z" THEN 2260
2550   GOTO 2520
2560 :
2570   REM ****END OF PROGRAM ******************************************
3000 :
3010   HGR : IF PEEK (2048) = 127 THEN 3020
3013   PRINT CHR$ (13) + CHR$ (4)"BLOAD SHAPES,A$0800"
3015   POKE 232,0:   POKE 233,8:   HCOLOR= 3:   SCALE= 1
.
.
.
```

Insert lines 3020-4090 from the GRAPH program here

```
.
.
5000 :
5010   REM **** START SUBROUTINE***************************************
5020   TEXT : HOME
5021   IF PEEK (104) = 64 THEN 5030
```

Figure A5.3. CURNLFIT listing for Apple II. (continued)

```
5022 P$ = CHR$ (13) + CHR$ (4) + "RUN CURNLFIT"
5023 POKE 103,0: POKE 104,64: POKE 16383,0: PRINT P$
5030 VTAB 1: HTAB 16: PRINT "CURNLFIT"
5040 VTAB 3: HTAB 8: PRINT "A PROGRAM FOR FITTING"
5050 HTAB 5: PRINT "NONLINEAR EQUATIONS TO DATA"
5060 VTAB 6: PRINT "PROGRAM INFORMATION IS FOUND IN THE BOOK"
5070 VTAB 7: HTAB 5: INVERSE
5080 PRINT "COMPUTER SIMULATION IN BIOLOGY:": HTAB 12: PRINT "A BASIC
     APPROACH": NORMAL
5090 PRINT " BY ROBERT E. KEEN & JAMES D. SPAIN"
5100 PRINT " (C) 1990 BY JOHN WILEY & SONS, INC"
5110 READ F$: RESTORE
5120 VTAB 12: PRINT "THIS PROGRAM IS SET TO READ DATA FROM"
5130 PRINT "A FILE NAMED": PRINT : HTAB 10: PRINT ">>>";F$: PRINT
5140 PRINT "IF THIS IS NOT THE NAME OF THE DISKFILE": PRINT "HOLDING THE
     DATA THEN QUIT THE PROGRAM"
5150 PRINT "& CHANGE THE FILENAME IN LINE 100": PRINT
5160 INPUT "ENTER 'P' TO PROCEED, 'Q' TO QUIT ";QM$
5170 IF QM$ = "Q" THEN HOME : TEXT : END
5180 IF QM$ = CHR$ (112) OR QM$ = CHR$ (113) THEN PRINT "CAPITAL LETTERS
     PLEASE"
5190 IF QM$ < > "P" THEN 5160
5200 D$ = CHR$ (13) + CHR$ (4): PRINT D$"VERIFY"F$: PRINT D$"OPEN"F$:
     PRINT D$"READ"F$
5210 INPUT N
5220 FOR J = 1 TO N: INPUT X(J),Y(J): NEXT J
5230 PRINT D$"CLOSE"F$
5240 DEF FN RO(X) = INT (10000 * X + 0.5) / 10000
5250 N$(1) = "USER'S EQN":E$(1) = "Y = F(X)"
5260 N$(2) = "HYPERBOLIC":E$(2) = "Y = (A*X) / (B+X)"
5270 N$(3) = "MOD. INVERSE":E$(3) = "Y = A / (B + X)"
5280 N$(4) = "EXPONENTIAL":E$(4) = "Y = A * EXP(B*X)"
5290 N$(5) = "EXP. RECIP.":E$(5) = "Y = A * EXP(B/X)"
5300 N$(6) = "MAXIMA FUNC.":E$(6) = "Y = A * X * EXP(B*X)"
5310 N$(7) = "EXP. SAT'N":E$(7) = "Y = A*[1 - EXP(B*X)]"
5320 N$(8) = "LOGISTIC":E$(8) = "Y = K/[1+A*EXP(B*X)]"
5330 N$(9) = "LOGARITHMIC":E$(9) = "Y = (A * X^B) + K"
5340 N$(10) = "SIGMOID":E$(10) = "Y = K / (1 + A * X^B)"
5350 RETURN
5400 :
5410 REM **** HARDCOPY SUBROUTINE ***************************************
5420 IF HL > 2 THEN 5460
5430 PRINT CHR$ (13) + CHR$ (4) + "PR#1"
5440 HP = 1
5450 ON HL GOTO 1470,2030
5460 REM SCREEN-DUMP FOR APPLE SILENTYPE PRINTER
5470 POKE - 12528,7: POKE - 12529,255: POKE - 12524,0
5480 PRINT CHR$ (13) + CHR$ (4) + "PR#1"
5490 PRINT CHR$ (17)
5500 PRINT CHR$ (13) + CHR$ (4) + "PR#0"
5510 HP = 0
5520 ON HL GOTO 1470,2030,2170,2310
5530 END
```

Figure A5.3. CURNLFIT listing for Apple II. (concluded)

APPENDIX 6

BASIC Programs

Listings of the GRAPH, CURFIT, POLYFIT, and CURNLFIT programs are given in Appendices 2, 3, 4, and 5. Entering these programs into a computer can be tedious and time-consuming. Individuals who wish to use these BASIC programs may obtain a diskette with versions of the programs for Apple II, IBM PC or Macintosh microcomputers. One diskette is available free of charge by mailing the card bound with this book to John Wiley & Sons, Inc., 605 Third Avenue, New York NY 10158-0012 USA. Instructors may distribute copies of the programs on disk for classroom use only; all other rights are reserved.

Inquiries about these programs should be directed to the authors at the Department of Biological Sciences, Michigan Technological University, Houghton MI 49931 USA, telephone 906-487-2346. Also obtainable from the authors is a manual for instructors using this book as a text. The manual contains BASIC listings and output for each exercise in the text, along with suggestions for formal instruction and coursework.

LITERATURE CITED

Alberghina, L. 1977. Dynamics of the cell cycle in mammalian cells. Journal of Theoretical Biology 69:633-643.

———. 1978. Modeling the control of cell growth. Simulation 31:37-41.

Anderson, N. 1974. A mathematical model for the growth of giant kelp. Simulation 22:97-105.

Anderson, R. M., H. C. Jackson, R. M. May, and A. M. Smith. 1981. Population dynamics of fox rabies in Europe. Nature 289:765-771.

Atkins, G. L. 1969. Multicompartment models for biological systems. Methuen, London, England.

Ayala, F. J. 1982. Population and evolutionary genetics: a primer. Benjamin/Cummings, Menlo Park, California, USA.

Bachelet, D., H. W. Hunt, J. K. Detling, and D. W. Hilbert. 1983. A simulation model of blue grama biomass dynamics, with special attention to translocational mechanism. Pages 457-446 in W. K. Lauenroth, G. V. Skogerboe, and M. Flug, editors. Analysis of ecological systems: State-of- the-art in ecological modelling. Developments in environmental modelling 5. Elsevier, New York, New York, USA.

Bacon, P. J. (editor). 1985. Population dynamics of rabies in wildlife. Academic Press, New York, New York, USA.

Bailey, N. T. J. 1957. The mathematical theory of epidemics. Hafner, New York, New York, USA.

———. 1964. The elements of stochastic processes with applications to the natural sciences. John Wiley and Sons, New York, New York, USA.

Baker, R., and G. T. Herman. 1972. Simulation of organisms using a developmental model. 2. The heterocyst formation problem in blue green algae. International Journal of Biomedical Computing 3:251-267.

Bakken, G. S., and T. H. Kunz. 1988. Microclimate methods. Pages 303-332 in T. H. Kunz, editor. Ecological and behavioral methods for the study of bats. Smithsonian Institution Press, Washington, D.C., USA.

Baly, E. C. C. 1935. The kinetics of photosynthesis. Proceedings of the Royal Society of London 117B:218-239.

Bartholomay, A. F. 1964. The general catalytic queue process. Pages 147-165 *in* J. Gurland, editor. Stochastic models in biology and medicine. University of Wisconsin Press, Madison, Wisconsin, USA.

Bartlett, M. S. 1960. Stochastic population models in ecology and epidemiology. Methuen, London, United Kingdom.

Bates, S. S., A. Tessier, P. G. C. Campbell, and J. Buffle. 1982. Zinc adsorption and transport by *Chlamydomonas variabilis* and *Scenedesmus subspicatus* (Chlorophyceae) grown in semi-continuous culture. Journal of Phycology 18:521-529.

Berg, H. C. 1983. Random walks in biology. Princeton University Press, Princeton, New Jersey, USA.

Bode, P. M., and H. R. Bode. 1984. Patterning in hydra. Pages 213-241 *in* G. M. Malacinski and S. V. Bryant, editors. Pattern formation. A primer in developmental biology. Macmillan, New York, New York, USA.

Bolin, B. 1981. Steady-state and response characteristics of a simple model of the carbon cycle. Pages 315-331 *in* B. Bolin, editor. Carbon cycle modelling. SCOPE 16. John Wiley and Sons, New York, New York, USA.

Booth, D. A., and F. M. Toates. 1974. A physiological control theory of food intake in the rat. Bulletin of the Psychonomic Society 3:442-444.

Boraas, M. 1983. Population dynamics of food-limited rotifers in two-stage chemostat culture. Limnology and Oceanography 28:546-563.

Braun, M. 1983. Differential equations and their applications. An introduction to applied mathematics. Third edition. Applied Mathematical Sciences. Volume 15. Springer-Verlag, New York, New York, USA.

Briggs, G. E., and J. B. S. Haldane. 1925. A note on the kinetics of enzyme action. Biochemical Journal 19:338-339.

Bronson, R. 1973. Modern introductory differential equations. McGraw-Hill, New York, New York, USA.

Butcher, J. C. 1964. On Runge-Kutta processes of higher order. Journal of the Australian Mathematics Society 4:179-185.

Caperon, J. 1968. Population growth response of *Isochrisis galbana* to nitrate variation in limiting concentration. Ecology 49:866-872.

Caperon, J., and J. Meyer. 1972. Nitrogen-limited growth of marine phytoplankton. Deep-Sea Research 19:601-632.

Carnahan, B., H. A. Luther, and J. O. Wilkes. 1969. Applied numerical methods. John Wiley and Sons, New York, New York, USA.

Carson, E. R., C. Cobelli, and L. Finkelstein. 1983. The mathematical modeling of metabolic and endocrine systems. Model formulation, identification and validation. Wiley-Interscience, New York, New York, USA.

Cartwright, M., and M. Husain. 1986. A model for the control of testosterone secretion. Journal of Theoretical Biology 123:239-250.

Chance, B. 1960. Analogue and digital representations of enzyme kinetics. Journal of Biological Chemistry 235:2440-2443.

Chance, B., D. Garfinkel, J. Higgins, and B. Hess. 1960. Metabolic control mechanisms. V. A solution of equations for glycolysis and respiration in ascites tumor cells. Journal of Biological Chemistry 235:2426-2439.

Charles-Edwards, D. A., D. Doley, and G. M. Rimmington. 1986. Modelling plant growth and development. Academic Press, Orlando, Florida, USA.

Chen, C. W. 1970. Concepts and utilities of ecologic model. Journal of the Sanitary Engineering Division, Proceedings of the American Society of Civil Engineers 96(SA5):1085-1098.

Cleland, W. W. 1970. Steady-state kinetics. Pages 1-65 in P. D. Boyer, editor. The enzymes. Third edition. Volume 2. Academic Press, New York, New York, USA.

Crank, J. 1956. The mathematics of diffusion. Clarendon Press, Oxford, England.

Crow, J. F., and M. Kimura. 1970. An introduction to population genetics theory. Harper and Row, New York, New York, USA.

Croxton, F. E., and D. J. Cowden. 1955. Applied general statistics. Prentice-Hall, Englewood Cliffs, New Jersey, USA.

Cullen, M. R. 1985. Linear models in biology. Linear systems analysis with biological applications. Ellis Horwood Ltd., Chichester, United Kingdom.

Cushing, D. H. 1977. The problems of stock and recruitment. Pages 116-133 in J. A. Gulland, editor. Fish population dynamics. Wiley-Interscience, New York, New York, USA.

Dahlberg, B. L., and R. C. Guettinger. 1956. The white-tailed deer in Wisconsin. Wisconsin Conservation Department Technical Wildlife Bulletin Number 14:1-282.

Davies, R. G. 1971. Computer programming in quantitative biology. Academic Press, New York, New York, USA.

Dawkins, R. 1986. The blind watchmaker. W. W. Norton, New York, New York, USA.

DeAngelis, D. L., R. A. Goldstein, and R. V. O'Neill. 1975. A model for trophic interaction. Ecology 56:881-892.

Deevey, E. S. 1947. Life tables for natural populations of animals. Quarterly Review of Biology 22:283-314.

DeMarais, L. T. 1985. Growth and respiration of two pleuronectiform juveniles from a western Mediterranean bay. Journal of Fish Biology 27:459-469.

Detwiler, R. P., and C. A. S. Hall. 1988. Tropical forests and the global carbon cycle. Science 239:42-47.

DiToro, D. M. 1980. Applicability of cellular equilibrium and Monod theory to phytoplankton growth kinetics. Ecological Modelling 8:201-218.

Dixon, R. 1981. The mathematical daisy. New Scientist 92:792-795.

Dowd, J. E., and D. S. Riggs. 1965. A comparison of estimates of Michaelis-Menten kinetic constants from various linear transformations. Journal of Biological Chemistry 240:863-869.

Droop, M. R. 1968. Vitamin B12 and marine ecology. IV. The kinetics, uptake and growth of *Monochrysis lotherei*. Journal of the Marine Biological Association of the United Kingdom 54:825-855.

———. 1973. Some thoughts on nutrient limitation in algae. Journal of Phycology 9:264-272.

Dugdale, R. C. 1967. Nutrient limitation in the sea: dynamics, identification and significance. Limnology and Oceanography 12:685-695.

Duggleby, R. G. 1981. A nonlinear regression program for small computers. Analytical Biochemistry 110:9-18.

Eisen, M. 1988. Mathematical methods and models in the biological sciences. Nonlinear and multidimensional theory. Prentice-Hall, Englewood Cliffs, New Jersey, USA.

Elias, H., and J. E. Pauly. 1966. Human microanatomy. F. A. Davis Company, Philadelphia, Pennsylvania, USA.

Elliot, E. T., R. G. Wiegert, and H. W. Hunt. 1983. Simulation of simple food chains. Pages 211-217 *in* W. K. Lauenroth, G. V. Skogerboe, and M. Flug, editors. Analysis of ecological systems: State-of- the-art in ecological modelling. Developments in environmental modelling 5. Elsevier, New York, New York, USA.

Emlen, J. M. 1973. Ecology: an evolutionary approach. Addison-Wesley, Reading, Massachusetts, USA.

Eyring, H., and D. W. Urry. 1965. Thermodynamics and chemical kinetics. Pages 57-95 *in* T. H. Waterman and H. J. Morowitz, editors. Theoretical and mathematical biology. Blaisdell, New York, New York, USA.

Fargo, W. S., and E. L. Bonjour. 1988. Developmental rate of the squash bug, *Anasa tristis* (Heteroptera: Coreidae), at constant temperatures. Environmental Entomology 17:926-929.

Fee, E. J. 1969. A numerical model for the estimation of photosynthetic production, integrated over time and depth, in natural waters. Limnology and Oceanography 14:906-911.

———. 1973. A numerical model for determining integral primary production and its application to Lake Michigan. Journal of the Fisheries Research Board of Canada 30:1447-1468.

Feller, W. 1968. An introduction to probability and its applications. Volume 1. Third Edition. John Wiley and Sons, New York, New York, USA.

Forrester, J. W. 1969. Urban dynamics. Massachusetts Institute of Technology Press, Cambridge, Massachusetts, USA.

Fretwell, S. 1972. Populations in a seasonal environment. Princeton University Press, Princeton, New Jersey, USA.

Fukuhara, O., and K. Takao. 1988. Growth and larval behavior of *Engraulis japonica* in captivity. Journal of Applied Ichthyology 4:158-167.

Gates, D. M. 1962. Energy exchange in the biosphere. Harper and Row, New York, New York, USA.

Gause, G. F. 1934. The struggle for existence. Williams and Wilkins, Baltimore, Maryland, USA.

Gause, G. F., and A. A. Witt. 1935. Behavior of mixed populations and the problem of natural selection. American Naturalist 69:596-609.

Gérardin, L. 1968. Bionics. McGraw-Hill, New York, New York, USA.

Gierer, A., and H. Meinhardt. 1972. A theory of biological pattern formation. Kybernetik 12:30-39.

Gillespie, P. A., and R. F. Vaccaro. 1978. A bacterial bioassay for measuring the copper-chelation capacity of seawater. Limnology and Oceanography 23:543-548.

Ginzburg, L. R., and E. M. Golenberg. 1985. Lectures in theoretical population biology. Prentice-Hall, Englewood Cliffs, New Jersey, USA.

Gleason, J. M. 1988. Statistical tests of the Apple IIe random number generator yield suggestions for generator seeding. Collegiate Microcomputing 6:108-112.

Goel, N. S., and N. Richter-Dyn. 1974. Stochastic models in biology. Academic Press, New York, New York, USA.

Gold, H. J. 1977. Mathematical modeling of biological systems. An introductory guidebook. Wiley-Interscience, New York, New York, USA.

Grabowski, C. T. 1983. Human reproduction and development. Saunders College Publishing, New York, New York, USA.

Grainger, J. N. R. 1959. The effect of constant and varying temperatures on the developing eggs of *Rana temporaria* L. Zoologischer Anzeiger 163:267-277.

Godfrey, K. 1983. Compartmental models and their application. Academic Press, New York, New York, USA.

Guyton, A. C. 1971. Basic human physiology. W. B. Saunders Company, Philadelphia, Pennsylvania, USA.

Hall, C. A. S. 1988. An assessment of several of the historically most influential theoretical models used in ecology and of the data provided in their support. Ecological Modelling 43:5-31.

———. 1977. Systems and models: terms and basic principles. Pages 5-36 in C. A. S. Hall and J. W. Day, Jr., editors. Ecosystem modelling in theory and practice: an introduction with case histories. John Wiley and Sons, New York, New York, USA.

Hamil, W. H., R. R. Williams, and C. Mackay. 1966. Principles of physical chemistry. Prentice-Hall, Englewood Cliffs, New Jersey, USA.

Hardin, G. 1966. Biology, its principles and implications. W. H. Freeman and Company, San Francisco, California, USA.

Herbert, D. 1958. Some principles of continuous culture. Pages 381-396 in G. Tunevall, editor. Recent progress in microbiology. Blackwell, Oxford, England.

Hoerl, A. E. 1954. Fitting curves to data. Pages 55-77 in J. H. Perry, editor. Chemical business handbook. McGraw-Hill, New York, New York, USA.

Holm, J., and S. Kellomaki. 1984. Comparison of growth strategies of two competing plant species in forest ground cover. Ecological Modelling 23:135-150.

Holmes, P. 1973. Letter. Science 180:1298-1299.

Holling, C. S. 1959. The components of predation as revealed by a study of small mammal predation of the European pine sawfly. Canadian Entomologist 91:293-320.

Holyoak, G. W., and R. C. Stones. 1971. Temperature regulation of the little brown bat, *Myotis lucifugus*, after acclimation at various ambient temperatures. Comparative Biochemistry and Physiology 39A:413-420.

Horn, H. S. 1975. Forest succession. Scientific American 232:90-98.

Horwitz, J. S., and D. C. Montgomery. 1974. A computer simulation model of a rubella epidemic. Computers in Biology and Medicine 4:189-198.

Hutchinson, G. E. 1957. A treatise on limnology. I. Geography, physics, and chemistry. John Wiley and Sons, New York, New York, USA.

———. 1973. Eutrophication. American Scientist 61:269-279.

———. 1978. An introduction to population ecology. Yale University Press, New Haven, Connecticut, USA.

Inagaki, S., and O. Yamashita. 1989. Delayed activation of glycogen synthase adjusted to the effective storage function of fat body in the premetamorphic instar of the silkworm, *Bombyx mori*. Comparative Biochemistry and Physiology 92B:81-86.

Ingham, P. W. 1988. The molecular genetics of embryonic pattern formation in *Drosophila*. Nature 335:25-34.

Irwin, J. O. 1964. The contributions of G. U. Yule and A. G. McKendrick to stochastic process methods in biology and medicine. Pages

147-165 *in* J. Gurland, editor. Stochastic models in biology and medicine. University of Wisconsin Press, Madison, Wisconsin, USA.

Jacob, F., and J. Monod. 1961. Genetic regulatory mechanisms in the synthesis of proteins. Journal of Molecular Biology 3:318-356.

James, M. L., G. M. Smith, and J. C. Wolford. 1977. Applied numerical methods for digital computation with FORTRAN and CSMP. Second edition. Harper and Row, New York, New York, USA.

Jenkins, D. H., and I. H. Bartlett. 1959. Michigan whitetails. Michigan Department of Conservation, Lansing, Michigan, USA.

Jenkins, S. H. 1988. The use and abuse of demographic models of population growth. Bulletin of the Ecological Society of America 69:201-207.

Johnson, F. H., H. Eyring, and M. J. Polissar. 1954. The kinetic basis of molecular biology. John Wiley and Sons, New York, New York, USA.

Jones, R. W. 1973. Principles of biological regulation: an introduction to feedback systems. Academic Press, New York, New York, USA.

Jones, R., and W. Hall. 1973. A simulation model for studying the population dynamics of some fish species. Pages 35-39 *in* M. S. Bartlett and R. Hiorns, editors. The mathematical theory of the dynamics of biological populations. Academic Press, New York, New York, USA.

Keen, R., and H. B. Hitchcock. 1980. Survival and longevity of the little brown bat (*Myotis lucifugus*) in southeastern Ontario. Journal of Mammalogy 61:1-7.

Keen, R., and R. Nassar. 1981. Confidence intervals for birth and death rates estimated with the egg-ratio technique for natural populations of zooplankton. Limnology and Oceanography 26:131-142.

Kelly, M. G., and G. M. Hornberger. 1973. Letter. Science 180:1299.

Kermack, W. D., and A. G. McKendrick. 1927. A contribution to the mathematical theory of epidemics. Proceedings of the Royal Society of London, Series A 115:700-721.

Kermack, W. D., and A. G. McKendrick. 1932. Contributions to the mathematical theory of epidemics. Proceedings of the Royal Society of London, Series A 138:55-83.

Knuth, D. E. 1981. The art of computer programming. Volume 2. Second edition. Addison-Wesley, Reading, Massachusetts, USA.

Kooloos, J. G. M., and G. A. Zweers. 1989. Mechanics of drinking in the mallard (*Anas platyrhynchos*, Anatidae). Journal of Morphology 199:327-347.

Kurta, A., and M. S. Fujita. 1988. Design and interpretation of laboratory thermoregulation studies. Pages 333-352 *in* T. H. Kunz, editor. Ecological and behavioral methods for the study of bats. Smithsonian Institution Press, Washington, D.C., USA.

Langmuir, I. 1916. Absorption of gases on solids. Journal of the American Chemical Society 38:2221-2236.

Lauwerier, H. A. 1984. Mathematical models of epidemics. Second printing. Mathematical Centre Tracts 138. Mathematical Centre, Amsterdam, The Netherlands.

Lehninger, A. L. 1975. Biochemistry. Worth Publishers, New York, New York, USA.

Leslie, P. H. 1945. The use of matrices in certain population mathematics. Biometrika 33:183-212.

Leslie, P. H., and J. C. Gower. 1960. Properties of a stochastic model for the predator-prey type of interaction between two species. Biometrika 47:219-301.

Levin, B. R. 1969. A model for selection in systems of species competition. Pages 237-275 in F. Heinmets, editor. Concepts and models of biomathematics. Simulation techniques and methods. Marcel Dekker, New York, New York, USA.

Li, C. C. 1976. First course in population genetics. Boxwood Press, Pacific Grove, California, USA.

Lindeman, R. L. 1942. The trophic-dynamic aspect of ecology. Ecology 23:399-418.

Logan, J. A. 1988. Toward an expert system for development of pest simulation models. Environmental Entomology 17:359- 376.

Lotka, A. J. 1925. Elements of physical biology. Williams and Wilkins, Baltimore, Maryland, USA. Reprinted in 1956 as Elements of mathematical biology. Dover, New York, New York, USA.

Lynch, M. 1989. The life history consequences of resource depression in *Daphnia pulex*. Ecology 70:246-256.

Machado, C. R., M. A. R. Garafalo, J. E. S. Roselino, I. C. Kettlehut, and R. H Migliorini. 1989. Effect of fasting on glucose turnover in a carnivorous fish (*Hoplias* sp.). American Journal of Physiology 256:R612-R615.

Macon, N. 1963. Numerical analysis. John Wiley and Sons, New York, New York, USA.

Mahler, H. R., and E. H. Cordes. 1966. Biological chemistry. Harper and Row, New York, New York, USA.

Malacinski, G. M., and S. V. Bryant (editors). 1984. Pattern formation. A primer in developmental biology. Macmillan, New York, New York, USA.

Maynard Smith, J. 1968. Mathematical ideas in biology. Cambridge University Press, New York, New York, USA.

McLay, C. 1970. A theory concerning the distance traveled by animals entering the drift of a stream. Journal of the Fisheries Research Board of Canada 27:359-370.

Meadows, D. H., D. L. Meadows, J. Randers, and W. W. Behrens. 1972. The limits to growth. Universe Books, New York, New York, USA.

Meinhardt, H. 1977. A model of pattern formation in insect embryogenesis. Journal of Cell Science 23:117-139.

———. 1982. Models of biological pattern formation. Academic Press, New York, New York, USA.

———. 1984. Models for pattern formation during development of higher organisms. Pages 47-72 in G. M. Malacinski and S. V. Bryant, editors. Pattern formation. A primer in developmental biology. Macmillan, New York, New York, USA.

Messenger, P. S., and N. E. Flitters. 1959. Effect of variable temperature environments on egg development of three species of fruit flies. Annals of the Entomological Society of America 52:191-204.

Milanese, M., and G. P. Molino. 1975. Structural identifiability of compartmental models and pathophysiological information from the kinetics of drugs. Mathematical Biosciences 26:175-190.

Milhorn, H. T. 1966. The application of control theory to physiological systems. W. B. Saunders Company, Philadelphia, Pennsylvania, USA.

Mitoma, C. 1985. Tests for hepatobiliary function. Pages 90-107 in C. A. Tyson and D. S. Sawhney, editors. Organ function tests in toxicity evaluation. Noyes Publications, Park Ridge, New Jersey, USA.

Miyazono, K., and C.-H. Heldin. 1989. Role for carbohydrate structures in TGF-$\beta1$ latency. Nature 338:158-160..

Molino, G. P., and M. Milanese. 1975. Structured analysis of compartmental models for the hepatic kinetics of drugs. Journal of Laboratory and Clinical Medicine 85:865-878.

Molino, G. P., M. Milanese, A. Villa, A. Cavanna, and G. P. Gaidano. 1978. Discrimination of hepatobiliary diseases by the evaluation of bromosulfophthalein blood kinetics. Journal of Laboratory and Clinical Medicine 91:396-408.

Moner, J. G. 1972. The effects of temperature and heavy water on cell division in heat-synchronized cells of Tetrahymena. Journal of Protozoology 19:382-385.

Monod, J. 1942. Recherches sur la croissance des cultures bacteriennes. Hermann, Paris, France.

Monod, J., J. Wyman, and J. P. Changeus. 1965. On the nature of allosteric transitions: a plausible model. Journal of Molecular Biology 12:88-96.

Motulsky, H. J., and L. A. Ransnas. 1987. Fitting curves to data using nonlinear regression: a practical and nonmathematical review. FASEB Journal 1:365-374.

Mukhtar, H., W. A. Khan, D. P. Bik, M. Das, and D. R. Bickers. 1989. Hepatic microsomal metabolism of leukotriene B4 in rats: biochemical characterization, effect of inducers, and age- and sex-dependent differences. Xenobiotica 19:151-159.

Mulholland, R. J., J. S. Read, and W. R. Emanuel. 1987. Asymptotic analysis of airborne fraction used to validate global carbon models. Ecological Modelling 36:139-152.

Murray, J. D. 1987. Modeling the spread of rabies. American Scientist 75:280-284.

Newsholme, E. A., and C. Start. 1976. Regulation in metabolism. John Wiley and Sons, London, England.

Novick, A., and L. Szilard. 1950a. Description of the chemostat. Science 112:715-716.

———. 1950b. Experiments with the chemostat on spontaneous mutations of bacteria. Proceedings of the National Academy of Sciences (USA) 36:708-719.

O'Brien, W. J. 1972. Limiting factors in phytoplankton algae: their meaning and measurement. Science 178:616-617.

———. 1973. Letter. Science 180:1299-1300.

Odum, H. T. 1957. Trophic structure and productivity of Silver Springs, Florida. Ecological Monographs 27:55-112.

Olinick, M. 1978. An introduction to mathematical models in the social and life sciences. Addison-Wesley, Reading, Massachusetts, USA.

O'Neill, R. V., R. A. Goldstein, H. H. Shugart, and J. B. Mankin. 1972. Terrestrial ecosystem energy model. U. S. IBP Eastern Deciduous Forest Biome Memo Report Number 72-19. Oak Ridge National Laboratory, Oak Ridge, Tennessee, USA.

Otis, D. L., K. P. Burnham, G. C. White, and D. R. Anderson. 1978. Statistical inference from capture data on closed animal populations. Wildlife Monographs 62:1-135.

Paris, O. H., and F. A. Pitelka. 1962. Population characteristics of the terrestrial isopod *Armadillidium vulgare* in California grassland. Ecology 43:229-248.

Park, T. 1948. Experimental studies of interspecies competition. 1. Competition between populations of the flour beetles, *Tribolium confusum* Duval and *Tribolium castaneum confusum* Herbst. Ecological Monographs 18:265-308.

Park, T., D. B. Mertz, W. Grodzinski, and T. Prus. 1965. Cannibalistic predation in populations of flour beetles. Physiological Zoology 38:289-321.

Patten, B. C. 1971. A primer for ecological modeling and simulation with analog and digital computers. Pages 3-121 *in* B. C. Patten, editor. Systems analysis and simulation in ecology. Volume 1. Academic Press, New York, New York, USA.

Pearl, R. 1927. The growth of populations. Quarterly Review of Biology 2:532-548.

Pearl, R., and J. R. Miner. 1935. Experimental studies on the duration of life. XIV. Comparative mortality. Quarterly Review of Biology 10:60-79.

Pearl, R., and L. J. Reed. 1920. On the rate of growth of the population of the United States since 1790 and its mathematical representation. Proceedings of the National Academy of Sciences (USA) 6:275-288.

Pielou, E. C. 1977. Mathematical ecology. John Wiley and Sons, New York, New York, USA.

Pollard, J. H. 1973. Mathematical models for the growth of human populations. Cambridge University Press, Cambridge, England.

Pollock, K. H. 1981. Capture-recapture models: a review of current methods, assumptions and experimental design. Pages 426-435 *in* D. J. Ralph and J. M. Scott, editors. Estimating the numbers of terrestrial birds. Studies in avian biology 6. Cooper Ornithological Society, Allen Press, Lawrence, Kansas, USA.

Poole, R. W. 1974. An introduction to quantitative ecology. McGraw-Hill, New York, New York, USA.

Powers, W. F., and R. P. Canale. 1975. Some applications of optimization techniques to water quality modeling and control. IEEE Transactions on Systems, Man and Cybernetics 5:312-321.

Prosser, C. L., and F. A. Brown, Jr. 1961. Comparative animal physiology. Second edition. W. B. Saunders Company, Philadelphia, Pennsylvania, USA.

Raaijmakers, J. G. W. 1987. Statistical analysis of the Michaelis-Menten equations. Biometrics 43:793-803.

Rai, L. C., M. J. P. Gaur, and H. D. Kumar. 1981. Phycology and heavy-metal pollution. Biological Reviews 56:99-151.

Rand Corporation. 1955. A million random digits with 100,000 normal deviates. The Free Press, Glencoe, Illinois, USA.

Ransom, R. 1981. Computers and embryos. Models in developmental biology. Wiley-Interscience, New York, New York, USA.

Rhee, G. Y. 1973. A continuous culture study of phosphate uptake, growth rate and polyphosphates in *Scenedesmus* spp. Journal of Phycology 9:495-506.

———. 1978. Effects of N:P atomic ratios and nitrate limitation on algal growth, cell composition, and nitrate uptake. Limnology and Oceanography 23:10-25.

Rhodes, D. G., J. J. Achs, L. Peterson, and D. Garfinkel. 1968. A method of calculating time-course behavior of multi-enzyme systems from the enzymatic equations. Computers and Biomedical Research 2:45-50.

Richter, O., and A. Betz. 1976. Simulation of biochemical pathways and its application to biology and medicine. Pages 181-197 *in* S. Levin, editor. Lecture notes in biomathematics 11. Springer- Verlag, New York, New York, USA.

Ricker, W. E. 1954. Stock and recruitment. Journal of the Fisheries Research Board of Canada 11:559-623.

Ricklefs, R. 1979. Ecology. Second edition. Chiron Press, Newton, Massachusetts, USA.

Rosenzweig, M. L. 1969. Why the prey curve has a hump. American Naturalist 103:81-87.

Rosenzweig, M. L., and R. H. MacArthur. 1963. Graphical representation and stability conditions of predator-prey interactions. American Naturalist 97:209-223.

Rotty, R. M. 1981. Data for global CO_2 production from fossil fuels and cement. Pages 121-125 *in* B. Bolin, editor. Carbon cycle modelling. SCOPE 16. John Wiley and Sons, New York, New York, USA.

Ryther, J. H. 1956. Photosynthesis in the ocean as a function of light intensity. Limnology and Oceanography 1:61-70.

———. 1959. Potential productivity of the sea. Science 130:602-608.

Sampson, J. R. 1984. Biological information processing. Current theory and computer simulation. Wiley-Interscience, New York, New York, USA.

Savageau, M. A. 1976. Biochemical systems analysis. Addison-Wesley, Reading, Massachusetts, USA.

Schaffer, W. M., and M. Kot. 1986. Differential systems in ecology and epidemiology. Pages 158-178 *in* A. V. Holden, editor. Chaos. Princeton University Press, Princeton, New Jersey, USA.

Schelske, C. L., E. F. Stoermer, and L. E. Feldt. 1971. Nutrients, phytoplankton productivity and species composition as influenced by upwelling in Lake Michigan. Pages 102-113 *in* Proceedings of the Fourteenth Conference on Great Lakes Research, International Association for Great Lakes Research, Ann Arbor, Michigan, USA.

Schuler, M. L., S. Leung, and C. C. Dick. 1979. A mathematical model for the growth of a single bacterial cell. Annals of the New York Academy of Science 77:35-55.

Searle, S. R. 1966. Matrix algebra for the biological sciences. John Wiley and Sons, New York, New York, USA.

Seber, G. A. F. 1982. The estimation of animal abundance. Second edition. Macmillan, New York, New York, USA.

Sharpe, P.J.H., and D. W. DeMichele. 1977. Reaction kinetics of poikilotherm development. Journal of Theoretical Biology 64:649-670.

Sheppard, C. W. 1948. The theory of transfers within a multi-compartment system using isotopic tracers. Journal of Applied Physics 19:70-76.

Sias, F.R. and T. G. Coleman. 1971. Digital simulation of biological systems using conversational languages. Simulation 16: 102-111.

Sirko, S., I. Bishai, and F. Coceani. 1989. Prostaglandin formation in the hypothalamus in vivo: effect of pyrogens. American Journal of Physiology 256:R616-R624.

Slobodkin, L. B. 1954. Population dynamics in *Daphnia obtusa* Kurz. Ecological Monographs 24:69-88.

Slobodkin, L. B. 1980. Growth and regulation of animal populations. Second edition. Dover, New York, New York, USA.

Smith, E. L. 1936. Photosynthesis in relation to light and carbon dioxide. Proceedings of the National Academy of Sciences (USA) 22:504.

Staub, J.-F., P. Brezillon, A. M. Perault-Staub, and G. Milhaud. 1981. Nonlinear modelling of calcium metabolism - first attempt in the calcium deficient rat. Transactions of the Institute of Measurement and Control (London) 3:89-97.

Steele, J. H. 1962. Environmental control of photosynthesis in the sea. Limnology and Oceanography 7:137-150.

Swartzman, G. L., and S. P. Kaluzny. 1987. Ecological simulation primer. Macmillan, New York, New York, USA.

Tanigoshi, L. K., and R. W. Browne. 1978. Influence of temperature on life table parameters of *Metaseiulus occidentalis* and *Tetranychus mcdanieli* (Acarina: Phytoseiidae, Tetranychidae). Annals of the Entomological Society of America 71:313-316.

Tanner, J. T. 1975. The stability and intrinsic growth rates of prey and predator populations. Ecology 56:855-867.

Tulasi, S. J., and J. V. Ramana Rao. 1989. Effects of organic and inorganic lead on the oxygen equilibrium curves of the fresh water field crab, *Barytelphusa guerini*. Bulletin of Environmental Contamination and Toxicology 42:247-253.

U. S. Bureau of the Census. 1986. Statistical abstract of the United States: 1987. 107th edition. United States Government Printing Office, Washington, D.C., USA.

Usher, M. B. 1972. Developments in the Leslie matrix model. Pages 29-60 *in* J. N. R. Jeffers, editor. Mathematical models in ecology. Blackwell, Oxford, England.

Vaccaro, R. F., F. Azam, and R. E. Hodson. 1977. Response of natural marine bacterial populations to copper: Controlled ecosystem pollution experiment. Bulletin of Marine Science 27:17-22.

Van Dover, C. D., P. J. S. Franks, and R. D. Ballard. 1987. Prediction of hydrothermal vent locations from distributions of brachyuran crabs. Limnology and Oceanography 32:1006-1010.

Vollenweider, R. A. 1965. Calculation models of photosynthesis-depth curves and some implications regarding day rate estimates in primary production measurements. Pages 425-457 in C. R. Goldman, editor. Primary productivity in aquatic environments. Supplement 18, Memorie Istituto Italiano di Idrobiologia. University of California Press, Berkeley, California, USA.

Vyhnalek, V. 1987. Interactions between algae and zooplankton in a continuous cultivation system. Ecological Modelling 39:33-43.

Wallace, B. 1972. Disease, sex, communication, behavior. Essays in social biology. Volume 3. Prentice-Hall, Englewood Cliffs, New Jersey, USA.

Walter, C. F. 1974. Some dynamic properties of linear, hyperbolic and sigmoidal multi-enzyme systems with feedback control. Journal of Theoretical Biology 44:219-240.

Walters, C. J. 1971. Systems ecology: the systems approach and mathematical models in ecology. Pages 276-292 in E. P. Odum. Fundamentals of ecology. Third edition. W. B. Saunders Company, Philadelphia, Pennsylvania, USA.

Waltman, P. 1974. Deterministic threshold models in the theory of epidemics. Lecture notes in biomathematics 1. Springer-Verlag, New York, New York, USA.

Wangersky, P. J. 1978. Lotka-Volterra population models. Annual Review of Ecology and Systematics 9:189-218.

Wangersky, P., and W. J. Cunningham. 1957. Time lag in population models. Cold Spring Harbor Symposium on Quantitative Biology 22:329-377.

Waters, J. 1966. Methods of numerical integration applied to a system having trivial function evaluations. Communications of the Association for Computing Machinery 9:293-310.

Watson, J. D. 1968. The double helix. Atheneum Press, New York, New York, USA.

Watt, K. E. F. 1968. Ecology and resource management. A quantitative approach. McGraw-Hill, New York, New York, USA.

Watt, K. E. F. 1970. The systems point of view in pest management. Pages 71-79 in R. L. Rabb and F. E. Guthrie, editors. Concepts of pest management. North Carolina State University Press, Raleigh, North Carolina, USA.

Watt, K. E. F. 1975. Critique and comparison of biome ecosystem modeling. Pages 139-152 in B. C. Patten, editor. Systems analysis and simulation in ecology. Volume 3. Academic Press, New York, New York, USA.

Webb, W. L., H. J. Schroeder, and L. A. Norris. 1975. Pesticide residue dynamics in a forest ecosystem: a compartment model. Simulation 24:161-169.

Wetzel, R. G. 1983. Limnology. Second edition. Saunders College Publishing, New York, New York, USA.

Wiebe, P. H. 1971. A computer model study on zooplankton patchiness and its effects on sampling error. Limnology and Oceanography 16:26-38.

Wiegert, R. G. 1975. Simulation models of ecosystems. Annual Review of Ecology and Systematics 6:311-338.

———. 1979. Population models: experimental tools for analysis of ecosystems. Pages 233-274 in D. J. Horn, G. R. Stairs, and R. D. Mitchell, editors. Analysis of ecological systems. Ohio State University Press, Columbus, Ohio, USA.

Wilby, O. K., and D. A. Ede. 1974. A model generating the pattern of skeletal elements in the embryonic chick limb. Pages 81-90 in Proceedings of the 1974 Conference on Biologically Motivated Automata Theory, Institute of Electrical and Electronics Engineers, New York, New York, USA.

Winkelman, J., D. C. Cannon, and S. L. Jacobs. 1974. Liver function tests, including bile pigments. Pages 1003-1009 in R. J. Henry, D. C. Cannon, and J. W. Winkelman, editors. Clinical chemistry: principles and techniques. Second edition. Harper and Row, New York, New York, USA.

Winter, M. E., B. S. Katcher and M. A. Koda-Kimble. 1980. Basic clinical pharmacokinetics. Applied Therapeutics, Spokane, Washington, USA.

Wright, S. 1922. Coefficients of inbreeding and relationship. American Naturalist 56:330-338.

Yates, R. A., and A. B. Pardee. 1956. Control of pyrimidine biosynthesis in *Escherichia coli* by a feed-back mechanism. Journal of Biological Chemistry 221:757-770.

Yearslay, J. R., and D. P. Lettenmaier. 1987. Model complexity and data worth: an assessment of changes in the global carbon budget. Ecological Modelling 39:201-226.

Zar, J. H. 1984. Biostatistical analysis. Second edition. Prentice-Hall, Englewood Cliffs, New Jersey, USA.

Zubay, G. 1988. Biochemistry. Second edition. Macmillan, New York, New York, USA.

INDEX

USE THIS
CARD TO
ORDER THE
COMPANION
SOFTWARE
DISKETTE